At a Glance

Color Atlas and Textbook of Human Anatomy

in 3 volumes

Volume 1:   Locomotor System
             by Werner Platzer

Volume 2:   Internal Organs
             by Helmut Leonhardt

Volume 3

Nervous System and Sensory Organs

by
Werner Kahle, M.D.
Professor Emeritus
Institute of Neurology
University of Frankfurt/Main
Frankfurt/Main, Germany

Michael Frotscher, M.D.
Professor
Anatomical Institute I
University of Freiburg
Freiburg, Germany

5th revised edition

179 color plates
Illustrations by Gerhard Spitzer

Thieme
Stuttgart · New York

Library of Congress Cataloging-in-Publication Data
is available from the publischer

1st German edition 1976
2nd German edition 1978
3rd German edition 1979
4th German edition 1982
5th German edition 1986
6th German edition 1991
7th German edition 2001

1st English edition 1978
2nd English edition 1984
3rd English edition 1986
4th English edition 1993

1st Dutch edition 1978
2nd Dutch edition 1981
3rd Dutch edition 1990
4th Dutch edition 2001
1st French edition 1979
2nd French edition 1993
1st Greek edition 1985
1st Hungarian edition 1996
1st Indonesian edition 1983
1st Italian edition 1979
2nd Italian edition 1987
3rd Italian edition 2001
1st Japanese edition 1979
2nd Japanese edition 1981
3rd Japanese edition 1984
4th Japanese edition 1990
1st Polish edition 1998
1st Serbo-Croatian edition 1991
1st Spanish edition 1977
2nd Spanish edition 1988
1st Turkish edition 1987

This book is an authorized and revised translation of the 7th German edition published and copyrighted 2001 by Georg Thieme Verlag, Stuttgart, Germany.
Title of the German edition: Taschenatlas der Anatomie, Band 3: Nervensystem und Sinnesorgane

Translated by
Ursula Vielkind, Ph. D., C. Tran.,
Dundas, Ontario, Canada

© 2003 Georg Thieme Verlag
Rüdigerstraße 14, D-70469 Stuttgart, Germany
http://www.thieme.de
Thieme New York, 333 Seventh Avenue,
New York, N.Y. 10001 U.S.A.
http://www.thieme.com

Cover design: Cyclus, Stuttgart

Typesetting by Druckhaus Götz GmbH,
71636 Ludwigsburg
Printed in Germany by Appl, Wemding

ISBN 3-13-533505-4 (GTV)

ISBN 1-58890-064-9 (TNY)   1 2 3 4 5

Preface to the 5th Edition of Volume 3

The number of students as well as colleagues in the field who have learned neuroanatomy according to volume 3 of the color atlas has been steadily increasing. Kahle's textbook has proved its worth. What should one do after taking on the job of carrying on with this text book, other than leaving as much as possible as it is? However, the rapid growth in our knowledge of neuroscience does not permit this. In just the last few years many new discoveries have been made that have shaped the way we view the structure and function of the nervous system. There was a need for updating and supplementing this knowledge. Hence, new sections have been added; for example, a section on modern methods of neuroanatomy, a section on neurotransmitter receptors, and an introduction to modern imaging procedures frequently used in the hospital. The Clinical Notes have been preserved and supplemented in order to provide a link to the clinical setting. The purpose was to provide the student not only with a solid knowledge of neuroanatomy but also with an important foundation of interdisciplinary neuroscience. Furthermore, the student is introduced to the clinical aspects of those fields in which neuroanatomy plays an important role. I sincerely hope that the use of modern multicolor printing has made it possible to present things more clearly and in a more uniform way. Thus, sensory pathways are now always presented in blue, motor pathways in red, paraympathetic fibers in green, and sympathetic fibers in yellow.

I wish to thank first and foremost Professor Gerhard Spitzer and Stephan Spitzer who took charge of the grapic design of the color atlas and provided their enormous experience also for the present edition. I thank Professor Jürgen Hennig and his co-workers at the radiodiagnostic division of the Medical School of the Albert Ludwig University of Freiburg, Germany, for their help with the new section on imaging procedures. Last but not least, I would like to thank Dr. André Diesel who took great care in screening the text for lack of clarity and who contributed significantly to the color scheme of the figures, al well as my secretary, Mrs. Regina Hummel, for her help with making the many corrections. My thanks go also to Mrs. Marianne Mauch and Dr. Jürgen Lüthje at Thieme Verlag, Stuttgart, for their generous advice and their patience.

Michael Frotscher
Fall 2002

Contents

The Nervous System

Introduction

The Nervous System—An Overall View

Development and Subdivision (A–D)

The nervous system serves information processing. In the most primitive forms of organization (**A**), this function is assumed by the **sensory cells** (**A–C1**) themselves. These cells are excited by stimuli coming from the environment; the excitation is conducted to a **muscle cell** (**A–C2**) through a cellular projection, or **process**. The simplest response to environmental stimuli is achieved in this way. (In humans, sensory cells that still have processes of their own are only found in the olfactory epithelium.) In more differentiated organisms (**B**), an additional cell is interposed between the sensory cell and the muscle cell – the nerve cell, or **neuron** (**BC3**) which takes on the transmission of messages. This cell can transmit the excitation to several muscle cells or to additional nerve cells, thus forming a **neural network** (**C**). A diffuse network of this type also runs through the human body and innervates all intestinal organs, blood vessels, and glands. It is called the **autonomic** (*visceral*, or *vegetative*) **nervous system** (ANS), and consists of two components which often have opposing functions: the **sympathetic nervous system** and the **parasympathetic nervous system**. The interaction of these two systems keeps the *interior organization* of the organism constant.

In vertebrates, the **somatic nervous system** developed in addition to the autonomic nervous system; it consists of the **central nervous system** (CNS; brain and spinal cord), and the **peripheral nervous system** (PNS; the nerves of head, trunk, and limbs). It is responsible for *conscious perception*, for *voluntary movement*, and for the processing of information (*integration*). Note that most textbooks include the peripheral nerves of the autonomic nervous system in the PNS.

The CNS develops from the *neural plate* (**D4**) of the ectoderm which then transforms into the *neural groove* (**D5**) and further into the *neural tube* (**D6**). The neural tube finally differentiates into the spinal cord (**D7**) and the brain (**D8**).

Functional Circuits (E, F)

The nervous system, the remaining organism, and the environment are functionally linked with each other. Stimuli from the environment (*exteroceptive stimuli*) (**E9**) are conducted by sensory cells (**E10**) via **sensory (afferent) nerves** (**E11**) to the CNS (**E12**). In response, there is a command from the CNS via **motor (efferent) nerves** (**E13**) to the muscles (**E14**). For control and regulation of the muscular response (**E15**), there is *internal feedback* from sensory cells in the muscles via sensory nerves (**E16**) to the CNS. This afferent tract does not transmit environmental stimuli but stimuli from within the body (*proprioceptive stimuli*). We therefore distinguish between **exteroceptive** and **proprioceptive sensitivities**.

However, the organism does not only respond to the environment; it also influences it spontaneously. In this case, too, there is a corresponding functional circuit: the action (**F17**) started by the brain via efferent nerves (**F13**) is registered by sensory organs (**F10**), which return the corresponding information via afferent nerves (**F11**) to the CNS (**F12**) (*reafference*, or *external feedback*). Depending on whether or not the result meets the desired target, the CNS sends out further stimulating or inhibiting signals (**F13**). Nervous activity is based on a vast number of such functional circuits.

In the same way as we distinguish between exteroceptive sensitivity (skin and mucosa) and proprioceptive sensitivity (receptors in muscles and tendons, autonomic sensory supply of the intestines), we can subdivide the motor system into an environment-oriented **ecotropic somatomotor system** (*striated, voluntary muscles*) and an **idiotropic visceromotor system** (*smooth intestinal muscles*).

A–C Models of primitive nervous systems (according to *Parker* and *Bethe*)

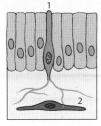

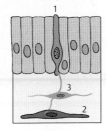

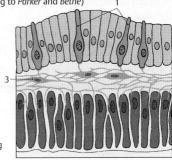

A Sensory cell with process to a muscle cell

B Nerve cell connecting a sensory cell and a muscle cell

C Diffuse neural network

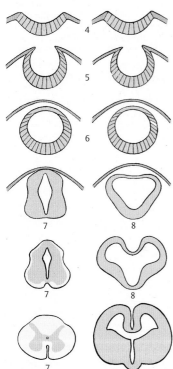

D Embryonic development of the central nervous system: spinal cord on the left, brain on the right

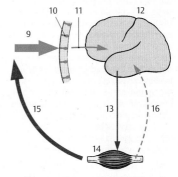

E Functional circuit: response of an organism to environmental stimuli

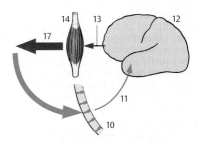

F Functional circuit: influence of an organism on its environment

Introduction

Position of the Nervous System in the Body (A, B)

The **central nervous system** (CNS) is divided into the brain, **encephalon** (**A1**), and the spinal cord (SC), **medulla spinalis** (**A2**). The brain in the cranial cavity is surrounded by a bony capsule; the spinal cord in the vertebral canal is enclosed by the bony vertebral column. Both are covered by meninges that enclose a cavity filled with a fluid, the **cerebrospinal fluid**. Thus, the CNS is protected from all sides by bony walls and the cushioning effect of a fluid (*fluid cushion*).

The **peripheral nervous system** (PNS) includes the *cranial nerves*, which emerge through holes (**foramina**) in the base of the skull, and the *spinal nerves*, which emerge through spaces between the vertebrae (**intervertebral foramina**) (**A3**). The peripheral nerves extend to muscles and skin areas. They form *nerve plexuses* before entering the limbs: the **brachial plexus** (**A4**) and the **lumbosacral plexus** (**A5**) in which the fibers of the spinal nerves intermingle; as a result, the nerves of the limbs contain portions of different spinal nerves (see pp. 70 and 86). At the entry points of the afferent nerve fibers lie **ganglia** (**A6**); these are small oval bodies containing sensory neurons.

When describing brain structures, terms like "top," "bottom," "front," and "back" are inaccurate, because we have to distinguish between different **axes of the brain** (**B**). Owing to the upright posture of humans, the neural tube is bent; the axis of the spinal cord runs almost vertically, while the axis of the forebrain (**Forel's axis**, orange) runs horizontally; the axis of the lower brain divisions (**Meinert's axis**, violet) runs obliquely. The positional terms relate to theses axes: the anterior end of the axis is called *oral* or *rostral* (*os*, mouth; *rostrum*, beak), the posterior end is called *caudal* (*cauda*, tail), the underside is called basal or *ventral* (*venter*, abdomen), and the upper side is called *dorsal* (*dorsum*, back).

The lower **brain divisions**, which merge into the spinal cord, are collectively called the *brain stem* (light gray) (**B7**). The anterior division is called the *forebrain* (gray) (**B8**).

The divisions of the brain stem, or **encephalic trunk**, have a common structural plan (consisting of *basal plate* and *alar plate*, like the spinal cord, see p. 13, C). Genuine *peripheral nerves* emerge from these divisions, as they do from the spinal cord. Like the spinal cord, they are supported by the *chorda dorsalis* during embryonic development. All these features distinguish the brain stem from the forebrain. The subdivision chosen here differs from the other classifications in which the diencephalon is viewed as part of the brain stem.

The forebrain, **prosencephalon**, consists of two parts, the *diencephalon* and the *telencephalon* or *cerebrum*. In the mature brain, the telencephalon forms the two hemispheres (*cerebral hemispheres*). The diencephalon lies between the two hemispheres.

A9 Cerebellum.

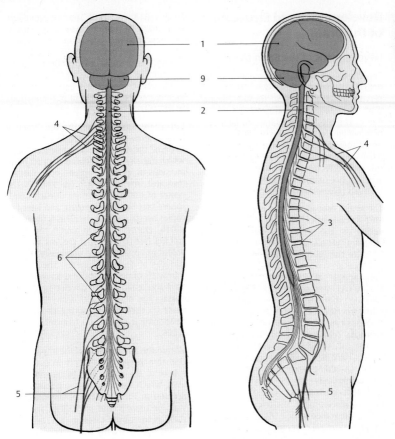

A Position of the central nervous system in the body

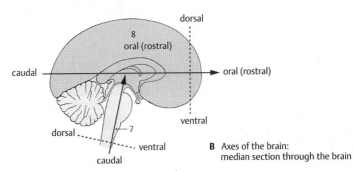

B Axes of the brain: median section through the brain

Development and Structure of the Brain

Development of the Brain (A–E)

The closure of the neural groove into the neural tube begins at the level of the upper cervical cord. From here, further closure runs in the oral direction up to the rostral end of the brain (*oral neuropore*, later the *terminal lamina*) and in the caudal direction up to the end of the spinal cord. Further developmental events in the CNS proceed in the same directions. Thus, the brain's divisions do not mature simultaneously but at intervals (*heterochronous maturation*).

The neural tube in the head region expands into several vesicles (p. 171, A). The rostral vesicle is the future forebrain, *prosencephalon* (yellow and red); the caudal vesicles are the future brain stem, *encephalic trunk* (blue). Two curvatures of the neural tube appear at this time: the cephalic flexure (**A1**) and the cervical flexure (**A2**). Although the brain stem still shows a uniform structure at this early stage, the future divisions can already be identified: **medulla oblongata** (elongated cord) (**A–D3**), **pons** (bridge of Varolius) (**A–D4**), **cerebellum** (**A–D5**, dark blue), and **mesencephalon** (midbrain) (**A–C6**, green). The brain stem is developmentally ahead of the prosencephalon; during the second month of human development, the telencephalon is still a thin-walled vesicle (**A**), whereas neurons have already differentiated in the brain stem (*emergence of cranial nerves*) (**A7**). The *optic vesicle* develops from the **diencephalon** (**AB8**, red) (p. 343, A) and forms the optic cup (**A9**). Anterior to it lies the *telencephalic vesicle* (**telencephalon**) (**A–D10**, yellow); initially, its anlage is unpaired (*impar telencephalon*), but it soon expands on both sides to form the two cerebral hemispheres.

During the third month, the prosencephalon enlarges (**B**). Telencephalon and diencephalon become separated by the *telodiencephalic sulcus* (**B11**). The anlage of the *olfactory bulb* (**B–D12**) has formed at the hemispheric vesicle, and the pituitary

anlage (**B13**) (p. 201 B) and the *mamillary eminence* (**B14**) have formed at the base of the diencephalon. A deep transverse sulcus (**B15**) is formed between the cerebellar anlage and the medulla oblongata as a result of the pontine flexure; the underside of the cerebellum comes to lie in apposition to the membrane-thin dorsal wall of the medulla (p. 283, E).

During the fourth month, the cerebral hemispheres begin to overgrow the other parts of the brain (**C**). The telencephalon, which initially lagged behind all other brain divisions in its development, now exhibits the most intense growth (p. 170, A). The center of the lateral surface of each hemisphere lags behind in growth and later becomes overlain with parts. This is the *insula* (**CD16**). During the sixth month, the insula still lies free (**D**). The first grooves and convolutions appear on the previously smooth surfaces of the hemispheres. The initially thin walls of neural tube and brain vesicles have thickened during development. They contain the neurons and nerve tracts that make up the brain substance proper. (For development of cerebral hemispheres, see p. 208.)

Within the anterior wall of the impar telencephalon, nerve fibers run from one hemisphere to the other. The *commissural systems*, which connect the two hemispheres, develop in this segment of the thickened wall, or *commissural plate*. The largest of them is the **corpus callosum** (**E**). The hemispheres grow mainly in the caudal direction; in parallel with their increase in size, the corpus callosum also expands in the caudal direction during its development and finally overlies the diencephalon.

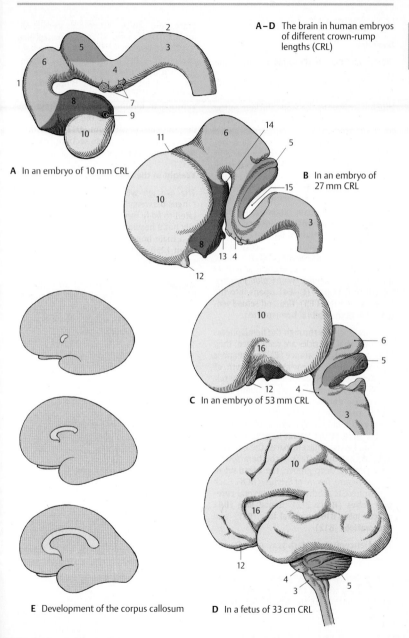

A–D The brain in human embryos of different crown-rump lengths (CRL)

A In an embryo of 10 mm CRL

B In an embryo of 27 mm CRL

C In an embryo of 53 mm CRL

D In a fetus of 33 cm CRL

E Development of the corpus callosum

Anatomy of the Brain (A – E)

Overview

The individual subdivisions of the brain contain cavities or ventricles of various shapes and widths. The *primary cavity of the neural tube and cerebral vesicle* becomes much narrower as the walls thicken. In the spinal cord of lower vertebrates, it survives as the *central canal*. In the human spinal cord, it becomes completely occluded (*obliterated*). In a cross section, only a few cells of the former lining of the spinal cord mark the site of the early central canal (**A1**). In the brain, the cavity survives and forms the **ventricular system** (p. 280) which is filled with a clear fluid, the *cerebrospinal fluid*. The **fourth ventricle** (**AD2**) is located in the segment of the medulla oblongata and the pons. After a narrowing of the cavity in the midbrain, the **third ventricle** (**CD3**) lies in the diencephalon. A passage on both sides of its lateral walls, the *interventricular foramen* (*foramen of Monro*) (**C–E4**), opens into the *lateral ventricles* (**CE5**) (**first** and **second ventricles**) of both cerebral hemispheres.

In frontal sections through the hemispheres (**C**), the **lateral ventricles** are seen twice; they have a curved appearance (**E**). This shape is caused by the crescent-shaped growth of the **hemispheres** (rotation of hemispheres, p. 208, C) which do not expand equally in all directions during development. In the middle of the semicircle is the *insula*. It lies deep in the lateral wall of the hemisphere on the floor of the *lateral fossa* (**C6**) and is overlain by the adjacent parts, the *opercula* (**C7**), so that the surface of the hemisphere shows only a deep groove, the **lateral sulcus** (*lateral fissure, fissure of Sylvius*) (**BC8**). Each hemisphere is subdivided into several **cerebral lobes** (**B**) (p. 212): **frontal lobe** (**B9**), **parietal lobe** (**B10**), **occipital lobe** (**B11**), and **temporal lobe** (**B12**).

The **diencephalon** (dark gray in **C**, **D**) and **brain stem** essentially become overlain by the cerebral hemispheres, thus rendered visible only at the base of the brain or in a longitudinal section through the brain. In a median section (**D**), the subdivisions of the brain stem can be recognized: **medulla oblongata** (**D13**), **pons** (**D14**), **mesencephalon** (**D15**), and **cerebellum** (**D16**). The fourth ventricle (**D2**) is seen in its longitudinal dimension. On its tentlike roof rests the cerebellum. The third ventricle (**D3**) is opened in its entire width. In its rostral section, the interventricular foramen (**D4**) leads into the lateral ventricle. Above the third ventricle lies the corpus callosum (**D17**); this fiber plate, seen here in cross section, connects the two hemispheres.

Weight of the Brain

The average weight of the human brain ranges between 1250 g and 1600 g. It is related to *body weight*: a heavier person usually has a heavier brain. The average weight of a male brain is 1350 g, that of a female brain 1250 g. By the age of 20, the brain is supposed to have reached its maximum weight. In old age, the brain usually loses weight owing to *age-related atrophy*. The weight of the brain does not indicate the intelligence of a person. Examination of the brains of prominent people ("elite brains") yielded the usual variations.

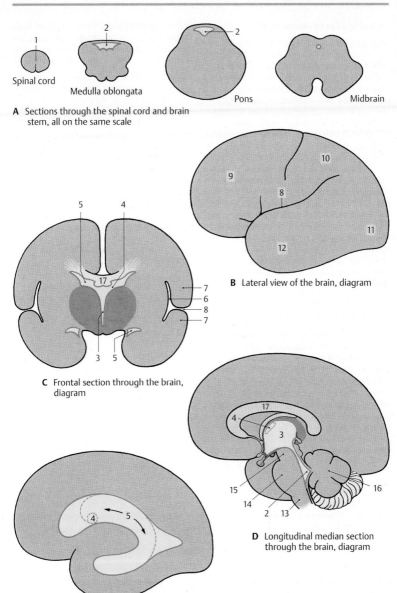

A Sections through the spinal cord and brain stem, all on the same scale

Spinal cord

Medulla oblongata

Pons

Midbrain

B Lateral view of the brain, diagram

C Frontal section through the brain, diagram

D Longitudinal median section through the brain, diagram

E Longitudinal paramedian section through the brain, diagram

Lateral and Dorsal Views (A, B)

The two cerebral hemispheres overlie all other parts of the brain; only the *cerebellum* (**A1**) and the *brain stem* (**A2**) are visible. The surface of the cerebral hemisphere is characterized by a large number of *grooves*, or *sulci*, and *convolutions*, or *gyri*. Beneath the surface of the relief of gyri lies the *cerebral cortex*, the highest nervous organ: consciousness, memory, thought processes, and voluntary activities all depend on the integrity of the cortex. The expansion of the cerebral cortex is increased through the formation of sulci and gyri. Only one-third of the cortex lies on the surface, while two-thirds lie in the depth of the gyri. As shown by the dorsal view (**B**), the hemispheres are separated by a deep groove, the **longitudinal cerebral fissure** (**B3**). On the lateral surface of the hemisphere lies the **lateral sulcus** (*sulcus of Sylvius*) (**A4**). A frontal section (pp. 9, 215, and 217) clearly shows that this is not a simple sulcus but a deep pit, the **lateral fossa**.

The anterior pole of the hemisphere is called the *frontal pole* (**A5**), the posterior one is called the *occipital pole* (**A6**). The cerebral hemisphere is subdivided into several lobes: the **frontal lobe** (**A7**) and the **parietal lobe** (**A9**), which are separated by the *central sulcus* (**A8**), the **occipital lobe** (**A10**), and the **temporal lobe** (**A11**). The central sulcus separates the **precentral gyrus** (**A12**) (region of voluntary movement) from the **postcentral gyrus** (**A13**) (region of sensitivity). Both together constitute the *central region*.

Median Section (C)

Between the hemispheres lies the **diencephalon** (**C14**); the **corpus callosum** (**C15**) above it connects the two hemispheres. The corpus callosum forms a fiber plate; its oral curvature encloses a thin wall segment of the hemisphere, the **septum pellucidum** (**C16**) (p. 221, B18). The *third ventricle* (**C17**) is opened. The adhesion of its two walls forms the **interthalamic adhesion** (**C18**). The **fornix** (**C19**) forms an arch above it. In the anterior wall of the third ventricle lies the **anterior commissure** (**C20**) (containing the crossing fibers of the olfactory brain); at its base lie

the decussation of the optic nerve, or **optic chiasm** (**C21**), the **hypophysis** (**C22**), and the paired **mamillary bodies** (**C23**); in the caudal wall lies the pineal gland, or **epiphysis** (**C24**).

The third ventricle is connected with the *lateral ventricle* of the hemisphere through the **interventricular foramen** (*foramen of Monro*) (**C25**); it turns caudally into the **cerebral aqueduct** (*aqueduct of Sylvius*) (**C26**) which passes through the midbrain and widens like a tent to form the *fourth ventricle* (**C27**) underneath the cerebellum. On the cut surface of the cerebellum (**C28**), the sulci and gyri form the *arbor vitae* ("tree of life"). Rostral to the cerebellum lies the quadrigeminal plate, or **tectal lamina** (**C29**), of the midbrain (a relay station for optic and acoustic tracts). The **pons** (**C30**) bulges at the base of the brain stem and turns into the elongated cord, or **medulla oblongata** (**C31**), which turns into the spinal cord.

C32 Choroid plexus.

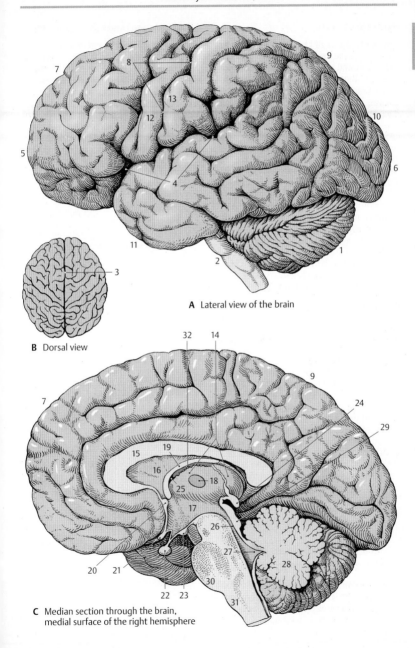

A Lateral view of the brain

B Dorsal view

C Median section through the brain,
medial surface of the right hemisphere

Introduction

Base of the Brain (A)

The basal aspect of the brain provides an overview of the brain stem, the ventral surfaces of frontal lobe (**A1**) and temporal lobe (**A2**), and the base of the diencephalon. The *longitudinal cerebral fissure* (**A3**) separates the two frontal lobes; at the basal surface of each hemisphere lies the *olfactory lobe* with the **olfactory bulb** (**A4**) and the **olfactory tract** (**A5**). The tract divides in the **olfactory trigone** (**A6**) into two *olfactory striae* which border the **anterior perforated substance** (**A7**); the latter is perforated by entering blood vessels. At the **optic chiasm** (**A8**), or decussation of the **optic nerves** (**A9**), the base of the diencephalon begins with the **hypophysis** (**A10**) and the **mamillary bodies** (**A11**). The **pons** (**A12**) bulges caudally and is followed by the **medulla oblongata** (**A13**). Numerous cranial nerves emerge from the brain stem. The cerebellum is divided into the medial, deeply-lying *vermis of the cerebellum* (**A14**) and the two *cerebellar hemispheres* (**A15**).

White and Gray Matter (B)

Upon dissecting the brain into slices, the white and gray matter, **substantia alba et grisea**, become visible on the cut surfaces. The gray matter represents a concentration of *neurons* and the white matter the fiber tracts, or *neuronal processes*, which appear light because of their white envelope, the myelin sheath. In the **spinal cord** (**B16**), the gray matter lies in the center and is enclosed by the bordering white matter (ascending and descending fiber tracts). In the **brain stem** (**B17**) and **diencephalon**, the distribution of gray and white matter varies. The gray areas are called *nuclei*. In the **telencephalon** (**B18**), the gray matter lies at the outer margin and forms the *cortex*, while the white matter lies inside. Thus, the distribution here is the reverse of that in the spinal cord.

The arrangement in the spinal cord represents a primitive state; it still exists in fish and amphibians where the neurons are in a *periventricular position* even in the telencephalon. The cerebral cortex represents the highest level of organization, which is fully developed only in mammals.

Subdivision into Longitudinal Zones (C)

During development, the neural tube is subdivided into longitudinal zones. The ventral half of the lateral wall, which differentiates early, is called the **basal plate** (**C19**) and represents the *origin of motor neurons*. The dorsal half, which develops later, is called the **alar plate** (**C20**) and represents the *origin of sensory neurons*. Between alar and basal plates lies a segment (**C21**) from which **autonomic neurons** originate. Thus, a structural plan of the CNS can be recognized in the spinal cord and brain stem, knowledge of which will aid in understanding the organization of various parts of the brain.

The derivatives of basal and alar plates are difficult to identify in diencephalon and telencephalon. Many authors therefore reject such a classification of the forebrain.

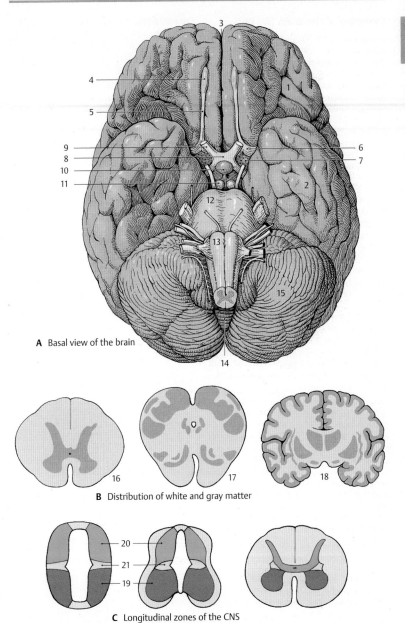

A Basal view of the brain

B Distribution of white and gray matter

C Longitudinal zones of the CNS

Evolution of the Brain (A – C)

In the course of evolution, the vertebrate brain developed into the organ of human intelligence. Since the ancestors are extinct, the developmental sequence can only be reconstructed by means of species that have retained a primitive brain structure. In *amphibians* and *reptiles*, the telencephalon (**A1**) appears as an appendix to the large olfactory bulb (**A2**); mesencephalon (**A3**) and diencephalon (**A4**) lie free at the surface. Already in primitive *mammals* (such as the hedgehog), however, the telencephalon expands over the rostral parts of the brain stem; in *lemurs*, it completely overlays the diencephalon and mesencephalon. Thus, the phylogenetic development of the brain essentially consists of a progressive enlargement of the telencephalon and a transfer of the highest integrative functions to this part of the brain. This is called **telencephalization**. Ancient primitive structures are still retained in the human brain and are intermingled with new, highly differentiated structures. Therefore, when we talk about new and old components of the human brain, we refer to the brain's evolution. The brain is neither a computer nor a thinking machine constructed according to rational principles; it is an organ that has evolved in countless variations over millions of years.

We can follow the **morphological evolution of the human brain** by means of casts made of fossil cranial cavities (**B**, **C**). The positive cast of the cranial cavity *(endocranial cast)* is a rough replica of the shape of the brain. When comparing the casts, the enlargement of the frontal and temporal lobes is striking. The changes from **Homo pekinensis** via **Neanderthal**, the inventor of sharp flint knifes, to **Cro-Magnon** (**B**), the creator of cave paintings, are obvious. However, there are no appreciable differences between Cro-Magnon and **present-day humans** (**C**).

During *phylogenesis* and *ontogenesis*, the individual brain divisions develop at different times. The parts serving the elementary vital functions develop early and are already formed in primitive vertebrates. The brain divisions for higher, more differentiated functions develop only late in higher mammals. During their expansion, they push the early-developed brain parts into a deeper location and bulge outward (they become *prominent*).

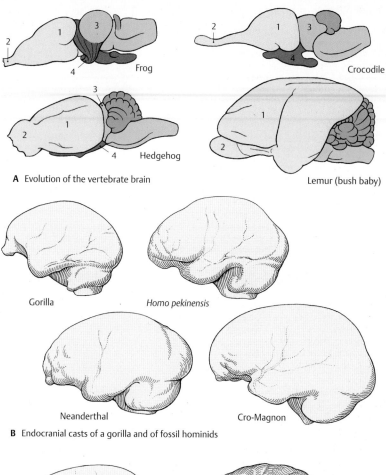

A Evolution of the vertebrate brain

Gorilla

Homo pekinensis

Neanderthal

Cro-Magnon

B Endocranial casts of a gorilla and of fossil hominids

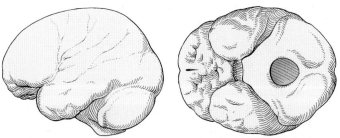

C Endocranial casts of *Homo sapiens*, lateral view and basal view

Basic Elements of the Nervous System

The Nerve Cell

The nervous tissue consists of **nerve cells** and **glial cells** which originate from the ectoderm (the latter are supporting and covering cells). Blood vessels and meninges do not belong to the nervous tissue; they are of mesodermal origin. The nerve cell (**ganglion cell** or **neuron**) is the functional unit of the nervous system. In its mature state, it is no longer able to divide, thus making proliferation and the replacement of old cells impossible. Very few nerve cells are formed after birth.

A neuron consists of the cell body, the *perikaryon* (**A1**), the processes, *dendrites* (**A2**), and one main process, the *axon* or *neurite* (**A – D3**).

The **perikaryon** is the *trophic center* of the cell, and processes that become separated from it degenerate. It contains the **cell nucleus** (**A4**) with a large, chromatin-rich **nucleolus** (**A5**) to which the **Barr body** (sex chromatin) (**A6**) is attached in females.

The **dendrites** enlarge the cell surface by branching. The processes of other neurons end here: the dendrites are the sites where *nerve impulses are received*. The processes of other neurons often end at small dendritic appendices, **spines** (thorns), which give the dendrites a rough appearance (**D**).

The **axon** conducts the nerve impulse and begins with the **axon hillock** (**AD7**), the site where *nerve impulses are generated*. At a certain distance from the perikaryon (*initial segment*) it becomes covered by the **myelin sheath** (**A8**), which consists of a lipid-containing substance (*myelin*). The axon gives off branches (**axon collaterals**) (**A9**) and finally ramifies in the terminal area (**A10**) to end with small end-feet (**axon terminals**, or **boutons**) on nerve cells or muscle cells. The bouton forms a synapse with the surface membrane of the next cell in line; it is here that impulse transmission to the other cell takes place.

Depending on the number of processes, we distinguish between **unipolar, bipolar,** or **multipolar neurons**. Most neurons are multi-polar. Some have short axons (*interneurons*), others have axons more than 1 m long (*projection neurons*).

A neuron cannot be visualized in its entirety by applying just one staining method. The different methods yield only partial images of neurons. The **cellular stain** (**Nissl's method**) shows nucleus and perikaryon (**B – D**). The latter, including the bases of the dendrites, is filled with clumps (**Nissl substance, tigroid bodies**) and may contain pigments (*melanin, lipofuscin*) (**D11**). The axon hillock is free of Nissl bodies. The Nissl substance is the light-microscopic equivalent of a well-developed *rough endoplasmatic reticulum*. Motor neurons possess a large perikaryon with coarse Nissl bodies, while sensory neurons are smaller and often contain only Nissl granules.

Impregnation with silver (**Golgi's method**) stains the entire cell including all neuronal processes; the cell appears as a brown-black silhouette (**B – D**). Other impregnation methods selectively stain the *axon terminals* (**E**), or the *neurofibrils* (**F**) running in parallel bundles through perikaryon and axon.

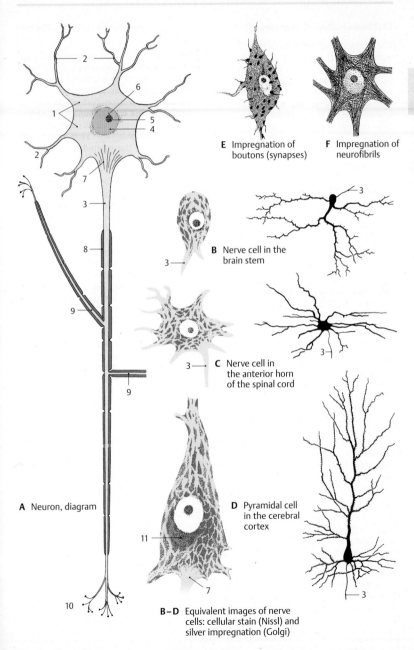

E Impregnation of boutons (synapses)

F Impregnation of neurofibrils

B Nerve cell in the brain stem

C Nerve cell in the anterior horn of the spinal cord

A Neuron, diagram

D Pyramidal cell in the cerebral cortex

B–D Equivalent images of nerve cells: cellular stain (Nissl) and silver impregnation (Golgi)

Methods in Neuroanatomy (A – E)

The availability of methods for studying the structure and function of cells, tissues, and organs is often the limiting factor in expanding our knowledge. Certain terms and interpretations can only be understood if the background of the method used is known. Therefore, the methods commonly used in neuroanatomy are presented here briefly.

Nerve cells and glial cells can be demonstrated in thin histological sections by various histological techniques. The **Nissl method** has proven helpful because of excellent visualization of the *rough endoplasmic reticulum* (p. 18), which is abundant in nerve cells. However, the different types of nerve cells are essentially characterized by their long processes, the dendrites and the axon, which are not stained by the Nissl method. For demonstration of as many of these processes as possible, thick sections (200 μm) are required. By using **silver impregnation** (**Golgi method**, p. 18), individual *nerve cells* with a large number of *processes* can be demonstrated in such thick sections. Recently, however, this 100-year-old, effective method has taken a back seat, because it is now possible to stain individual nerve cells by filling them with a dye using **recording electrodes** (**A**). The advantage of this technique is that electrical signals can be recorded from the neuron in question at the same time. In addition to visualization by *light microscopy*, the intracellularly stained or Golgi-impregnated nerve cells can subsequently be examined by *electron microscopy* to show the synaptic contacts of these neurons.

An important characteristic of nerve cells is their specific *neurotransmitter* or *messenger substance* by which communication with other nerve cells is achieved. By means of **immunocytochemistry** and the use of antibodies against the messenger substances themselves, or against *neurotransmitter-synthesizing enzymes*, it is possible to visualize nerve cells that produce a specific transmitter (**B**). Again, these immunocytochemically stained nerve cells and their processes can subsequently be examined by electron microscopy.

The longest processes of nerve cells, the axons (which can be up to 1 m long in humans), cannot be traced to their target area in histological sections. In order to demonstrate the axonal projections of neurons to different brain regions, *axonal transport* (p. 28, D) is utilized. By means of anterograde and retrograde axonal transport, substances are transported from the nerve cell body to the axon terminal and from the axon terminal back to the nerve cell body. Very long fiber connections can be visualized (**C – E**) by means of **tracers** (e.g., fluorescent dyes) that are injected either into the target area or into the region containing the cell bodies of the corresponding population of neurons; the tracers are then taken up by the axon terminals or by the cell bodies of the projection neurons, respectively. When using **retrograde transport** (**C**), the tracer is injected into the assumed target area. If the assumed connecting tracts exist, the tracer will accumulate in the cell bodies. By means of retrograde transport and the use of different fluorescent dyes (**D**), different *projection zones* of one and the same neuron can be demonstrated. When using **anterograde transport** (**E**), the tracer is injected into the region of the cell bodies of projecting neurons. Labeled axon terminals will be visible in the assumed *target zone* if the labeled neurons indeed project to this area.

Tissue cultures of nerve cells are being employed to an increasing extent for studying the processes of development and regeneration of nerve cells, and also for studying the effects of pharmaceuticals.

Basic Elements

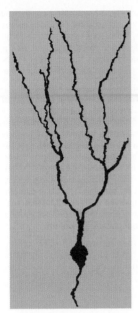

A Visualization of a neuron by means of an intracellularly injected marker

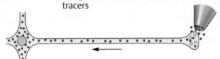

C–E Visualization of projections by means of retrograde and anterograde axonal transport of tracers

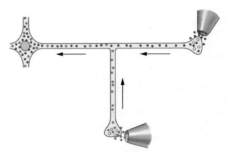

C Retrograde transport

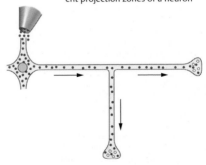

D Retrograde transport from different projection zones of a neuron

E Anterograde transport to different projection zones of a neuron

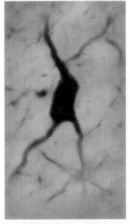

B Immunocytochemical visualization of a cholinergic neuron using an antibody against choline acetyltransferase

Ultrastructure of the Nerve Cell (A–C)

Electron micrographs show the **cell nucleus** (**A–C1**) to be enclosed by a *double-layered membrane* (**A2**). It contains the *nuclear pores* (**BC3**) that probably open only temporarily. The *karyoplasm* of the nucleus contains finely dispersed *chromatin granules*, which consist of DNA and proteins. The **nucleolus** (**A–C4**), a spongiform area of the nucleus made up of a dense granular component and a loose filamentous component, consists of RNA and proteins.

In the **cytoplasm**, the Nissl bodies appear as **rough endoplasmic reticulum** (**A–C5**), a lamellar system of membranes that enclose flattened, intercommunicating cisternae (**BC6**). Attached to the cytoplasmic side of the membranes are the protein-synthesizing **ribosomes** (**BC7**). To maintain the long axon (up to 1 m long), it is essential that the cell has an extremely high rate of protein synthesis (structural metabolism). Ribosome-free membranes form the agranular or **smooth endoplasmic reticulum** (**C8**). The rough endoplasmic reticulum communicates with the *perinuclear space* (**BC9**) and with the *marginal cisternae* (**A10**) below the cell surface. Marginal cisternae are often found at sites where boutons or glial cell processes are attached. The cytoplasm is crossed by **neurofilaments** and **neurotubules** (**A–C11**) that are arranged into long parallel bundles inside the axon. The neurotubules correspond to the microtubules of other cells.

The transport of substances takes place along neurofilaments and neurotubules (p. 28, D). *Neurofibrils* are the light-microscopic equivalent of densely packed neurotubules.

The neuron contains a large number of **mitochondria** (**A–C12**). These are enclosed in a double membrane; the inner membrane shows projections (*cristae*) (**C13**) into the inner space (*matrix*). The mitochondria are of various shapes (short and plump in the perikaryon, long and slender in the dendrites and the axon) and move constantly along fixed cytoplasmic paths between the

Nissl bodies. The mitochondria are the site of cellular respiration and, hence, of energy generation. Numerous enzymes are localized in the inner membrane and in the matrix, among others the enzymes of the *citric acid cycle* and *respiratory-chain (oxidative) phosphorylation*.

The **Golgi complex** consists of a number of **dictyosomes** (**A–C14**), which are stacks of flattened, noncommunicating cisternae. The dictyosome has a *forming side* (*cis* face) (**C15**) and a *maturing side* (*trans* face) (**C16**). The forming side receives transport vesicles from the endoplasmic reticulum. At the margins of the maturing side, *Golgi vesicles* are formed by budding. The Golgi complex is mainly involved in the modification (e.g., glycosylation, phosphorylation) of proteins from the endoplasmic reticulum.

The numerous **lysosomes** (**A–C17**) contain various enzymes (e.g., esterases, proteases) and are mainly involved in intracellular digestion.

A18 Pigment.

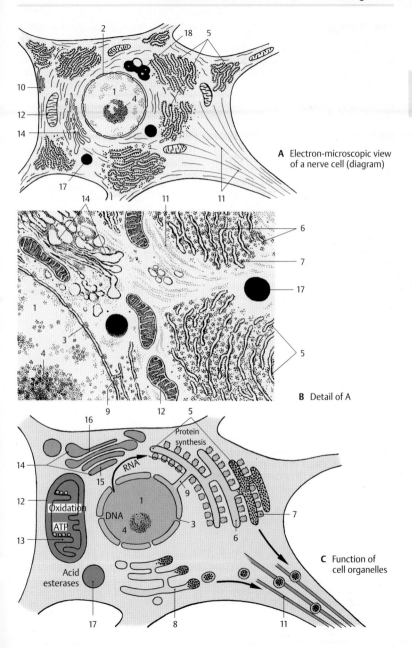

A Electron-microscopic view of a nerve cell (diagram)

B Detail of A

C Function of cell organelles

The Synapse

The axon ends with numerous small knob-like swellings, the **axon terminals** or **boutons**. Together with the apposed membrane of the next neuron, the bouton forms the **synapse** where excitation is transmitted from one neuron to another.

The synapse consists of the **presynaptic component** (*bouton*) (**AB1**) with the *presynaptic membrane* (**BC2**), the *synaptic cleft* (**B3**), and the **postsynaptic component** with the *postsynaptic membrane* (**BC4**) of the next neuron. The bouton is free of neurofilaments and neurotubules but contains mitochondria and small, mainly clear **vesicles** (**BC5**) which are clustered near the presynaptic membrane (*active zone*). The synaptic cleft contains filamentous material and communicates with the extracellular space. The presynaptic and postsynaptic membranes exhibit dense **zones of apposition**, which resemble those found at various cell junctions (*zonulae* or *maculae adherentes*, adherent junctions or desmosomes). In asymmetric synapses (see below), the density of the postsynaptic membrane (**B6**) is more prominent than the presynaptic density.

Synapses can be classified according to their *localization*, their *structure*, and their *function*, or according to the *neurotransmitter substances* they contain.

Localization (A)

The boutons may be apposed to dendrites (**AC7**) of the receptor neuron (**axodendritic synapses**) (**A8, C**), to small projections of the dendritic membrane, spines (**axospinous synapses**) (**A9**), to the perikaryon (**axosomatic synapses**) (**A10**), or to the initial segment of the axon (**axoaxonal synapses**) (**A11**). Large neurons are occupied by thousands of boutons.

Structure (B)

Depending on the width of the synaptic cleft and the properties of the apposing membranes, two types of synapses, type I

and type II, can be distinguished according to Gray. In *type I synapses*, the synaptic cleft is wider and the density of the postsynaptic membrane is more pronounced (**asymmetric synapse**). In *type II synapses*, the synaptic cleft is narrower and the postsynaptic density is about the same as the presynaptic density (**symmetric synapse**).

Function (C)

There are **excitatory** and **inhibitory** synapses. The majority of the excitatory synapses are found at the dendrites, often at the heads of the spines (**A9**). Most of the inhibitory synapses are found at the perikaryon or at the axon hillock, where excitation is generated and can be most effectively suppressed. While synaptic vesicles are usually round, some boutons contain oval or elongated vesicles (**C12**). They are characteristic of inhibitory synapses. *Asymmetric synapses* (type I) are often excitatory, whereas *symmetric synapses* (type II) are mostly inhibitory.

C13 Mitochondria.

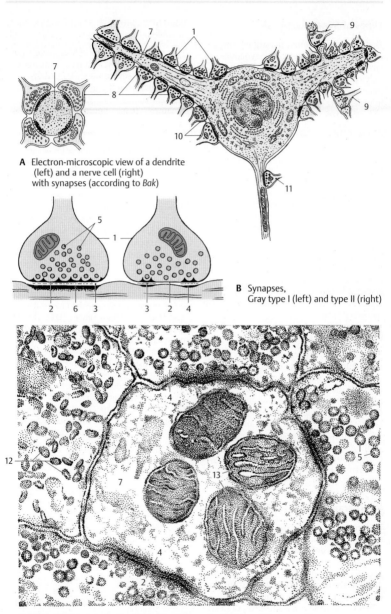

A Electron-microscopic view of a dendrite (left) and a nerve cell (right) with synapses (according to *Bak*)

B Synapses, Gray type I (left) and type II (right)

C Electron microscopic view of a cross section of a dendrite with surrounding synapses (diagram according to *Uchizono*)

Types of Synapses (A, B)

There are numerous variations on the simple form of synapses. The synaptic contact between parallel axons and dendrites is called **parallel contact** or *bouton en passant* (**A1**). Many dendrites have thornlike projections (spines) that form a **spinous synapse** (**A2**) with the bouton. On the apical dendrites of some pyramidal cells, the terminal swelling of the axon encloses the entire spine, which may be relatively large and branched, bearing numerous synaptic contacts (**complex synapse**) (**B**). Several axons and dendrites can join to form **glomerulus-like complexes** in which the different synaptic elements are closely intertwined. They probably affect each other in terms of fine-tuning (modulating) the transmission of impulses.

Each brain division has characteristic forms of synapses. *Gray type I and II synapses* are predominantly found in the cerebral cortex, *glomerulus-like complexes* are found in the cerebellar cortex, in the thalamus, and in the spinal cord.

Electrical synapses

Adjacent cells can communicate through pores (tunnel proteins), called **gap junctions**. Cells linked by gap junctions are electrically coupled; this facilitates the transmission of impulses from one cell to another (e.g., in smooth muscles, p. 303, B8). Gap junctions in neurons are therefore also called *electrical synapses* in contradistinction to the *chemical synapses*, which release neurotransmitters. Electrical coupling via gap junctions occurs not only between neurons but also between glial cells.

Neurotransmitters (C, D)

Transmission of impulses at the chemical synapses is mediated by neurotransmitters. The most widely distributed transmitter substances in the nervous system are **acetylcholine** (ACh), **glutamate**, **gamma-aminobutyric acid** (GABA), and **glycine**. Glutamate is the most common excitatory transmitter, GABA is a transmitter of the inhibitory synapses in the brain, and glycine is an inhibitory transmitter in the spinal cord.

The *catecholamines* **norepinephrine** (NE) and **dopamine** (DA) also act as transmitters, and so does **serotonin** (5-HT). Many **neuropeptides** act not only as hormones in the bloodstream but also as transmitters in the synapses (e.g., neurotensin, cholecystokinin, somatostatin).

The transmitters are produced in the perikaryon and stored in the vesicles of the axon terminals. Often only the enzymes required for transmitter synthesis are produced in the perikaryon, while the transmitter substances themselves are synthesized in the boutons. The *small and clear vesicles* are thought to carry glutamate and ACh, the *elongated vesicles* of the inhibitory synapses carry GABA, while norepinephrine and dopamine are present in the *granular vesicles* (**C**).

Most vesicles are located near the presynaptic membrane, the density of which can be demonstrated by special procedures as a grid with hexagonal spaces (**D3**). The vesicles pass through these spaces to reach the presynaptic membrane and, upon excitation, empty their content into the synaptic cleft by fusing with the presynaptic membrane (*omega figure*) (**D4**). The transmitter substances are delivered in certain quanta, the morphological equivalents of which are the vesicles. Some of the transmitter molecules return into the bouton by *reuptake* (**D5**).

D6 Axonal filaments.

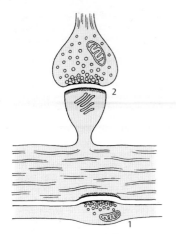

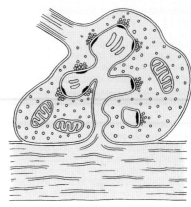

A Parallel contact (1) and spinous synapse (2)

B Complex synapse

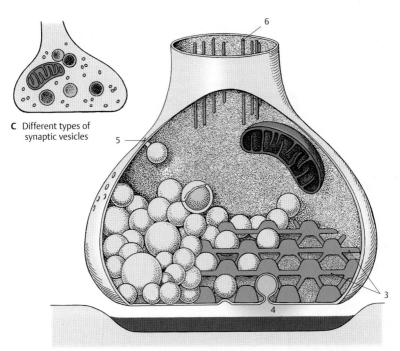

C Different types of synaptic vesicles

D Model of a synapse (according to *Akert, Pfenniger, Sandri* and *Moor*)

Neurotransmitters (continued) (A–C)

Many neurons, perhaps most of them, produce more than one transmitter substance. Nevertheless, they are classified according to their functionally most important neurotransmitter as *glutamatergic, cholinergic, catecholaminergic* (noradrenergic and dopaminergic), *serotoninergic,* and *peptidergic neurons.* The **catecholaminergic** and **serotoninergic neurons** can be identified by fluorescence microscopy because their transmitters show a green-yellow fluorescence following exposure to *formalin vapor* (**A**, **B**). It is thus possible to trace the axon and to recognize the perikaryon and the outline of its nonfluorescent nucleus. The fluorescence is very faint in the axon, more distinct in the perikaryon, and most intense in the axon terminals. It is here that the highest concentration of transmitters occurs. **Cholinergic neurons** can be demonstrated by a histochemical assay for **acetylcholinesterase**, the enzyme required for the degradation of acetylcholine. Since this enzyme is also produced by noncholinergic neurons, the proper assay is by immunocytochemistry using antibodies against **choline acetyltransferase**, the acetylcholine-synthesizing enzyme. Other transmitters and neuropeptides can also be demonstrated by immunocytochemistry (**C**). It has been shown by double-labeling that many neuropeptides are produced together with classical neurotransmitters within the same neuron. So far, the functional significance of *cotransmission,* i.e., the release of different transmitters by the same neuron, has been studied in detail only on some neurons of the autonomic nervous system.

Axonal Transport (D, E)

The transmitter substances or their synthesizing enzymes are produced in the perikaryon and must be transported to the axon terminal. The microtubules of the neuron, **neurotubules** (**D1**), play a key role in this transport mechanism. If they are destroyed by applying the mitotic poison *colchicine,* the **intra-axonal transport** stops. This rapid transport of material is energy-dependent and takes place in vesicles that are moved along the microtubules by **motor proteins**. The retrograde transport (in the direction of the cell body and toward the minus end of the microtubules) is mediated by *dynein* (**D2**), while the anterograde transport (in the direction of the axon terminal and toward the plus end of the microtubules) is mediated by *kinesin* (**D3**). The transporting vesicles are endowed with several motor proteins, the ATP-binding heads of which interact with the surface of the microtubule in an alternating and reversible fashion. This results in ATP being hydrolyzed, and the released energy is converted into molecular movement that causes the vesicles to roll along the microtubules in the target direction. The velocity of the rapid intra-axonal transport has been calculated at 200–400 mm per day. Proteins, viruses, and toxins reach the perikaryon by retrograde transport from the axon terminals.

In addition to the rapid intra-axonal transport, there is also a continuous **flow of axoplasm** which is much slower, namely, 1–5 mm per day. It can be demonstrated by ligating a single axon (**E**); proximal to the constricted site, the axoplasm is held back and the axon shows swelling.

The anterograde and retrograde transport mechanisms are used in neuroanatomy to study connecting tracts (see p. 20).

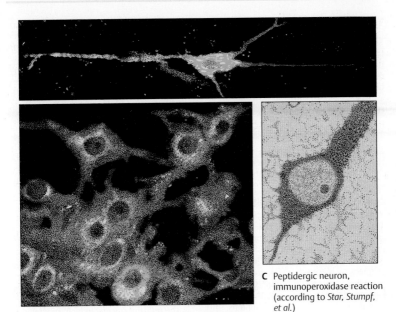

C Peptidergic neuron, immunoperoxidase reaction (according to *Star, Stumpf, et al.*)

A, B Catecholaminergic neurons in the brain stem, fluorescence microscopic views (according to *Dahlström and Fuxe*)

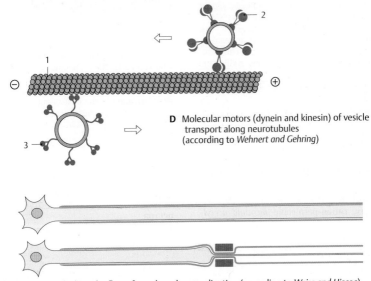

D Molecular motors (dynein and kinesin) of vesicle transport along neurotubules (according to *Wehnert and Gehring*)

E Blocking the flow of axoplasm by axon ligation (according to *Weiss and Hiscoe*)

Transmitter Receptors (A, B)

There are two categories of neurotransmitter receptors: *ligand-gated ion channels* and *transmitter receptors* coupled to an intracellular guanosine triphosphate-(GTP-) binding protein (*G protein*).

Ligand-gated Ion Channels

Ligand-gated ion channels consist of different subunits (**A1**) that are inserted into the cell membrane (**A2**). Binding of the neurotransmitter to the specific receptor causes the channel to become permeable to certain ions (**B**).

Excitatory amino acid receptors. Receptors for the excitatory transmitter *glutamate* are classified according to the synthetic ligands binding to them. There are three types of *glutamate-gated ion channels*: the AMPA (aminohydroxymethylisoxazolepropionic acid) receptor (**C3**), the NMDA (*N*-methyl-D-aspartate) receptor (**C4**), and the kainate receptor. Binding to the AMPA receptor causes an influx of sodium ions, thus leading to depolarization of the cell. Similarly, activation of the NMDA receptor causes an influx of both Na^+ and Ca^{2+}. Under conditions of resting potential, the NMDA receptor is blocked by magnesium; the magnesium blockade is lifted by depolarization (through AMPA receptors). This temporal shift in activities of the AMPA and NMDA receptors results in a graduated response of the postsynaptic neurons to the neurotransmitter glutamate.

Inhibitory GABA and glycine receptors. GABA is the most common inhibitory transmitter in the brain, and glycine in the spinal cord. Both receptors are ligand-gated ion channels that cause the influx of chloride ions when activated. The cell thus becomes hyperpolarized and inhibited.

Ligand-gated ion channels include the excitatory, cation-permeable *nicotinic acetylcholine receptor* and the *serotonin (5-HT) receptor*.

Receptors Coupled to G Protein

Most neurotransmitters do not bind to ligand-gated channels but to receptors coupled to G protein. The main difference between the two types of receptors is in the *speed of the synaptic response*. In the case of ligand-gated ion channels, the activation causes a rapid synaptic potential lasting only for milliseconds. Activation of G protein-coupled receptors results in responses that last seconds or minutes. G proteins regulate enzymes that produce intracellular messenger substances. These have an effect on ion channels or, via regulatory proteins, on the expression of genes.

Synaptic Transmission (C)

The synaptic transmission is essentially characterized by three processes:

1 Conversion of the *action potential* arriving at the axon terminal into a chemical signal. Depolarization results in the opening of calcium channels (**C5**) and in the *influx of calcium*, which, mediated by certain proteins, causes fusion of synaptic vesicles (**C6**) with the presynaptic membrane and *release of the transmitter* into the cleft (**C7**).

2 The released transmitter binds to *specific receptors*.

3 In the case of ligand-gated ion channels, this results in their opening for certain ions. In the case of glutamatergic receptors, the influx of Na^+ or Ca^{2+} causes depolarization of the postsynaptic membrane (*excitatory postsynaptic potential*, EPSP). In the case of GABA and glycine receptors, the influx of Cl^- causes hyperpolarization of the postsynaptic membrane (*inhibitory postsynaptic potential*, IPSP). The activation of G protein-coupled receptors results in a long-lasting response that may finally lead to a *change in gene expression* in the postsynaptic neuron.

A, B Ligand-gated ion channel (adapted from *Kandel, Schwartz and Jessel*)

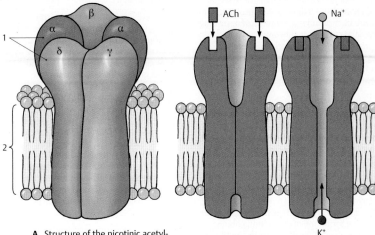

A Structure of the nicotinic acetylcholine receptor

B Binding of acetylcholine (ACh)
opens the channel for Na⁺ and K⁺

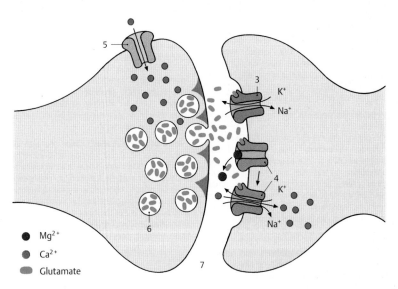

- Mg²⁺
- Ca²⁺
- Glutamate

C Synaptic transmission at a glutamatergic synapse

Neuronal Systems

Groups of neurons that have the same transmitter and the axons of which often form dense fiber bundles are described according to their transmitter substance as *cholinergic, noradrenergic, dopaminergic, serotoninergic, GABAergic,* or *peptidergic systems.* The impulse can be transmitted to neurons of the same type or to neurons with a different neurotransmitter. In the sympathetic nervous system, for example, the neurons in the spinal cord are cholinergic; however, transmission in the peripheral ganglia switches to *noradrenergic* neurons (p. 297, C).

Noradrenergic, dopaminergic, and serotoninergic neurons are located in the *brain stem.* **Noradrenergic neurons** form the *locus caeruleus* (**A1**) (p. 100, B28; p. 132, D18) and cell groups in the lateral part of the *reticular formation* of the medulla oblongata and the pons; their fibers project to the hypothalamus, to the limbic system, diffusely into the neocortex, and to the anterior horn and lateral horn of the spinal cord. **Serotoninergic neurons** lie in the *raphe nuclei* (**A2**) (p. 108, B28), especially in the *posterior raphe nucleus* (**A3**); their fibers project to the hypothalamus, to the olfactory epithelium, and to the limbic system. **Dopaminergic neurons** make up the compact part of the *substantia nigra* (**A4**) (p. 134, A17; p. 136, AB1) from where the *nigrostriatal fibers* extend to the striatum.

Peptidergic neurons are found in phylogenetically older brain regions, namely, in the *central gray of the midbrain* (**A5**), in the *reticular formation* (**A6**), in the *hypothalamus* (**A7**), in the *olfactory bulb*, in the *habenular nucleus* (**A8**), *interpeduncular nucleus* (**A9**), and *solitary nucleus* (**A10**). Numerous peptidergic neurons have also been demonstrated in the *cerebral cortex*, in the *thalamus*, in the *striatum*, and in the *cerebellum*. The significance of the different peptides is still largely unclear. It is assumed that they act as **cotransmitters** and have a modulating function. Many of these peptides are found in other organs as well, such as the digestive system (e.g., vasoactive intestinal polypeptide, somatostatin, cholecystokinin).

Glutamate is often the transmitter of projection neurons with long axons. **Glutamatergic neurons** are, for example, the **projection neurons of the cerebral cortex**, the pyramidal cells (p. 240, C; p. 242, A1 and B11). **GABAergic inhibitory neurons** are often classified according to the target structures on which they form inhibitory synapses. GABAergic *basket cells*, which form synapses with cell bodies, are distinguished from *axo-axonal cells*. The latter develop inhibitory synapses at the beginning of the axon (initial segment) of a projection neuron. GABAergic neurons often form local circuits (**interneurons**). They often contain peptides (see above) and calcium-binding proteins apart from GABA as the classic transmitter.

Cholinergic neurons are found in the *brain stem* and also in the *basal forebrain*. As in the case of catecholaminergic neurons, far-reaching projections originate from circumscribed groups of cells, for example, in the *basal nucleus* (**B11**) and in certain *septal nuclei* (**B12**) that supply, via fibers in the *cingulate gyrus* (limbic gyrus) (**B13**) and in the *fornix* (**B14**), respectively, large regions of the *neocortex* and the *hippocampus* (**B15**). These ascending cholinergic projections from the basal forebrain are thought to be associated with **processes of learning and memory**. They are affected in **Alzheimer's disease** which is accompanied by disturbed learning and memory.

Synthesis, degradation, and storage of transmitter substances can be influenced by *pharmaceuticals*. An excess or deficiency in transmitters can be created in the nerve cells, leading to changes in motor or mental activity. Changing transmitter synthesis and degradation is not the only way neuropharmaceuticals can influence the synaptic transmission; they can also act on the receptors as transmitter agonists or antagonists.

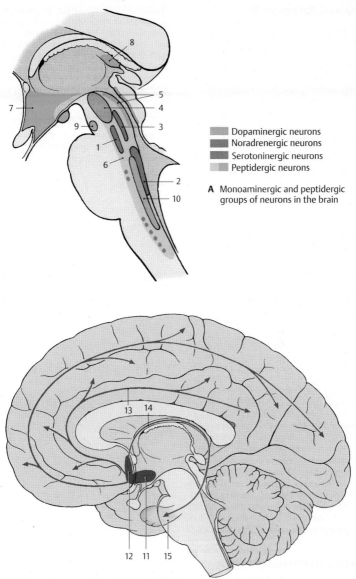

Dopaminergic neurons
Noradrenergic neurons
Serotoninergic neurons
Peptidergic neurons

A Monoaminergic and peptidergic groups of neurons in the brain

B Cholinergic groups of neurons in the basal forebrain, including the cholinergic innervation of cerebral cortex and hippocampus

Neuronal Circuits (A – D)

The nerve cells and their processes form a network (**A**) that is not a continuum of nerve fibers (*continuity theory*) but consists of countless individual elements, the neurons (*neuron theory*). As the basic building block of the nervous system, the neuron represents a structural, genetic, trophic, and functional unit.

The neurons in the network are interconnected in a specific way (*neuronal circuits*). The connections for the inhibition of excitation are as important as those for the transmission of excitation, for they are the ones through which the continuous influx of impulses is restricted and selected: important signals are transmitted, unimportant ones are suppressed.

Postsynaptic inhibition does not inhibit synaptic transmission but subsequent discharge of the postsynaptic neuron.

Inhibitory GABAergic neurons can be integrated into the neuronal circuit in different ways. In the case of **recurrent (feedback) inhibition** (**B**), an axon collateral of the excitatory projection neuron (green) activates the inhibitory cell (red), which in turn inhibits the projection neuron via a recurrent collateral. In the case of **feed-forward inhibition** (**C**), the inhibitory interneuron is not activated by a recurrent collateral of the excitatory cell but by excitatory afferents from another brain region. The effect on the projection neuron, however, is the same; activation of the inhibitory GABAergic neurons leads to inhibition of the projection neurons. In case of **disinhibition** (**D**), an inhibitory interneuron is again activated by an excitatory afferent. The target cell of this interneuron, however, is another inhibitory interneuron. Activation of the first interneuron through the excitatory afferent therefore means an increased inhibition of the second interneuron, which now cannot exert an inhibitory effect on the next projection neuron. The inhibitory effect is removed (disinhibition).

A single nerve cell in the brain receives a large number of connections (**convergence**).

One pyramidal cell of the cerebral cortex can establish over 10 000 synaptic connections with other nerve cells. In turn, such a cell itself creates numerous connections with many other nerve cells by numerous axon collaterals (**divergence**). The *spatial* and *temporal summation* of excitatory and inhibitory inputs of a cell decides at a given moment whether the cell is depolarized and generates an action potential, which then runs along the axon and leads to excitation of subsequent neurons in the series. If the inhibitory inputs predominate, the discharge of neurons does not take place.

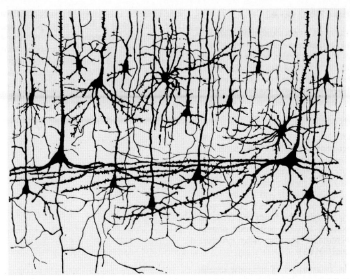

A Neuronal network in the cerebral cortex, silver impregnation (according to *Cajal*)

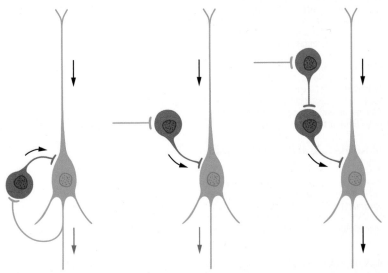

B Recurrent (feedback) inhibition **C** Feedforward inhibition **D** Disinhibition

B–D Synaptic relay principles of inhibitory neurons.

The Nerve Fiber (A – G)

The axon (**AFG1**) is surrounded by a sheath: in *unmyelinated nerve fibers* by the cytoplasm of the sheath cells, and in *myelinated nerve fibers* by the **myelin sheath** (**ABG2**). Axon and sheath together are called the **nerve fiber**. The myelin sheath begins behind the initial segment of the axon and ends just before the terminal ramification. It consists of **myelin**, a lipoprotein produced by the sheath cells. The sheath cells in the *CNS* are **oligodendrocytes**; in the *peripheral nerves* they are **Schwann cells**, which originate from the neural crest (p. 62, C2). The myelin sheath of fresh, unfixed nerve fibers appears highly refractile and without structure. Its lipid content makes it birefringent in polarized light. The lipids are removed upon fixation, and the denatured protein scaffold remains as a gridlike structure (*neurokeratin*) (**D3**).

At regular intervals (1 – 3 mm), the myelin sheath is interrupted by deep constrictions, the **nodes of Ranvier** (**AB4 F**). The segment between two nodes of Ranvier in peripheral nerves, the **internode** or **interannular segment** (**F**), corresponds to the expansion of one sheath cell. The cell nucleus (**ADF5**) and perinuclear cytoplasm form a slight bulge on the myelin sheath in the middle of the internode. Cytoplasm is also contained in oblique indentations, the **Schmidt–Lanterman incisures** (**C**, **F6**) (see also p. 40, A4). The margins of the sheath cells define the node of Ranvier at which axon collaterals (**E**) may branch off or synapses may occur.

Ultrastructure of the Myelin Sheath (G)

The electron micrograph shows the axon enclosed by a plasma membrane, the **axolemma**; it is surrounded by a series of regularly spaced, concentric dark and light lines (*period lines*). The width of each lamella from one dark line to the next measures 120 Å on average (1 Å = 0.1 nm), with the dark line taking up 30 Å and the light line 90 Å. As seen at higher magnification, the light lines are subdivided by a thin irregular line resembling a string of pearls (**G7**). We thus distinguish a dense **major period line** and a fainter **intraperiod line**. Studies using polarized light and X-rays have shown that the myelin sheath is made up of alternating layers of protein and lipid molecules. Accordingly, the dark lines (major period line and intraperiod line) are regarded as layers of protein molecules and the light lines as layers of lipid molecules.

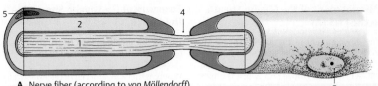

A Nerve fiber (according to *von Möllendorff*)

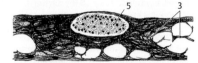

B Node of Ranvier, osmium stain (diagram)

C Schmidt–Lanterman incisures

D Perikaryon of a Schwann cell

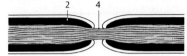

E Axonal branching

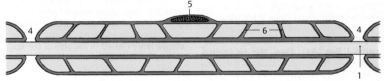

F Internode (according to *Cajal*)

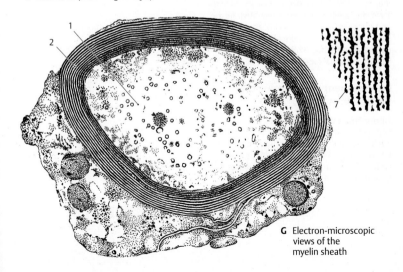

G Electron-microscopic views of the myelin sheath

Development of the Myelin Sheath in the PNS (A)

The development of the myelin sheath affords an insight into the structure of its spiraling lamellae. The cell body of the **Schwann cell** (**A1**) forms a groove into which the **axon** (**A2**) becomes embedded. The groove deepens and its margins approach each other and finally meet. In this way, a duplication of the cell membrane is formed, the **mesaxon** (**A3**), which wraps around the axon like a spiral as the Schwann cell migrates around the encircled axon.

The term *mesaxon* is based on the term mesenterium, a thin duplication that is formed as a suspension band by the peritoneum and encloses the intestine. In a similar way, the Schwann cell forms a duplication and envelops the axon. Like all plasma membranes, the cell membrane of the Schwann cell consists of an outer and an inner dense layer of protein and a light lipid layer between them. Upon membrane duplication, the two outer protein layers come into apposition first and fuse to form the **intraperiod line** (**A4**). Thus, the six-layered membrane duplication becomes the five-layered **myelin lamella**. With further encircling, the inner protein layers of the cell membrane make contact as well and fuse to form the dense **major period line** (**A5**). At the end of the process, the start of the duplication lies inside the myelin sheath, the **internal mesaxon** (**AB6**), while the end lies outside, the **external mesaxon** (**7** in **A, B**).

Development of Unmyelinated Nerve Fibers (A)

Unmyelinated nerve fibers (**A8**) are enveloped by sheath cells, each of which encircles several axons. The margins of the grooves may also form a membrane duplication (*mesaxon*) but without fusion of the membrane layers.

Structure of the Myelin Sheath in the CNS (B, C)

The myelin sheath in the CNS (**B**) shows distinct differences as compared to the myelin sheath of the peripheral nerves (see p. 41, B). Whereas the Schwann cell in the PNS myelinates only one axon, an **oligodendrocyte** (**B9**) in the CNS myelinates several axons and will later remain connected with several internodes via cytoplasmic bridges. The extent and shape of the cell becomes clear when visualizing the internodes as being unfurled (**C**). The mechanism of the myelination process is unknown. The external mesaxon forms an *external bulge* (**B10**) starting from the cytoplasmic bridge. The myelin lamellae terminate at the *paranodal region* (**B11**) (node of Ranvier). As seen in the longitudinal section, the innermost lamella terminates first and the outermost lamella covers the remaining endings, terminating directly at the node of Ranvier. At the ends of the lamellae, the dense major period lines widen into *pockets filled with cytoplasm* (**B12**). The axon of the central nerve fiber is completely exposed in the area of the node of Ranvier. There are no Schmidt–Lanterman incisures in the CNS.

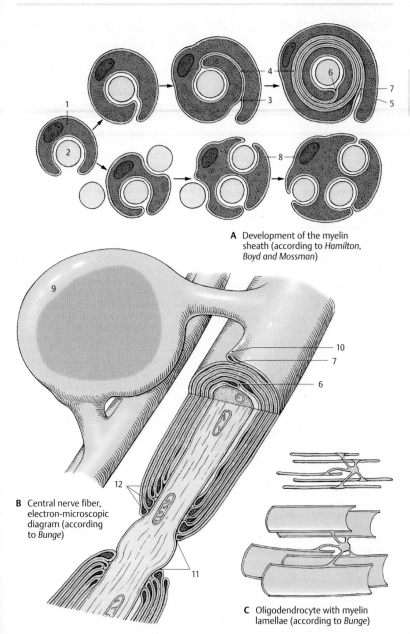

A Development of the myelin sheath (according to *Hamilton, Boyd and Mossman*)

Basic Elements

B Central nerve fiber, electron-microscopic diagram (according to *Bunge*)

C Oligodendrocyte with myelin lamellae (according to *Bunge*)

Peripheral Nerve (A–D)

The **myelin sheath** of peripheral nerve fibers is surrounded by the cytoplasm of the **Schwann cell** (**A1**). The outer cell membrane borders on a **basal lamina** (**AB2**), which envelops the entire peripheral nerve fiber. The nucleus of the Schwann cell (**A3**) is depicted in cross section. The **Schmidt–Lanterman clefts** (**A4**) are depicted in longitudinal section as cytoplasmic crevices of the major period lines. In the three-dimensional reconstruction, they appear as spirals in which the cytoplasm communicates between the inside and outside. At the **node of Ranvier** (**B5**), the Schwann cell processes (**AB6**) slide over the paranodal region and over the axon (**ABD7**). They interdigitate and thus form a dense envelope around the node of Ranvier.

Differences between the structures of the myelin sheaths in CNS and PNS are illustrated in **B**.

There is a regular relationship between the circumference of the axon, the thickness of its myelin sheath, the distance between the nodes of Ranvier, and the **conduction velocity** of a nerve fiber. The larger the circumference of an axon, the thicker the enclosing myelin sheath and the longer the internodes. When myelinated nerve fibers are still growing (e.g., in the nerves of the limbs), the internodes are growing in length. The longer the internodes, the faster the conduction velocity of the fiber. We distinguish between myelinated, poorly myelinated, and unmyelinated nerve fibers, also referred to as A, B, and C fibers. The **myelinated A fibers** have an axonal diameter of 3–20 μm and a conduction velocity of up to 120 m/s; the **poorly myelinated B fibers** are up to 3 μm in diameter and have a conduction velocity of up to 15 m/s. Conduction velocity is the slowest in the **unmyelinated C fibers** (up to 2 m/s); we are dealing here with a *continuous spread of excitation*. By contrast, conduction in myelinated nerves is saltatory, that is, it takes place in jumps. The morphological basis of *saltatory conduction* is the alternation of myelinated internodes and unmyelinated nodes of Ranvier; the current inside the axon jumps from one node to the next, and the current circuit is closed each time at the nodes through changes in the permeability of the axolemma (triggered by voltage-gated ion channels). This mode of conduction is much faster and requires less energy than the continuous spread of excitation.

The peripheral nerve fiber is surrounded by longitudinal collagenous connective-tissue fibrils; together with the basal membrane, they form the **endoneural sheath**. The nerve fibers are embedded in a loose connective tissue, the **endoneurium** (**D8**). A variable number of nerve fibers is collected into *bundles* or *fascicles* (**C10**) by the **perineurium** (**CD9**) which consists mainly of circular fibers. The innermost layer of the perineurium is formed by *endothelial cells* that enclose the endoneural space in several thin layers. The perineural endothelial cells possess a basal membrane at their perineural and endoneural surfaces and are joined together by *zonulae occludentes* (tight junctions). They represent a barrier between nerve and surrounding tissue, similarly to the endothelial cells of cerebral capillaries (p. 45 E). The mechanical strength of the peripheral nerve is based on its content of *circular elastic fibers*. In the nerves of the limbs, the perineurium is reinforced in the joint regions. The **epineurium** (**CD11**) borders on the perineurium; its inner layers form concentric lamellae as well. They change into *loose connective tissue* containing *fat cells* (**D12**), *blood vessels*, and *lymph vessels*.

D13 Cell nuclei of Schwann cells.
D14 Capillaries.

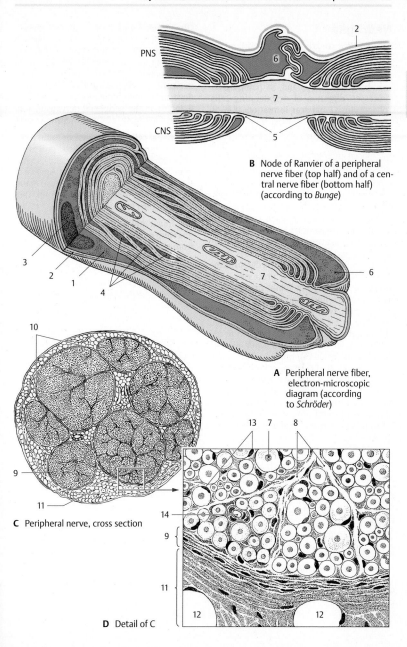

PNS

CNS

B Node of Ranvier of a peripheral nerve fiber (top half) and of a central nerve fiber (bottom half) (according to *Bunge*)

A Peripheral nerve fiber, electron-microscopic diagram (according to *Schröder*)

C Peripheral nerve, cross section

D Detail of C

Neuroglia

The **neuroglia** (*glia*, glue) is the *supporting and covering tissue of the CNS* and has all the functions of connective tissue: support, metabolite exchange, and, in pathological processes, digestion of degenerating cells (phagocytosis) and scar formation. It is of *ectodermal origin*. After Nissl staining, only cell nuclei and cytoplasm are visible; visualization of cellular processes is achieved only by special impregnation methods and by immunocytochemistry. We distinguish three different types of neuroglia: *astroglia* (*macroglia*), *oligodendroglia*, and *microglia* (**A**).

Astrocytes have a large, clear cell nucleus and numerous processes which give the cell a starlike appearance (**A**, **C**). There are **protoplasmic astrocytes** with few processes (usually present in gray matter) and **fibrous astrocytes** with numerous long processes (predominantly present in white matter). The latter produce fibers and contain glial filaments in cell body and processes. They form glial scars after damage to the brain tissue. Astrocytes are regarded as supporting elements, since they form a three-dimensional scaffold. On the outer surface of the brain, the scaffold thickens to form a dense fiber felt, the **glial limiting membrane**, which forms the outer limit of the ectodermal tissue against the mesenchymal meninges. Astrocytes extend processes to blood vessels and play a role in metabolite exchange (p. 45 A, B).

In addition, astrocytes play a decisive part in *maintaining the interior environment* of the CNS, particularly the *ion balance*. Potassium ions released upon excitation of groups of neurons are removed from the extracellular space via the network of astrocyte processes. Astrocytes probably also take up CO_2 released by nerve cells and thus keep the interstitial pH at a constant value of 7.3. Astrocyte processes enclose the synapses and seal off the synaptic cleft. They also take up neurotransmitters (uptake and release of GABA via astrocytes have been demonstrated).

Oligodendrocytes have a smaller, darker cell nucleus and only a few, sparsely branched processes. In the gray matter, they accompany neurons (**satellite cells**) (**B**). In the white matter, they lie in rows between the nerve fibers (**intrafascicular glia**). They produce and maintain the *myelin sheath* (p. 39, B and C). In the peripheral nervous system the myelin sheath is formed by **Schwann cells**.

Microglial cells have an oval or rodlike cell nucleus and short, branched processes. They exhibit ameboid mobility and can migrate within the brain tissue. In response to tissue destruction, they phagocytose material (**scavenger cells**) and round up into spheres (**gitter cell**).

The frequently expressed view that microglia were not derived from the ectoderm but from the mesoderm (*mesoglia*) is not supported by evidence.

Fibrous astrocyte   Protoplasmic astrocyte   Oligodendroglia   Microglia

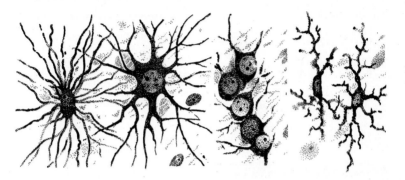

A Equivalent images of neuroglia: Nissl staining (top row), silver impregnation (bottom row)

B Oligodendrocytes as satellites of a nerve cell

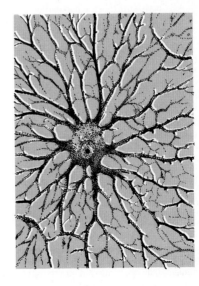

C Astrocyte in tissue culture

Blood Vessels

Cerebral blood vessels are of *mesodermal* origin. They grow during development from the mesodermal coverings into the brain tissue. In histological preparations, they are mostly surrounded by a narrow empty cleft (*Virchow–Robin space*, perivascular space), an artifact caused by tissue shrinkage during histological preparation. Arteries and arterioles are of the *elastic type*, that is, their muscles are poorly developed and their contractility is limited. The **capillaries** exhibit a nonfenestrated *closed endothelium* and a *closed basal membrane*. There are *no lymph vessels* in the CNS.

Astrocyte processes extend to the capillaries and widen into **perivascular glial feet** (**AB1**). In electron micrographs, capillaries are completely covered by perivascular feet. The capillary wall consists of **endothelial cells** (**BE2**) which overlap at their margins like roof tiles and are joint together by *zonulae occludentes* (tight junctions). The capillary is enclosed by the **basal lamina** (**BE3**) and the **astrocyte covering** (**BE4**). The latter can be compared to the *glial limiting membrane* (p. 42); both structures separate the ectodermal tissue of the CNS from the adjacent mesodermal tissue.

The sealing of the brain tissue from the rest of the body manifests itself in the **blood–brain barrier**, a *selective barrier* for numerous substances that are prevented from penetrating from the bloodstream through the capillary wall into the brain tissue.

This barrier was first demonstrated by Goldmann's experiments using *trypan blue*. If the dye is injected intravenously into experimental animals (**Goldmann's first experiment**) (**C**), almost all organs stain blue, but the brain and spinal cord remain unstained. Minor blue staining is only found in the *gray tubercle* (**C5**), the *postremal area*, and the *spinal ganglia*. The *choroid plexus* (**C6**) and the *dura* (**C7**) show a distinct blue staining. The same pattern is observed in cases of jaundice in humans; the bile pigment stains all organs yellow, only the CNS remains unstained. If the dye is injected into the space

of the cerebrospinal fluid (**Goldmann's second experiment**) (**D**), brain and spinal cord are diffusely stained on the surface, while the rest of the body remains unstained. Thus, there exists a barrier between the CSF and the blood but not between the CSF and the CNS. We therefore distinguish between a blood–brain barrier and a **blood–cerebrospinal fluid barrier**. The two barriers behave in different ways.

The site of the blood–brain barrier is the **capillary endothelium** (**E**) (see also vol. 2); in the brain, it forms a closed wall without fenestration. By contrast, the capillary walls of many other organs (liver, kidney [**E8**]) exhibit prominent fenestration which allows for extensive metabolite exchange. The barrier effect has been demonstrated for numerous substances in studies using isotopes. The barrier may result in a complete blockade or in a delay of penetration. Whether or not drugs can penetrate this barrier has major practical implications.

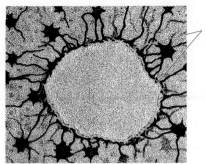

A Blood vessel surrounded by astrocytes, silver impregnation

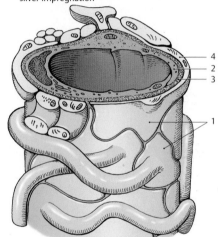

B Blood vessel with perivascular glial feet (diagram according to *Wolff*)

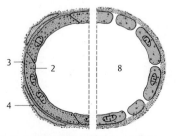

E Brain capillary (left) and kidney capillary (right), diagram based on electron-microscopic findings

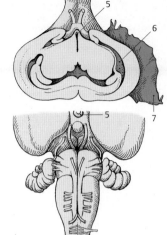

C Goldmann's first experiment

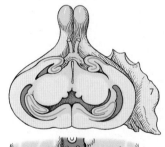

D Goldmann's second experiment

C–D Blood-brain barrier in the rabbit (according to *Spatz*)

Spinal Cord and Spinal Nerves

Overview

The *spinal cord*, or **spinal medulla**, is located in the channel of the spinal column, the **vertebral canal**, and is surrounded by the *cerebrospinal fluid*. It has two spindle-shaped swellings: one in the neck region, the **cervical enlargement** (**C1**), and one in the lumbar region, the **lumbar enlargement** (**C2**). At the lower end, the spinal cord tapers into the **medullary cone** (**BC3**) and ends as a thin thread, the **terminal filament** (**C4**). The **anterior median fissure** at the ventral side and the **posterior median sulcus** (**BC5**) at the dorsal side mark the boundaries between the two symmetrical halves of the spinal cord. Nerve fibers enter dorsolaterally and emerge ventrolaterally at both sides of the spinal cord and unite to form the *dorsal roots*, **posterior roots**, and the *ventral roots*, **anterior roots**. The roots join to form short nerve trunks of 1 cm in length, the **spinal nerves**. Intercalated into the posterior roots are **spinal ganglia** (**B6**) containing *sensory nerve cells*. Only the posterior roots of the first cervical spinal nerves do not have a spinal ganglion, or only a rudimentary one.

In humans, there are *31 pairs of spinal nerves* which emerge through the *intervertebral foramina* from the *vertebral canal*. Each spinal nerve pair supplies one body segment. The spinal cord itself is unsegmented. The impression of segmentation is created by the bundling of nerve fibers emerging from the foramina (p. 66).

The spinal nerves are subdivided into **cervical nerves**, **thoracic nerves**, **lumbar nerves**, **sacral nerves**, and **coccygeal nerves** (**A**). There are

- *8 pairs of cervical nerves* (C1 – C8) (the first pair emerges between occipital bone and atlas)
- *12 pairs of thoracic nerves* (T1 – T12) (the first pair emerges between the first and second thoracic vertebrae
- *5 pairs of lumbar nerves* (L1 – L5) (the first pair emerges between the first and second lumbar vertebrae)
- *5 pairs of sacral nerves* (S1 – S5) (the first pair emerges through the upper sacral foramina)

- *one pair of coccygeal nerves* (emerging between the first and second coccygeal vertebrae)

Spinal cord and vertebral canal are initially of the same length so that each spinal nerve emerges from the foramen lying at its own level. During development, however, the vertebral column increases much more in length than does the spinal cord. As a result, the lower end of the spinal cord moves further up in relation to the surrounding vertebrae. In the newborn, the lower end of the spinal cord lies at the level of the third lumbar vertebra, and in the adult, at the level of the first lumbar or twelfth thoracic vertebra. Thus, the spinal nerves no longer emerge at their levels of origin; instead, their roots run down a certain distance within the vertebral canal to their foramen where they emerge. The more caudally the roots originate from the spinal cord, the longer their run within the vertebral canal. The levels where the spinal nerves emerge are therefore no longer identical with the corresponding levels of the spinal cord.

From the medullary cone (**BC3**) onward, the vertebral canal contains only a dense mass of descending spinal roots, known as the **cauda equina** (*tail of a horse*) (**B7**).

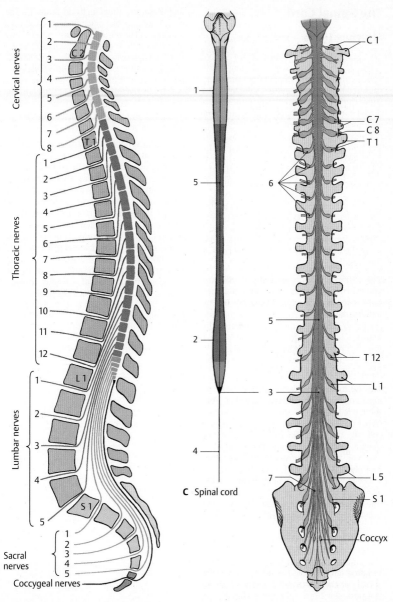

A Lateral view of the spinal nerves

C Spinal cord

B Dorsal view of the spinal ganglia

Spinal Cord *(left margin)*

The Spinal Cord

Structure (A, B)

The **gray matter**, **substantia grisea** (nerve cells), appears in transverse section of the spinal cord as a butterfly configuration surrounded by the *white matter*, **substantia alba** (fiber tracts). We distinguish on either side a *dorsal horn* (**posterior horn**) (**AB1**) and a *ventral horn* (**anterior horn**) (**AB2**). Both form columns in the longitudinal dimension of the spinal cord, the **anterior column** and the **posterior column**. Between them lies the *central intermediate substance* (**A3**) with the obliterated central canal (**A4**). In the thoracic spinal cord, the *lateral horn* (**AB5**) is interposed between the anterior and posterior horns. The **lateral posterior sulcus** (**A6**) is the site where the posterior root fibers (**AB7**) enter. The anterior root fibers (**AB8**) leave the anterior side of the spinal cord as fine bundles.

The posterior horn is *derived from the alar plate* (origin of sensory neurons) and contains neurons of the afferent system (**B**). The anterior horn is *derived from the basal plate* (origin of motor neurons) and contains the anterior horn cells, the efferent fibers of which run to the muscles. The lateral horn contains autonomic nerve cells of the *sympathetic nervous system* (p. 292).

The **white matter** is subdivided into the *dorsal column*, **or posterior funiculus** (**A9**), which reaches from the *posterior septum* (**A10**) to the *posterior horn*, the *lateral column*, or **lateral funiculus** (**A11**), which reaches from the posterior horn to the anterior root, and the *ventral column*, or **anterior funiculus** (**A12**), which reaches from the anterior root to the anterior fissure (**A13**). The latter two form the **anterolateral column**. The **white commissure** (**A14**) connects the two halves of the spinal cord.

Reflex Arcs (C – G)

The afferent fibers of the posterior root, which originate from the nerve cells of the spinal ganglion, transmit sensory signals to the posterior horn cells of the spinal cord, and these pass them on to the brain (**C**). The relay may also take place in the medulla oblongata. However, the afferent fibers may also run to the anterior horn cells and transmit the signal directly to these cells. The resulting muscle reaction is called *reflex*, the underlying neuronal circuit is called *reflex arc* (**D**). In general, the afferent fibers do not run directly to the motor neuron (*monosynaptic reflex arc*) but via interneurons that are interposed (*multisynaptic reflex arc*) (**E**).

The monosynaptic intrinsic reflex (*stretch reflex*) and the multisynaptic extrinsic reflex (*withdrawal reflex*) are of clinical importance. In the **stretch reflex** (**F**), the muscle is briefly stretched by a tap on its tendon. Stimulation of the muscle receptors (p. 314) results in a momentary contraction of the muscle as a counter reaction. The reflex involves only a few neurons at any level of the spinal cord. In the **withdrawal reflex** (**G**), skin receptors are stimulated (pain); the withdrawal movement is brought about by the coordinated action of several muscle groups. The signal spreads through several levels of the spinal cord and involves many interneurons.

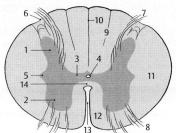

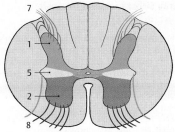

A Cross section of spinal cord

B Longitudinal zones of spinal cord

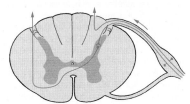

C Afferent fibers (ascending tracts)

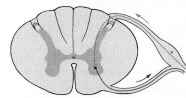

D Monosynaptic reflex arc

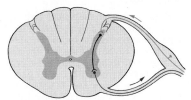

E Multisynaptic reflex arc

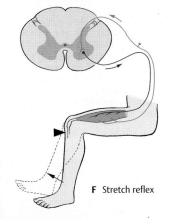

F Stretch reflex

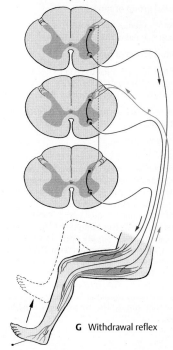

G Withdrawal reflex

Gray Substance and Intrinsic System (A–E)

The **posterior horn** is formed by the **nucleus proprius** (**A1**), the major portion of the posterior horn from which the **dorsal nucleus** (*Clarke's nucleus*) (**A2**) is set apart. The **gelatinous substance** (*Rolando's substance*) (**A3**) borders dorsally on the nucleus proprius. On it sits like a cap the end of the posterior horn, the marginal zone (**nucleus posteromarginalis**) (**A4**). The posterior horn is separated from the surface of the spinal cord through the **posterolateral tract** (*Lissauer's tract*) (**A5**). Between posterior horn and anterior horn lies the *intermediate gray matter* (**A6**) and lateral to it the **lateral horn** (**A7**). The border to the white matter between posterior horn and lateral horn is diffuse (*reticular formation*) (**A8**).

In the **anterior horn**, the motor neurons are arranged in groups of nuclei.

Medial group of nuclei

- Anteromedial nucleus (**A9**)
- Posteromedial nucleus (**A10**)

Lateral group of nuclei

- Anterolateral nucleus (**A11**)
- Posterolateral nucleus (**A12**)
- Retroposterolateral nucleus (**A13**)

Central group of nuclei in the cervical spinal cord

- Phrenic nucleus
- Accessory nucleus

For example, in the cervical spinal cord (**B**), the anterior horn is subdivided somatotopically so that the neurons of the medial group of nuclei supply neck and back muscles, intercostal, and abdominal muscles (**B14**). The neurons of the anterolateral nucleus supply the muscles of shoulder girdle and upper arm (**B15**), and the neurons of the posterolateral nucleus supply the muscles of lower arm and hand (**B16**). Finally, the retroposterolateral nucleus contains particularly large motor neurons that supply the small finger muscles (**B17**).

The neurons for the extensor muscles (**B18**) lie in the anterior field of the anterior horn, and those for the flexor muscles (**B20**) lie posterior to them. The somatotopic subdivisions do not occupy a single plane in the anterior horn but are spread over a certain height in such a way that the neurons for the shoulder girdle lie at a higher level, below them those for the upper arm, and still deeper those for the lower arm and hand. Diagram (**C**) illustrates the nerve supply to all body muscles.

To bring about an orderly movement during contraction of a muscle group, there must be simultaneous relaxation of the corresponding antagonists. This is achieved through inhibition of the corresponding anterior horn cells (**D**). For example, if an impulse is passed on by a neuron of the *extensor muscles* (**D18**), it is simultaneously transmitted by an axon collateral to inhibitory interneurons, the **Renshaw cells** (**D19**), which then inhibit the neurons of the *flexor muscles* (**D20**).

Intrinsic system of the spinal cord (E). Other interneurons mediate the spread of impulses over several levels, either on the same side or on the opposite side. Their ascending and descending fibers run in *basic bundles*, **fasciculi proprii** (**E21**), which border directly on the gray matter. In general, the ascending and descending fibers reach only one or two root levels. However, the fasciculi proprii also contain long fibers connecting the cervical spinal cord and the lumbar spinal cord (as shown in cats and monkeys). These fibers transmit excitatory and inhibitory impulses to anterior horn motor cells, a fact that is thought to be important for coordinated movement of the anterior and posterior extremities during locomotion. About half of the posterolateral tract (Lissauer's tract) (**E5**) consists of fibers of the intrinsic system.

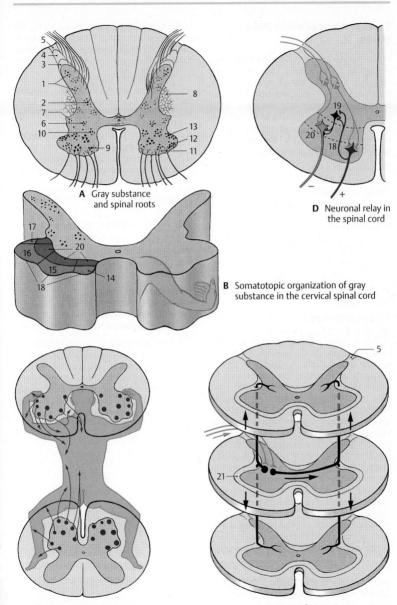

A Gray substance and spinal roots

D Neuronal relay in the spinal cord

B Somatotopic organization of gray substance in the cervical spinal cord

C Somatotopic organization of gray substance, overview (according to *Bossy*)

E Fasciculi proprii

Cross Sections of the Spinal Cord (A – D)

Cross sections at different levels (left, myelin stain; right, cellular stain) vary considerably. In the regions of cervical enlargement and lumbar enlargement, the cross-sectional area is larger than in the rest of the spinal cord; it is largest at the C4 – C5 and L4 – L5 levels. In both swellings, the numerous nerves that supply the extremities cause an increase in gray matter.

The **white matter** is most extensive in the cervical region and diminishes gradually in caudal direction; the ascending sensory tracts increase in number from the sacral to the cervical region as more fibers are added, while the descending motor tracts decrease from the cervical to the sacral regions as fibers terminate at various levels.

The butterfly configuration of the **gray matter** changes in shape at the various levels, and so does the **posterolateral tract** (*Lissauer's tract*) (**A – D1**).

The **posterior horn** is narrow in the cervical spinal cord; its tip ends in the cap-shaped marginal zone (**nucleus posteromarginalis**) (**A2**). The lateral angle between the posterior and anterior horn is occupied by the reticular formation (**AD3**). The gelatinous substance (*Rolando's substance*) (**A – D4**) contains small, mostly peptidergic neurons where posterior root fibers of various calibers terminate; it also contains descending fibers from the brain stem (raphe nuclei, p. 108, B28; reticular formation, p. 146). Unmyelinated processes of neurons ascend or descend for one to four root levels within the posterolateral tract (*Lissauer's tract*) and then reenter into the gelatinous substance. Some of the processes run within the lateral spinothalamic tract to the thalamus (p. 324). The fibers of proprioceptive sensibility in the muscles (muscle spindles) terminate in the **posterior thoracic nucleus** (*dorsal nucleus of Clarke*) (**AB5**) where the tracts to the cerebellum begin. The reduced gray matter of the thoracic spinal cord has a slender posterior horn with a prominent dorsal nucleus. In the plump posterior horn

of the lumbar and sacral spinal cords, the gelatinous substance (**CD4**) is much enlarged and borders dorsally on the narrow band of the marginal zone (**CD2**).

The **lateral horn** forms in the thoracic spinal cord the **lateral intermediate substance** (**B6**). It contains sympathetic nerve fibers mainly for the vasomotor system, the efferent fibers of which emerge via the anterior root. Sympathetic neurons also lie medially in the **intermediomedial nucleus** (**B7**). In the sacral spinal cord, parasympathetic neurons form the *intermediolateral nucleus* und *intermediomedial nucleus* (**D8**).

The **anterior horn** expands in the cervical spinal cord and contains several nuclei with large motor neurons, all of which are cholinergic.

Medial group of nuclei
- Anteromedial nucleus (**A9**)
- Posteromedial nucleus (**A10**)

Lateral group of nuclei
- Anterolateral nucleus (**A11**)
- Posterolateral nucleus (**A12**)
- Retroposterolateral nucleus (**A13**)

In the region supplying the upper limbs, the anterior horn is far more differentiated than in the thoracic spinal cord where only a few cell groups can be identified. The expanded, plump anterior horn of the lumbar and sacral spinal cords, which supplies the lower limbs, again contains several groups of nuclei.

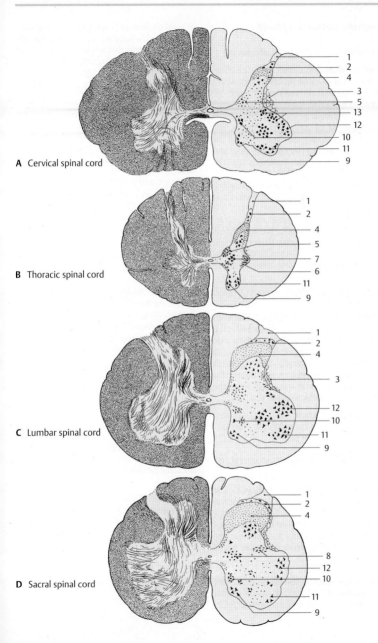

A Cervical spinal cord

B Thoracic spinal cord

C Lumbar spinal cord

D Sacral spinal cord

Spinal Cord

Spinal Cord

Ascending Pathways (A – D)

Tracts of the Anterolateral Funiculus (A)

Lateral spinothalamic tract (A1). The afferent, poorly myelinated posterior root fibers (**A2**) (first neuron of sensory pathway) bifurcate in the *posterolateral tract* (Lissauer's tract) and terminate at the cells of the *gelatinous substance* of the posterior horn. The fibers of the tract originate here, cross in the *white commissure* to the opposite side, and ascend in the lateral funiculus to the thalamus (second neuron). The pathway transmits *pain and temperature sensation, exteroceptive* and *proprioceptive impulses.* It is somatotopically subdivided; sacral (S) and lumbar (L) fibers are located dorsolaterally, while thoracic (T) and cervical (C) fibers are located ventromedially. Fibers for pain sensation probably lie superficially, while those for temperature sensation lie more deeply.

Anterior spinothalamic tract (A3). The afferent fibers (**A4**) (first neuron) bifurcate into ascending and descending branches and terminate at posterior horn cells, the fibers of which cross to the opposite side and ascend in the anterior funiculus to the thalamus (second neuron). They transmit *crude touch and pressure sensations.* Together with the lateral tract, they form the pathway of **protopathic sensibility** (p. 324).

The **spinotectal tract** (**A5**) carries pain fibers to the roof of the midbrain (contraction of pupils when in pain).

Pathways of the Posterior Funiculus (C, D)

Fasciculus gracilis (of Goll) (**C6**) and **fasciculus cuneatus** (of Burdach) (**C7**). The thick heavily myelinated fibers ascend without relay in the ipsilateral posterior funiculi. They belong to the first neuron of the sensory pathway and terminate at the nerve cells of the posterior funiculus nuclei (second neuron) (p. 140, B5, B6). They transmit exteroceptive and proprioceptive impulses of the **epicritic sensibility** (*exteroceptive,* information on localization and quality of tactile sensation; *proprioceptive,* infor-

mation on limb position and body posture). The posterior funiculi are somatotopically subdivided; the sacral fibers lie medially, followed laterally by the lumbar and thoracic fibers (*fasciculus gracilis*). The fibers from T3 to C2 lie laterally and form the *fasciculus cuneatus.*

Short ascending collaterals (**C8**) branch from the ascending fibers. They terminate at the posterior horn cells and form compact bundles, namely, the *comma tract of Schultze* (**D9**) in the cervical spinal cord, *Flechsig's oval field* (**D10**) in the thoracic spinal cord, and the *Phillippe – Gombault triangle* (**D11**) in the sacral spinal cord.

Cerebellar Pathways of the Lateral Funiculus (B)

Posterior spinocerebellar tract (Flechsig's tract) (B12). The afferent posterior horn fibers (first neuron) terminate at the cells of the *dorsal nucleus of Clarke* (**B13**) from where the tract (second neuron) originates. It runs along the margin of the ipsilateral lateral funiculus to the cerebellum and transmits mainly *proprioceptive impulses* (from joints, tendons, muscle spindles).

Anterior spinocerebellar tract (Gowers' tract) (B14). The cells of origin lie in the posterior horn. Their fibers (second neuron) ascend ipsilaterally as well as contralaterally along the anterolateral margin of the spinal cord to the cerebellum, to which they transmit *exteroceptive and proprioceptive impulses.* Both cerebellar pathways are somatotopically subdivided; the sacral fibers lie dorsally, the lumbar and thoracic fibers ventrally.

The **spino-olivary tract** (**B15**) and **vestibulospinal tract** (**B16**) arise from the posterior horn cells of the cervical spinal cord; they transmit mainly *proprioceptive impulses* to the inferior olive of the opposite side and to the vestibular nuclei.

A – C17 Neurons in the spinal ganglion (first neuron) (p. 71, A7).

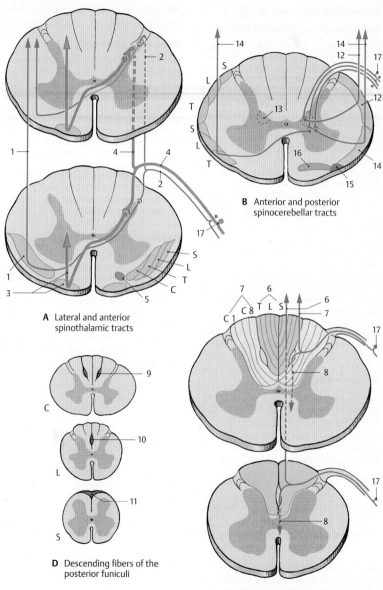

A Lateral and anterior spinothalamic tracts

B Anterior and posterior spinocerebellar tracts

C Fasciculus gracilis, fasciculus cuneatus

D Descending fibers of the posterior funiculi

Spinal Cord

Spinal Cord

Descending Pathways (A–C)

Corticospinal Tract, Pyramidal Tract (A)

The fibers of the pyramidal pathway originate mostly from the precentral gyrus and the cortex in front of it (areas 4 and 6) (p. 308, A1, A2). Furthermore, some of the fibers are thought to be derived from the cortical regions of the parietal lobe. Eighty percent of all fibers cross in the lower medulla oblongata to the collateral side (**pyramidal decussation**) (**A1**) and run as **lateral corticospinal tract** (**A2**) in the *lateral funiculus*. The remaining fibers run uncrossed as **anterior corticospinal tract** (**A3**) in the *anterior funiculus* and cross only at the level of their termination. More than half of the pyramidal tract fibers terminate in the cervical spinal cord to supply the upper limb, and one-fourth terminate in the lumbar spinal cord to supply the lower limb. The lateral funiculus is subdivided somatotopically, with the fibers for the lower limb lying at the periphery and those for the trunk and upper limb lying further inside. Most of the fibers terminate on interneurons that transmit impulses for the voluntary motor system to the anterior horn cells. However, the fibers not only conduct excitation to the anterior horn cells but also mediate cortical inhibition via interneurons (p. 308, p. 316).

Extrapyramidal Tracts (B)

The extrapyramidal tracts comprise descending systems from the brain stem that influence the motor system (p. 310):

- **Vestibulospinal tract** (**B4**) (*balance, muscle tone*)
- **Anterior and lateral reticulospinal tract** (**B5**) from the pons
- **Lateral reticulospinal tract** (**B6**) from the medulla oblongata
- **Tegmentospinal** tract (**B7**) from the midbrain

The **rubrospinal tract** (**B8**) (in humans largely replaced by the tegmentospinal tract) and the **tectospinal tract** (**B9**) terminate in the cervical spinal cord and influence only the *differentiated motor system of the head and upper limb*. The **medial longitudinal fascicle**

(**B10**) contains various fiber systems of the brain stem (p. 142).

Autonomic Pathways (C)

The autonomic pathways consist of poorly myelinated or unmyelinated fibers and rarely form compact bundles. The **parependymal tract** (**C11**) runs along both sides of the central canal. Its ascending and descending fibers can be traced back up to the diencephalon (hypothalamus) and are thought to transmit impulses for *genital function, micturition,* and *defecation*. Anterior to the lateral pyramidal tract runs the descending pathway for *vasoconstriction* and *sweating* (*Foerster*) (**C12**) with a somatotopic subdivision corresponding to that of the lateral pyramidal tract.

Visualization of Pathways (D–E)

The various pathway systems cannot be identified on normal transverse sections of the spinal cord. Only under special circumstances (in experimental transection, in spinal cord injury, or during *development* when tracts become myelinated at different times), can they be distinguished from each sother, such as the late myelinating pyramidal tract (**D2**). In case *of injuries*, distal fibers separated from the perikaryon degenerate so that their area in the spinal cord becomes visible, such as the fasciculus gracilis (**E13**).

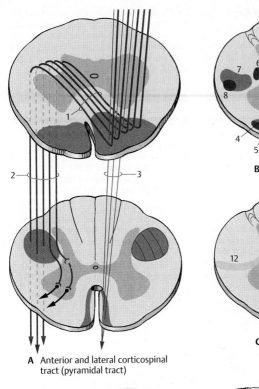

A Anterior and lateral corticospinal tract (pyramidal tract)

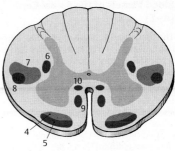

B Descending pathways

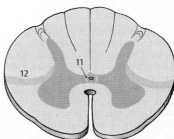

C Autonomic pathways

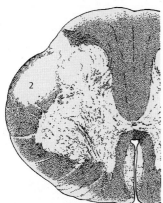

D Unmyelinated pyramidal tract in the newborn

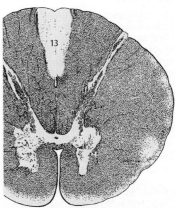

E Degeneration of the fasciculus gracilis after injury to the spinal cord

Blood Vessels of the Spinal Cord (A–E)

The spinal cord is supplied with blood from two sources, the *vertebral arteries* and the *segmental arteries* (*intercostal arteries* and *lumbar arteries*).

Vertebral arteries (A1). Before they unite, they give off two thin *posterior spinal arteries* that form a network of small arteries along the posterior surface of the spinal cord. At the level of the pyramidal decussation, two additional branches of the vertebral arteries join to form the *anterior spinal artery* (**AD2**) which runs along the anterior surface of the spinal cord at the entrance to the anterior sulcus.

Segmental arteries (C3). Their posterior branches (**C4**) and the vertebral arteries give off **spinal branches** (**C5**) which enter through the intervertebral foramina and divide at the spinal roots into dorsal and ventral branches to supply the spinal roots and the spinal meninges. Of the 31 spinal arteries, only 8 to 10 extend to the spinal cord and contribute to its blood supply. The levels at which the radicular arteries approach the spinal cord vary, and so do the sizes of the vessels. The largest vessel approaches the spinal cord at the level of the lumbar enlargement between T12 and L3 (**large radicular artery**) (**A6**).

The **anterior spinal artery** is widest at the level of the cervical and lumbar enlargements. Its diameter is much reduced in the mid-thoracic region of the spinal cord. As this region is also the border area between two supplying radicular arteries, this segment of the spinal cord is especially at risk in case of circulatory problems (**A**, arrow). Depending on the variation of the radicular arteries, this may also apply to other segments of the spinal cord.

The anterior spinal artery gives off numerous small arteries into the anterior sulcus, the **sulcocommissural arteries** (**D7**). In the cervical and thoracic spinal cords, they turn alternately to the left and right halves of the spinal cord; in the lumbar and sacral spinal cords, they divide into two branches. In ad-

dition, anastomoses arise between the anterior and posterior spinal arteries, so that the spinal cord is surrounded by a vascular ring (**vasocorona**) (**D8**) from where vessels radiate into the white matter. Injection of tracers revealed that the gray matter is much more vascularized than the white matter (**D**).

Areas of blood supply (E). The anterior spinal artery supplies the anterior horns, the bases of the posterior horns, and the largest part of the anterior lateral funiculi (**E9**). The posterior funiculi and the remaining parts of the posterior horns are supplied by the **posterior spinal arteries** (**E10**). The marginal zone of the anterior lateral funiculus is supplied by the plexus of the vasocorona (**E11**).

The **spinal veins** (**B**) form a network in which one *anterior spinal vein* and two *posterior spinal veins* stand out. The efferent veins run along the spinal roots and open into the *epidural venous plexus* (see vol. 2). The spinal veins lack valves prior to their penetration through the dura.

C12 Aorta.

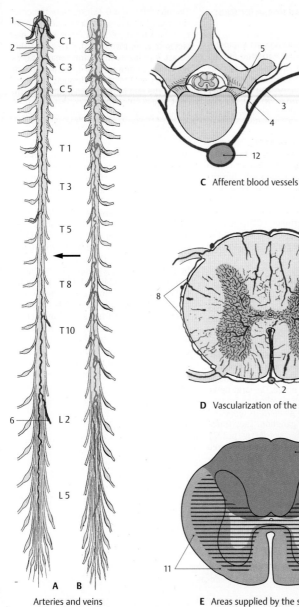

A B Arteries and veins of the spinal cord

C Afferent blood vessels

D Vascularization of the spinal cord

E Areas supplied by the spinal cord arteries (according to *Gillilan*)

Spinal Ganglion and Posterior Root (A–H)

The posterior spinal root contains a spindle-shaped bulge, the **spinal ganglion** (**A**), an accumulation of cell bodies of sensory neurons; their bifurcated processes send one branch to the periphery and the other branch to the spinal cord (p. 70, A7). They lie as cell clusters or as cell rows between the bundles of nerve fibers.

Development of the ganglia (C). The cells originate from the lateral zone of the **neural plate** (**C1**); however, they do not participate in the formation of the neural tube but remain at both sides as the **neural crest** (**C2**). Hence, the spinal ganglia can be regarded as gray matter of the spinal cord that became translocated to the periphery. Other derivatives of the neural crest are the cells of the autonomic ganglia, the paraganglia, and the adrenal medulla.

From the capsule (**A3**) of the spinal ganglion, which merges into the perineurium of the spinal nerve, connective tissue extends to the interior and forms a sheath around each neuron (*endoganglionic connective tissue*) (**B4**). The innermost sheath, however, is formed by ectodermal *satellite cells* (**BE5**) and is surrounded by a basal membrane comparable to that around the Schwann cells of the peripheral nerve. The large nerve cells (**B6, E**) with their myelinated process conglomerated into a *glomerulus* represent only one-third of the ganglion. They transmit impulses of *epicritic sensibility* (p. 322). The remainder consists of medium-sized and small ganglion cells with poorly myelinated or unmyelinated nerve fibers which are thought to conduct pain signals and sensations from the intestine. There are also some multipolar nerve cells.

Development of the ganglion cells (D). The spinal ganglion cells are initially *bipolar cells*. During development, however, the two processes fuse to form a single trunk which then bifurcates in a T-shaped manner. The cells are therefore called *pseudounipolar nerve cells*.

The posterior root is thicker than the anterior root. It contains fibers of various calibers, two-thirds of them being poorly myelinated or unmyelinated fibers. The thin poorly myelinated and unmyelinated fibers, which transmit impulses of the *protopathic sensibility* (p. 324), enter through the lateral part of the root into the spinal cord (**F7**). The thick myelinated fibers transmit impulses of the *epicritic sensibility* and enter through the median part of the root into the spinal cord (**F8**).

At the entrance into the spinal cord, there is a narrow zone where the myelin sheaths are very thin so that the fibers appear unmyelinated. This zone is regarded as the *boundary between the central and the peripheral nervous systems* (**Redlich–Obersteiner zone**) (**G**). In the electron-microscopic image (**H**), however, this boundary does not exactly coincide with the Redlich–Obersteiner zone. For each axon, the boundary is marked by the last node of Ranvier prior to the entrance into the spinal cord. Up to this point, the peripheral myelin sheath is surrounded by a *basal membrane* (blue in **H**). The next internode no longer has a basal membrane. For unmyelinated fibers, the boundary is also marked by the basal membrane of the enveloping Schwann cell. Thus, the basal membranes around the spinal cord form a boundary that is only penetrated by the axons.

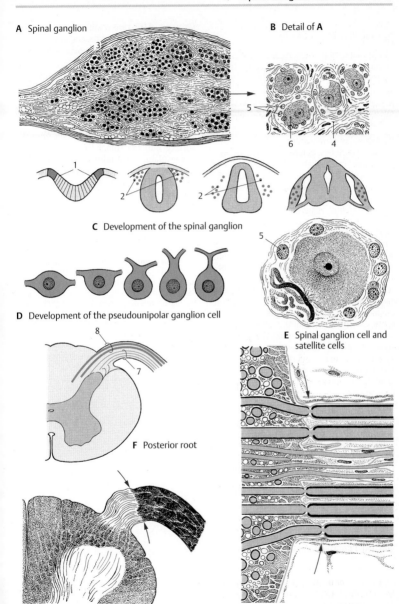

A Spinal ganglion

B Detail of **A**

C Development of the spinal ganglion

D Development of the pseudounipolar ganglion cell

E Spinal ganglion cell and satellite cells

F Posterior root

G Redlich–Obersteiner zone

H Posterior root, electron-microscopic diagram (according to *Andres*)

Spinal Meninges (A – E)

The spinal cord in the vertebral canal is surrounded by the following connective tissue membranes: the *tough spinal meninx* (*pachymeninx*), or **spinal dura mater** (**A1**), and the *soft spinal meninx* (*leptomeninx*) consisting of the **spinal arachnoidea** (**A2**) and the **spinal pia mater** (**A3**).

The spinal dura mater forms the outermost sheath which is separated from the periosteum-like lining of the vertebral canal, the **endorhachis** (**A4**), by the **epidural space** (**A5**). The space is filled with adipose tissue and contains an extensive venous plexus, the *internal vertebral venous plexus* (see vol. 2). The dura mater forms caudally the *dural sac* (**B6**), enveloping the *cauda equina* (**B7**), and finally extends together with the terminal filament as a thin cord up to the periosteum of the coccyx (*dural terminal filament*) (**B8**). Only at the oral end at the *foramen magnum* (occipital bone) is the dural sac attached to bone. The epidural space forms a resilient cushion for the dural sac, which moves together with the vertebral column and the head. Bending the head pulls the dural sac upward, causing mechanical stress on the spinal cord; when bending the head forward, roots and blood vessels are stretched (**D9**), when bending the head backward, they are compressed (**D10**).

The **arachnoidea** borders closely onto the inner surface of the dura mater. It forms the boundary of the **subarachnoidal space** (**AC11**), which is filled with *cerebrospinal fluid* (CSF). Between the inner surface of the dura and the arachnoidea lies a capillary cleft, the *subdural space*, which widens into a real space only under pathological conditions (subdural bleeding). Dura and arachnoidea accompany the spinal roots (**AC12**), pass with them through the intervertebral foramina, and also envelope the spinal ganglia (**AC13**). The funnel-like root sleeves contain CSF in their proximal portions. The dura then turns into the *epineurium* (**A14**), and the arachnoidea into the *perineurium* (**A15**) of the spinal nerves. The part of the root leaving the vertebral canal, the *radicular nerve* (**A16**), runs obliquely downward in the cervical and lumbosacral regions and obliquely upward in the midthoracic region (**C**).

The **spinal pia mater** borders directly onto the marginal glial layer of the spinal cord. This represents the boundary between mesodermal envelopes and ectodermal nerve tissue. The pia mater contains numerous small blood vessels that penetrate from the surface into the spinal cord. A connective tissue plate, the **denticulate ligament** (**A17**), extends on both sides of the spinal cord from the pia to the dura and is attached to the latter by individual pointed processes. The ligament extends from the cervical spinal cord to the midlumbar spinal cord, thus keeping the spinal cord, which floats in the CSF, in position.

▥ Clinical Note: Under sterile conditions, cerebrospinal fluid may be safely withdrawn for examination from the lower segment of the dural sac that contains only the fibers of the cauda equina. For this purpose, with the patient bending over, a needle is deeply inserted between the processes of the second to fifth lumbar vertebrae until CSF begins to drop (*lumbar puncture*) (**E**).

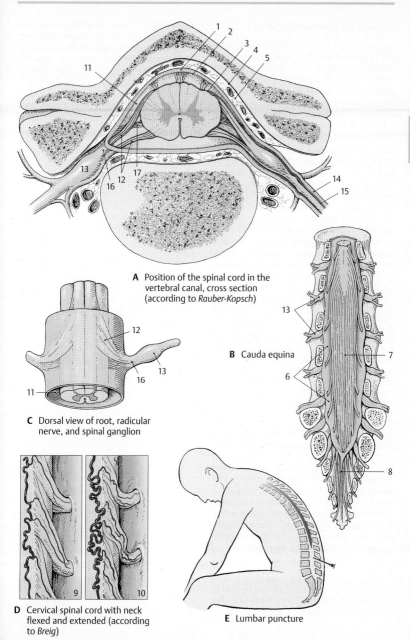

Spinal Cord

A Position of the spinal cord in the vertebral canal, cross section (according to *Rauber-Kopsch*)

B Cauda equina

C Dorsal view of root, radicular nerve, and spinal ganglion

D Cervical spinal cord with neck flexed and extended (according to *Breig*)

E Lumbar puncture

Segmental Innervation (A–C)

The vertebrate body, with the exception of the head, is originally subdivided into *segments* or *metameres*. The vertebrae, ribs, and intercostal muscles can be regarded as remnants of such a segmentation in humans. Metamerism concerns only tissues of the mesoderm (**myotomes, sclerotomes**) but not derivatives of the ectoderm. Thus, there are no spinal cord segments, only the levels at which the individual spinal roots enter and emerge. However, the spinal fibers join to form the spinal nerves as they emerge through the metameric intervertebral foramina, thus creating an *apparent secondary segmentation*. The sensory fibers of the spinal nerves supply stripe-shaped zones of the skin, called **dermatomes** in analogy to myotomes and sclerotomes. This, too, is a secondary segmentation and reflects the innervation of each dermatome by a single posterior root (segmental innervation).

Clinical Note: The dermatomes play an important role in the *diagnosis and localization of spinal cord injuries.* Loss of sensibility in certain dermatomes indicates a specific level of injury in the spinal cord. Simplified reference points are the line through the nipples, regarded as the boundary between T4 and T5, and the groin, regarded as the boundary between L1 and L2. The first cervical spinal nerve has no sensory representation on the body surface, for the spinal ganglion of its posterior root is absent or rudimentary.

There are slightly different segmental boundaries for various modalities, such as touch and pain, and for sweating and piloerection. The diagram (**A**) was designed according to the *decrease in sensibility* (**hypoesthesia**) resulting from disk prolapse; it shows how the dermatomes extending around the trunk become elongated in the limbs. They may even lose their continuity with the midline (C7, L5). They become translocated to the distal limb areas during embryonic development when the limbs are budding (**C**).

The dermatomes overlap like roof tiles, as illustrated by the shift in boundaries that have been determined according to the expanded areas in case of *posterior root pain* (hypersensitivity to pain, **hyperalgesia**) (**B**). The loss of a single posterior root cannot be demonstrated for touch sensation, since the corresponding dermatome is also supplied by the neighboring posterior roots. The dermatomes for pain and temperature sensation are narrower, and the loss of a posterior root can still be demonstrated when these modalities are tested.

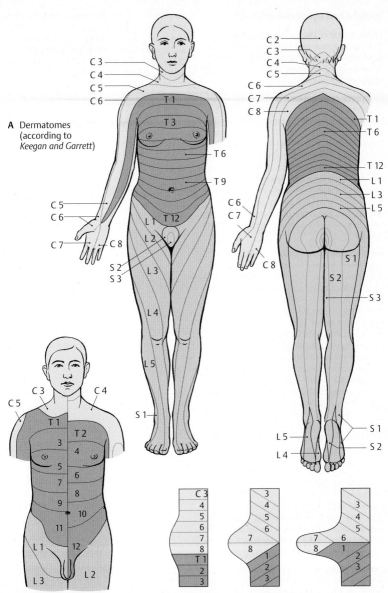

A Dermatomes (according to *Keegan and Garrett*)

B Overlap of dermatomes (according to *Förster*)

C Development of dermatomes in the upper limb (according to *Bolk*)

Spinal Cord

Spinal Cord Syndromes (A – C)

The anatomy of the spinal cord causes very specific patterns of functional deficiencies after injury; depending on the site of lesion, different pathways and therefore different functions are lost.

Complete transection (**A**) cuts off all descending motor pathways, causing *complete paralysis* below the injured level. At the same time, it interrupts all ascending pathways, causing a complete *loss of all sensations*. If the lesion is above the sacral spinal cord, it results in the loss of voluntary control over urination and defecation. If the lesion lies above the lumbar enlargement, both lower limbs are paralyzed (**paraplegia**), and if it lies above the cervical enlargement, both upper limbs are also paralyzed (**tetraplegia**).

Hemisection of the spinal cord (**B**) results in the **Brown – Séquard's syndrome**. For example, hemisection on the left interrupts the lateral and anterior corticospinal tracts (**B1**) and results in left-sided *paralysis*. Transection of the vasomotor pathway causes ipsilateral vasomotor paralysis. Transection of the posterior funiculi (**B2**) and the cerebellar lateral funiculi (**B3**) leads to severely *impaired deep sensibility* (posture sensation). On the same side as the lesion, there is also **hyperesthesia** (touch is perceived as pain). This is thought to be caused by a loss of epicritic sensibility (posterior funiculi) with retention of the protopathic sensibility (crossing pathways of the anterior corticospinal tract ascend contralaterally) (**B4**). Finally, there is *dissociated anesthesia* on the intact right side from the lesion downward; while touch sensation is hardly impaired, pain and temperature sensations are lost (ipsilateral interruption of the crossing pathway of the anterior corticospinal tract) (**B5**). The anesthetic zone (**B6**) above the transection on the affected side is attributed to destruction of the posterior root entrance zone at the level of the spinal cord lesion.

Central injury (**C**) to the gray substance of the spinal cord also causes *dissociated anesthesia* at the corresponding levels. The epicritic sensibility transmitted via the ipsilateral posterior funiculi (**C2**) is retained. However, pain and temperature sensations are lost (**analgesia** and **thermoanesthesia**), because their fibers, which cross through the white commissure, are interrupted (**C5**).

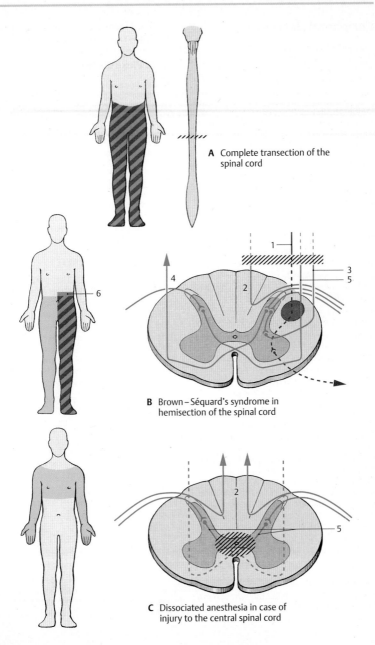

A Complete transection of the spinal cord

B Brown–Séquard's syndrome in hemisection of the spinal cord

C Dissociated anesthesia in case of injury to the central spinal cord

Spinal Cord

Peripheral Nerves

The **peripheral nerves** may contain four different types of fibers:

- **Somatomotor** (efferent) fibers (**A1**) for striated muscles
- **Somatosensory** (afferent) fibers (**A2**) for skin sensibility
- **Visceromotor** fibers (**A3**) for smooth muscles
- **Viscerosensory** fibers (**A4**) for inner organs

The spinal nerves usually contain several types of fibers; they are *mixed nerves*.

The different fibers have the following pathways. The *somatomotor fibers* pass from the anterior horn cells (**A5**) through the anterior root (**A6**); the *somatosensory* and *viscerosensory fibers* originate from the nerve cells of the spinal ganglia (**A7**); and the *visceromotor fibers* of the lateral horn cells (**A8**) pass mostly through the anterior root. Anterior and posterior roots (**A9**) join to form the spinal nerve (**A10**), which contains all types of fibers. This short nerve trunk then divides into four branches:

- The *meningeal branch* (**A11**), a recurrent sensory branch extending to the spinal meninges
- The *posterior branch* (**A12**)
- The *anterior branch* (**A13**)
- The *communicating branch* (**A14**)

The **posterior branch** supplies motor fibers to the deep (autochthonous) muscles of the back and sensory fibers to the skin areas on both sides of the vertebral column (p. 84). The **anterior branch** supplies motor fibers to the muscles of the anterior and lateral walls of the trunk and to the muscles of the limbs; it also supplies sensory fibers to the corresponding skin areas. The **communicating branch** connects with the sympathetic chain ganglion (**A15**) (autonomic nervous system, p. 292). It usually forms two independent communicating branches, the *white communicating branch* (**A16**) (myelinated) and the *gray communicating branch* (**A17**) (unmyelinated). The visceromotor fibers pass via the white branch to the sympathetic chain ganglion, where they are relayed to neurons, the axons of which partly reenter

the spinal nerve as *postganglionic fibers* (p. 297 **A5**) via the gray branch.

Nerve Plexusus (B)

At the level of the limbs, the *anterior branches* of the spinal nerves form networks (**plexusus**) in which fibers are exchanged. The resulting nerve trunks, which then extend to the periphery, possess a newly organized supply of fibers derived from different spinal nerves.

Cervical plexus (p. 72). The *plexus of the neck* is formed by the anterior branch of the first four spinal nerves. The following nerves originate here: the lesser occipital nerve (**B18**), the greater auricular nerve (**B19**), the transverse nerve of the neck (**B20**), the supraclavicular nerves (**B21**), the phrenic nerve (**B22**), and also the roots of the deep cervical ansa (**B23**).

Brachial plexus (p. 74). The *plexus of the arm* is formed by the anterior branches of spinal nerves C5 to C8 and by a part of the T1 nerve. We distinguish between a section lying above the clavicle, the **supraclavicular part,** and a section lying below the clavicle, the **infraclavicular part.** The anterior branches pass through the scalene gap into the posterior cervical triangle, where they form *three primary trunks* above the clavicle:

- The **superior trunk** (**B24**) (C5, C6)
- The **medial trunk** (**B25**) (C7)
- The **inferior trunk** (**B26**) (C8, T1)

The nerves that originate here form the supraclavicular part (p. 74). Below the clavicle, *three secondary cords* form; they are named according to their position relative to the axillary artery (**B27**):

- The **lateral cord** (**B28**) (p. 74) (from the anterior branches of the superior and medial trunks)
- The **medial cord** (**B29**) (p. 78) (from the anterior branch of the inferior trunk)
- The **posterior cord** (**B30**) (p. 80) (from the dorsal branches of the three trunks)

The lateral cord gives rise to the *musculocutaneous nerve* (**B31**). The remaining fibers together with fibers of the medial cord form

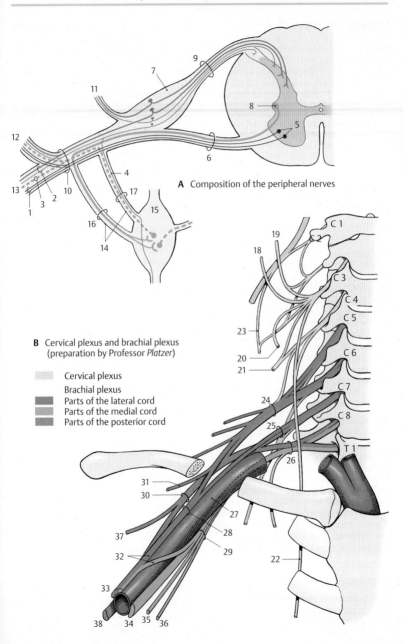

A Composition of the peripheral nerves

B Cervical plexus and brachial plexus
(preparation by Professor *Platzer*)

Cervical plexus
Brachial plexus
Parts of the lateral cord
Parts of the medial cord
Parts of the posterior cord

Spinal Cord

Spinal Cord

the **median loop** (**B32**) (p. 76, AC1) and unite to form the *median nerve* (**B33**). The medial cord gives rise to the *ulnar nerve* (**B34**), the *medial cutaneous nerve of the forearm* (**B35**), and the *medial cutaneous nerve of the arm* (**B36**). The posterior cord gives off the *axillary nerve* (**B37**) and continues as the *radial nerve* (**B38**).

Cervical Plexus (C1 – C4) **(A – D)**

Innervation of the muscles (A). Short nerves run from the **anterior branches** directly to the deep neck muscles, namely, the anterior (**A1**) and lateral (**A2**) rectus capitis muscles, the long muscle of the head, and the long muscle of the neck (**A3**). From the anterior branch of C4, nerves run to the upper part of the anterior scalene muscle (**A4**) and to the medial scalene muscle (**A5**).

The anterior branches of C1 – C3 form the deep cervical ansa (**C6**): fibers from C1 and C2 temporarily appose the hypoglossal nerve (**AC7**) and then leave it as the **superior root** (anterior) (**AC8**); the fibers for the thyrohyoid muscle (**A9**) and the geniohyoid muscle then continue with the hypoglossal nerve. The superior root combines with the **inferior root** (posterior) (**AC10**) (C2, C3) to form the cervical ansa, from where branches run to supply the infrahyoid muscles, namely, the omohyoid muscle (**A11**), the sternothyroid muscle (**A12**), and sternohyoid muscle (**A13**).

Innervation of the skin (B, C). The sensory nerves of the plexus pass behind the sternocleidomastoid muscle through the fascia, where they form the **punctum nervosum** (**B14**). From here they spread over head, neck, and shoulder; the **lesser occipital nerve** (**BC15**) (C2, C3) extends to the occiput, the **greater auricular nerve** (**BC16**) (C3) into the area surrounding the ear (auricula, mastoid process, region of the mandibular angle). The **transverse nerve of the neck** (**BC17**) (C3) supplies the upper neck region up to the chin, while the **supraclavicular nerves** (**BC18**) (C3, C4) supply the subclavicular fossa and the shoulder region.

Area innervated by the phrenic nerve (C, D). The **phrenic nerve** (**CD19**) (C3, C4) contains fibers of the fourth, and often also of the third, spinal nerve. It crosses the anterior scalene muscle and enters into the superior thoracic aperture in front of the subclavian artery. It extends through the mediastinum to the diaphragm and, on its way, gives off fine branches for sensory supply to the pericardium, the **pericardiac branches** (**D20**). At the surface of the diaphragm, it branches and supplies all muscles of the diaphragm (**D21**). Fine branches provide the sensory fibers for the membranes bordering on the diaphragm, that is, cranially the pleura and caudally the peritoneum of the diaphragm and the peritoneal covering of the upper intestinal organs.

▬ **Clinical Note:** Injury to the cervical spinal cord or its roots at the C3 – C5 levels results in paralysis of the diaphragm and in reduced respiration. In case of paralysis of the thoracic muscles, on the other hand, respiration can still be maintained by the cervical spinal cord via the phrenic nerve.

Posterior Branches (C1 – C8)

The *dorsal branches of the cervical nerves*, or **posterior branches**, supply motor fibers to neck muscles belonging to the autochthonous muscles of the back and sensory fibers to the skin of the neck.

The posterior branch of the first cervical nerve is exclusively motor and runs as *suboccipital nerve* to the small muscles in the region of occiput, atlas, and axis.

The *greater occipital nerve* runs from the second cervical nerve to the occiput and supplies its skin up to the vertex (p. 84 **D4**).

The posterior branch of the third cervical spinal nerve, the *third occipital nerve*, supplies sensory fibers to the neck region.

The remaining posterior branches of the cervical spinal nerves supply sensory fibers to the skin area bordering caudally and motor fibers to the autochthonous back muscles of this region.

Innervation of the skin (B). Autonomic zone (dark blue) and maximum zone (light blue).

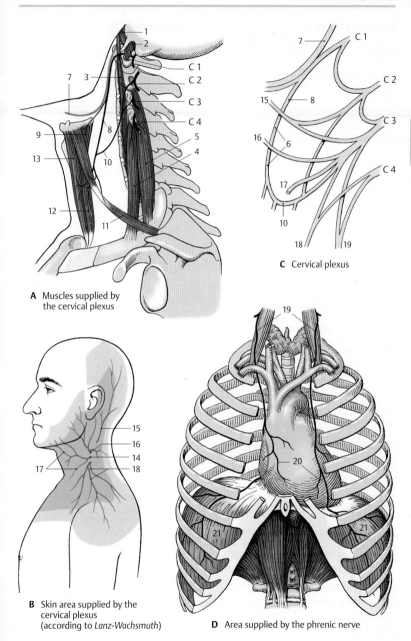

A Muscles supplied by
the cervical plexus

C Cervical plexus

B Skin area supplied by the
cervical plexus
(according to *Lanz-Wachsmuth*)

D Area supplied by the phrenic nerve

Brachial Plexus (C5 – T1)

Peripheral sensory innervation. The innervation of the skin by peripheral nerves originating from the plexus differs from the segmental innervation (p. 66). The regions supplied by individual nerves overlap at their margins. The region innervated by a single nerve is called the *autonomic zone* (dark blue), and the total area supplied by the nerve including the area cosupplied by adjacent nerves is called the *maximum zone* (light blue).

▓▓ **Clinical Note:** Interruption of a nerve causes complete insensibility (*anesthesia*) in the autonomic zone but only a decreased sensibility (*hypoesthesia*) in the maximum zones.

Supraclavicular Part (A – C)

The supraclavicular part gives rise to motor nerves that innervate the *muscles of the shoulder girdle*.

The following nerves run to the *posterior and lateral surfaces of the thorax*: the **dorsal scapular nerve** (**A1**) (C5) to the scapular muscle (**C2**) and to the lesser (**C3**) and greater (**C4**) rhomboid muscles; the **long thoracic nerve** (**A5**) (C5 – C7), the branches of which terminate at the lateral thoracic wall in the peaks of the anterior serratus muscle (**B6**); and the **thoracodorsal nerve** (**A7**) (C7, C8), which supplies the latissimus dorsi muscle (**C8**). The muscles of the shoulder blade are innervated at the posterior surface of the shoulder blade (supraspinous muscle [**C9**] and infraspinous muscle [**C10**]) by the **suprascapular nerve** (**A11**) (C5, C6), and at the anterior surface by the **subscapular nerve** (**A12**) (C5 – C7), which extend to the subscapular muscle and the greater teres muscle (**C13**).

The following nerves reach the *anterior surface of the thorax*: the **subclavius nerve** (**A14**) (C4 – C6) (to the subclavius muscle [**B15**]), the **lateral pectoral nerve** (**A16**) (C5 – C7) and the **medial pectoral nerve** (**A17**) (C7 – T1), which supply the greater (**B18**) and lesser (**B19**) pectoral muscles.

▓▓ **Clinical Note:** Injury to the supraclavicular part leads to paralysis of the muscles of the shoulder girdle and makes it impossible to raise the arm. This type of *upper brachial plexus paralysis* (*Erb's palsy*) may be caused by dislocation of the shoulder joint during birth, or through improper positioning of the arm during anesthesia. Injury to the infraclavicular part of the brachial plexus results in *lower brachial plexus paralysis* (*Klumpke 's palsy*), which predominantly involves the small muscles of the hand and possibly also the flexor muscles of the forearm.

Infraclavicular Part (D – F)

Three main trunks of the anterior branches, the *superior, middle,* and *inferior trunks of the brachial plexus,* give rise to three cords, the *lateral, middle,* and *posterior fascicles;* they are named according to their position relative to the axillary artery.

Lateral Fascicle

The lateral fascicle gives rise to the *musculocutaneous nerve* and the *median nerve.*

Musculocutaneous nerve (C5 – C7) (**D – F**). The nerve passes through the coracobrachial muscle and runs between the biceps muscle and the brachial muscle down to the elbow. It gives off branches (**E20**) to the flexor muscles of the upper arm, namely, to the coracobrachial muscle (**D21**), to the short head (**D22**) and long head (**D23**) of the biceps muscle of the arm, and to the brachial muscle (**D24**).

The sensory fibers of the nerves come to the surface through the fascia at the elbow and supply the skin in the lateral region of the forearm as **lateral cutaneous nerve of the forearm** (**D – F25**). Injury to this nerve causes loss of sensibility in a small zone of the elbow; diminished sensibility extends to the middle of the forearm.

Innervation of the skin (F). Autonomic zone (dark blue) and maximum zone (light blue).

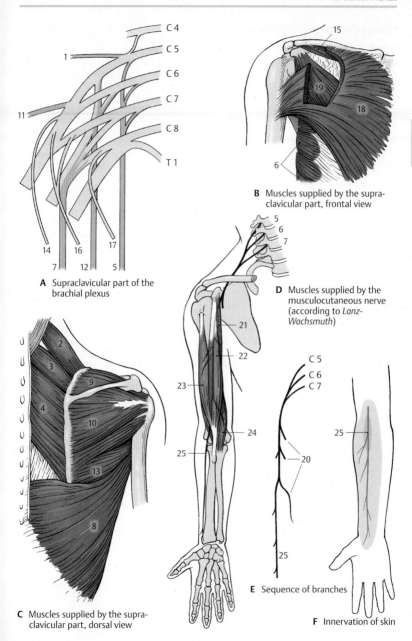

A Supraclavicular part of the brachial plexus

B Muscles supplied by the supraclavicular part, frontal view

C Muscles supplied by the supraclavicular part, dorsal view

D Muscles supplied by the musculocutaneous nerve (according to *Lanz-Wachsmuth*)

E Sequence of branches

F Innervation of skin

Infraclavicular Part (continued)

Lateral Fascicle (continued) (A–D)

Median nerve (C6–T1). Parts of the lateral and medial fascicles form the **median loop** (**AC1**) at the anterior surface of the axillary artery and join to form the *median nerve*.

The nerve extends in the medial bicipital sulcus along the surface of the brachial artery to the elbow, where it passes between the two heads of the round pronator muscle to the forearm. It runs between the superficial flexor muscle of the fingers and the deep flexor muscle of the fingers to the wrist. Prior to its passage through the *carpal tunnel*, it lies superficially between the tendons of the radial flexor muscle of the wrist and the long palmar muscle. In the carpal tunnel, it ramifies into its terminal branches.

The **muscular branches** (**C2**) of the nerve supply the *pronator muscles* and most of the *flexor muscles of the forearm*, namely, the round pronator muscle (**A3**), the radial flexor muscle of the wrist (**A4**), the long palmar muscle (**A5**), and the superficial flexor muscle of the fingers with radial head (**A6**) and humeroulnar head (**A7**). In the elbow, the **anterior interosseous nerve of the forearm** (**AC8**) branches off and runs along the interosseous membrane to the quadrate pronator muscle (**A9**). It gives off branches to the long flexor muscle of the thumb (**A10**) and to the radial part of the deep flexor muscle of the fingers.

In the lower third of the forearm, the sensory **palmar branch of the median nerve** (**A–C11**) branches off to the skin of the ball of the thumb (thenar eminence), to the radial side of the wrist, and to the palm.

After passing through the carpal tunnel, the median nerve divides into three branches: the **common palmar digital nerves I–III** (**A–C12**), each of which bifurcates at the level of the metacarpophalangeal joints into two **proper palmar digital nerves** (**A–C13**). From the *first common palmar digital nerve*, a branch extends to the thenar eminence (short abductor muscle of thumb [**A14**], su-

perficial head of the short flexor muscle of thumb [**A15**], and opposing muscle of thumb [**A16**]). The *common palmar digital nerves* supply the lumbrical muscles I–III (**A17**). They run to the interdigits and bifurcate in such a way that each pair of *proper palmar digital nerves* provides sensory fibers to the lateral aspects of an interdigit. Thus, the first pair of nerves supplies the ulnar side of the thumb and the radial side of the index finger, the second pair supplies the ulnar side of the index finger and the radial side of the middle finger, and the third pair supplies the ulnar side of the middle finger and the radial side of the ring finger. The area innervated by the proper palmar digital nerves on the posterior side includes the end phalanx of the thumb as well as the distal and middle phalanges of the fingers (**B**).

The median nerve gives off branches to the periosteum, elbow joint, radiocarpal joint, and mediocarpal joint. At the level of the wrist, there is normally an anastomosis with the ulnar nerve.

▬▬ **Clinical Note:** After injury to the nerve, pronation of the forearm is no longer possible and flexion is severely restricted. As to the hand, the thumb, index finger and middle finger can no longer be flexed at the end and middle phalanges, resulting in a characteristic feature of median paralysis, the so-called **hand of oath** (**D**). On passing the carpal tunnel, the nerve can be injured by pressure in older persons (*carpal tunnel syndrome*).

Innervation of the skin (B). Autonomic zone (dark blue) and maximum zone (light blue).

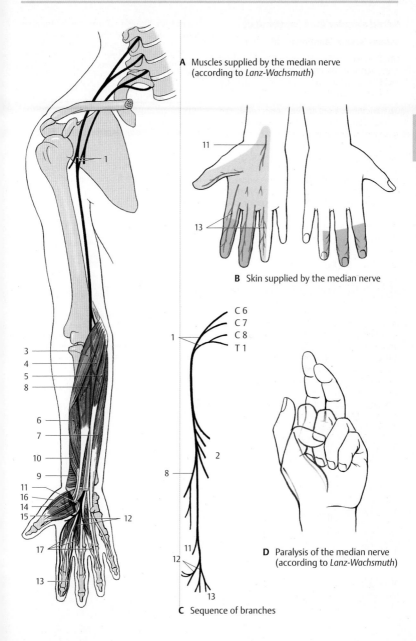

A Muscles supplied by the median nerve (according to *Lanz-Wachsmuth*)

B Skin supplied by the median nerve

C 6
C 7
C 8
T 1

C Sequence of branches

D Paralysis of the median nerve (according to *Lanz-Wachsmuth*)

Infraclavicular Part (continued)

Medial Fascicle (A–D)

Ulnar nerve (C8–T1). Initially, the ulnar nerve runs in the *upper arm* in the medial bicipital sulcus without giving off any branches.

On the ulnar side of the upper arm, the nerve runs down behind the medial intermuscular septum, being covered by the medial head of the triceps muscle. It crosses the elbow joint on the extensor side in a bony groove, the sulcus for the ulnar nerve, at the medial epicondyle of the humerus. Here, the nerve can be palpated, and the pressure causes an electrifying pain radiating into the ulnar side of the hand. The nerve then passes between the two heads of the ulnar flexor muscle of the wrist to the flexor side of the forearm and runs beneath this muscle down to the wrist. It does not pass through the carpal tunnel but extends over the flexor retinaculum to the palm of the hand, where it divides into a *superficial branch* and a *deep branch*.

In the *forearm*, the nerve gives off branches (**C1**) to the ulnar flexor muscle of the wrist (**A2**) and to the ulnar half of the deep flexor muscle of the fingers (**A3**). A sensory branch, the **dorsal branch of the ulnar nerve** (**BC4**), branches off in the middle of the forearm and runs to the ulnar side of the back of the hand where it supplies the skin. As for the rest of the back of the hand, its area of innervation overlaps with that of the radial nerve. Another sensory branch, the **palmar branch of the ulnar nerve** (**BC5**), branches off in the distal third of the forearm. It extends to the palm and supplies the skin of the hypothenar eminence.

The **superficial branch** runs as **common palmar digital nerve IV** (**BC6**) toward the interdigital space between ring finger and little finger and divides into the **proper palmar digital nerves** (**BC7**), which supply sensory fibers to the volar aspects of the little finger and the ulnar side of the ring finger and reach to the distal phalanges on the extensor side of both fingers. There is a connection to a branch of the median nerve, called the communicating branch of the median nerve with the ulnar nerve (**C8**).

The **deep branch** (**AC9**) sinks into the depth of the palm and curves toward the thenar eminence. It gives off branches for all muscles of the hypothenar eminence (**C10**) (abductor muscle of fifth finger [**A11**], short flexor muscle of fifth finger [**A12**], opposing muscle of fifth finger [**A13**]), for all dorsal and palmar interosseous muscles (**A14**), for the lumbrical muscles III and IV (**A15**), and finally, at the thenar eminence, for the abductor muscle of thumb (**A16**) and the deep head of the short flexor muscle of thumb (**A17**).

▦ **Clinical Note:** Injury to the ulnar nerve causes the formation of a so-called **clawhand** (**D**), where the fingers are extended in the metacarpophalangeal joints but flexed in the proximal and distal interphalangeal joints. This characteristic posture of the fingers is caused by paralysis of the interosseous muscles and lumbrical muscles, which flex the phalanges in the metacarpophalangeal joints but extend them in the proximal and distal interphalangeal joints. Failure of the flexor muscles causes the fingers to remain in this posture due to the now predominant extensor muscles. Since the little finger and the adductors of the thumb are paralyzed, thumb and little finger can no longer touch each other.

Innervation of the skin (B). Autonomic zone (dark blue) and maximum zone (light blue).

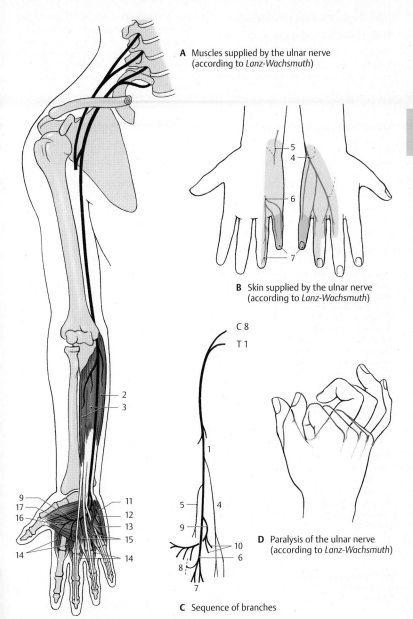

A Muscles supplied by the ulnar nerve
(according to *Lanz-Wachsmuth*)

B Skin supplied by the ulnar nerve
(according to *Lanz-Wachsmuth*)

C 8
T 1

D Paralysis of the ulnar nerve
(according to *Lanz-Wachsmuth*)

C Sequence of branches

Spinal Cord

Infraclavicular Part (continued)

Medial Fascicle (continued) (A – C)

In addition to the ulnar nerve, the medial fascicle gives rise to the *medial cutaneous nerve of the arm* and the *medial cutaneous nerve of the forearm*; both are exclusively sensory nerves supplying the skin on the median side of the arm.

Medial cutaneous nerve of the arm (C8 – T1) **(A, B).** The nerve approaches the anterior surface of the upper arm below the axillary fossa. Here it ramifies and supplies the skin of the medial aspect between axilla and elbow joint. It reaches to the flexor side with its anterior branches and to the extensor side of the upper arm with its posterior branches. Frequently, there are anastomoses to the intercostobrachial nerve.

Medial cutaneous nerve of the forearm (C8 – T1) **(A, C).** The nerve runs below the fascia on the ulnar side of the forearm and passes in the lower third through the fascia with two branches, the **anterior branch** (**AC1**) and the **ulnar branch** (**AC2**). The anterior branch supplies the medial flexor side of the forearm almost up to the midline, and the ulnar branch supplies the upper region of the medial extensor side almost up to the midline. The area innervated by the medial cutaneous nerve of the forearm extends slightly to the upper arm and to the hand.

Posterior Fascicle (D, F)

The posterior fascicle gives rise to the *axillary nerve* and the *radial nerve*.

Axillary nerve (C5 – C6). This runs deep inside the axilla and across the capsule of the shoulder joint around the surgical neck on the back of the humerus. It passes through the lateral axillary gap and extends beneath the deltoid muscle to the anterior margin of the latter.

Before the nerve trunk passes through the lateral axillary gap, it gives off a motor branch (**DF3**) to the lesser teres muscle (**D4**), which also passes through the lateral axillary gap. At the same level, the *superior lateral cutaneous nerve of the arm* (**D – F5**)

branches off and reaches the skin at the posterior margin of the deltoid muscle, where it supplies the skin of the lateral aspects of shoulder and upper arm. From the nerve trunk extending beneath the deltoid muscle to the front, numerous branches (**D6**) to the deltoid muscle (**D7**) branch off and supply its various parts.

▬▬ **Clinical Note:** As a result of its location on the capsule of the shoulder joint, the nerve can be injured by dislocation of the humerus or by humeral neck fracture. This causes anesthesia in the skin area over the deltoid muscle.

Innervation of the skin (B, C, E). Autonomic zone (dark blue) and maximum zone (light blue).

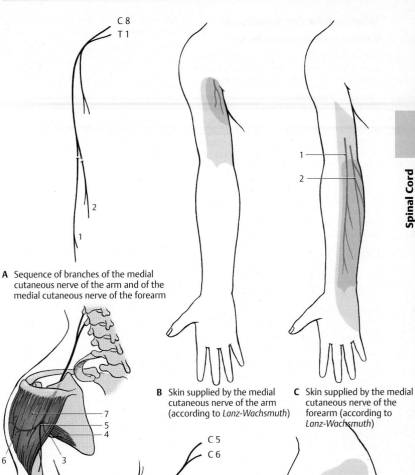

Spinal Cord

A Sequence of branches of the medial cutaneous nerve of the arm and of the medial cutaneous nerve of the forearm

B Skin supplied by the medial cutaneous nerve of the arm (according to *Lanz-Wachsmuth*)

C Skin supplied by the medial cutaneous nerve of the forearm (according to *Lanz-Wachsmuth*)

F Sequence of branches

D Muscles supplied by the axillary nerve

E Skin supplied by the axillary nerve (according to *Lanz-Wachsmuth*)

Infraclavicular Part (continued)

Posterior Fascicle (continued) (A–D)

Radial nerve (C5–C8) **(A–C).** The main nerve of the posterior cord supplies the *extensor muscles of upper arm and forearm.*

The nerve trunk extends from the axilla into the proximal third of the medial bicipital sulcus and then spirals around the dorsal surface of the humerus, to which it is directly apposed in the *sulcus of the radial nerve.* In the distal third of the upper arm, it passes to the flexor side between brachial muscle and brachioradialis muscle. In the sulcus of the radial nerve, the nerve can easily be injured by pressure or by bone fractures because of its proximity to the bone. The nerve crosses the elbow joint on the flexor side and divides at the level of the head of radius into two terminal branches, the *superficial branch* and the *deep branch.* The superficial branch continues in the forearm on the medial surface of the brachioradial muscle and then runs in the lower third between brachioradial muscle and radius to the extensor side in order to reach the back of the hand. The deep branch obliquely penetrates the supinator muscle, gives off numerous muscular branches, and extends as the thin *posterior interosseous nerve of the forearm* to the wrist.

For the upper arm, the radial nerve gives off the **posterior cutaneous nerve of the arm** (**A–C1**), which supplies a skin area on the extensor side of the upper arm with sensory fibers, and the **inferior lateral cutaneous nerve of the arm** (**A–C2**). In the middle third of the upper arm, it gives off *muscular branches* (**AC3**) for the long head, the lateral head, and the medial head of the triceps muscle (**A4**). The branch for the medial head gives off also the branch for the anconeus muscle (**A5**).

The *posterior cutaneous nerve of the forearm* (**A–C6**) branches off in the region of the upper arm; it supplies a strip of skin on the radial extensor side of the forearm. At the level of the lateral epicondyle, *muscular branches* (**C7**) extend to the brachioradial muscle (**A8**) and to the long radial extensor muscle of the wrist (**A9**). The nerve trunk then ramifies into its two major branches in the forearm.

At the back of the hand, the **superficial branch** (**A–C10**) gives off the **dorsal digital nerves** (**A–C11**); they supply sensory fibers to the radial back of the hand, the extensor side of the thumb, the proximal phalanges of index and middle fingers, and the radial half of the extensor side of the ring finger. The ulnar communicating branch of the radial nerve connects with the ulnar nerve (**C12**).

The **deep branch** (**AC13**) gives off *muscular branches* to the short radial extensor muscle of the wrist (**A14**) and to the supinator muscle, while passing through the supinator muscle. Thereafter, it gives off motor branches to the hand extensor muscles, namely, to the common extensor muscle of the fingers (**A15**), the extensor muscle of the little finger (**A16**), the ulnar extensor muscle of the wrist (**A17**), the long abductor muscle of the thumb (**A18**), and the short extensor muscle of the thumb (**A19**). Finally, the terminal branch of the deep branch, the **posterior interosseous nerve**, gives off branches to the long extensor muscle of the thumb (**A20**) and to the extensor muscle of the index finger (**A21**).

The nerve sends sensory branches to the shoulder joint and wrist.

▬ **Clinical Note:** Injury to the main nerve trunk in the area of the upper arm results in paralysis of the extensor muscles. This mainly affects the hand, leading to the so-called **wristdrop** (**D**) characteristic for radial paralysis: extension is possible neither in the wrist nor in the fingers, thus making the hand drop down limply.

Innervation of the skin (B). Autonomic zone (dark blue) and maximum zone (light blue).

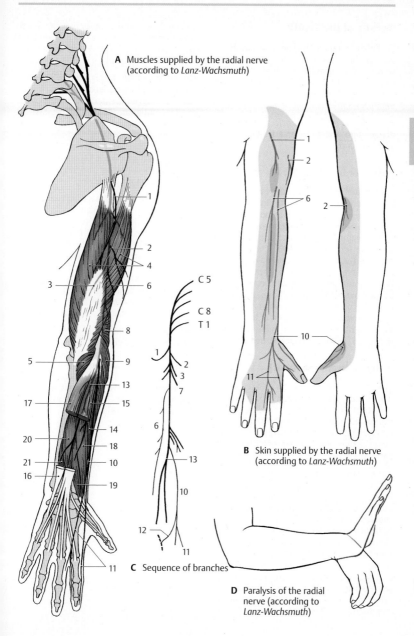

A Muscles supplied by the radial nerve
(according to *Lanz-Wachsmuth*)

C 5

C 8
T 1

B Skin supplied by the radial nerve
(according to *Lanz-Wachsmuth*)

C Sequence of branches

D Paralysis of the radial
nerve (according to
Lanz-Wachsmuth)

Nerves of the Trunk

In the trunk region, the original metamerism of the body can still be recognized through the arrangement of the ribs and their intercostal muscles. The thoracic nerves, too, fit in well with this segmental organization.

Each of the *twelve thoracic spinal nerves* divides into a **posterior branch** (**A1**) and an **anterior branch** (**A2**).

Posterior Branches (A, D)

Each posterior branch divides into a *medial* and a *lateral branch*. Both supply motor fibers to the deep autochthonous back muscles. Sensory innervation of the back comes mainly from the lateral branches of the posterior branches (**AD3**). The area supplied by the posterior branches of cervical spinal nerves expands widely and includes the occiput (**greater occipital nerve**) (**D4**). In the lumbar region, sensory innervation of the back comes from the posterior branches of the lumbar spinal nerves L1–L3 and the sacral spinal nerves S1–S3 (**superior cluneal nerves [D5]** and **medial cluneal nerves [D6]**).

Anterior Branches (A–D)

The anterior branches of the spinal thoracic nerves run as **intercostal nerves** between the ribs, initially on the inner surface of the thorax and later within the internal intercostal muscles. We distinguish between an upper group and a lower group of intercostal nerves.

The **nerves of the upper group** (T1–T6) run up to the sternum and supply the intercostal muscles (**C7**), the superior and inferior posterior serrate muscles, and the transverse thoracic muscle. They give off sensory branches to the skin of the thorax, namely, the **lateral cutaneous branches** (**AD8**) at the anterior margin of the anterior serrate muscle, which further divide into anterior and posterior branches, and the **anterior cutaneous branches** (**AD9**) close to the sternum, which also divide into anterior and posterior branches. The lateral and medial cutaneous branches of anterior branches 4–6, which extend to the area of the mammary gland, are referred to as *lateral and medial mammary branches*.

The **nerves of the lower group** (T7–T12), the intercostal segments of which no longer end at the sternum, extend across the costal cartilages up to the white line. They take an increasingly oblique downward path and supply the muscles of the abdominal wall (abdominal transverse muscle [**C10**], external [**C11**] and internal [**C12**] abdominal oblique muscles, rectus abdominis muscle [**C13**] and pyramidal muscle).

Special features. Intercostal nerve 1 participates in forming the brachial plexus and sends only a thin branch to the intercostal space. Intercostal nerve 2 (and often 3 as well) gives off its *lateral cutaneous branch* to the upper arm (**intercostobrachial nerve**) (**B14**), where it connects with the *medial cutaneous nerve of the arm*. The last intercostal nerve running beneath the twelfth rib is referred to as the **subcostal nerve;** it runs obliquely downward across the iliac crest.

The inguinal region and hip region receive their sensory innervation from the uppermost branches of the lumbar plexus, namely, from the **iliohypogastric nerve** (**D15**) (lateral branch and anterior branch), from the **ilioinguinal nerve** (**D16**), and from the **genitofemoral nerve** (genital branch [**D17**], femoral branch [**D18**]).

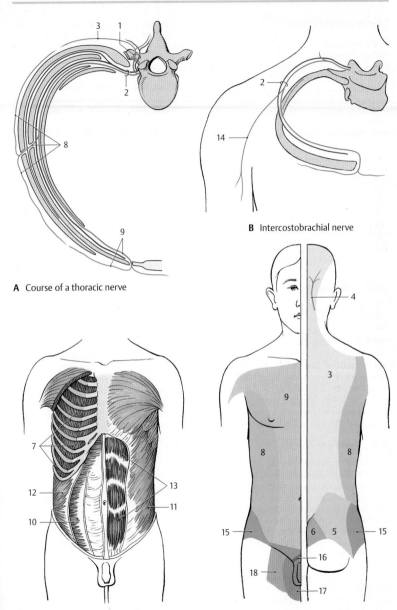

A Course of a thoracic nerve

B Intercostobrachial nerve

C Muscles supplied by the intercostal nerves

D Innervation of the skin of the trunk

Lumbosacral Plexus (A)

The lumbosacral plexus is formed by the *anterior branches* of the lumbar and sacral spinal nerves. Its branches provide sensory and motor innervation to the lower limb. The branches of L1–L3 and part of L4 form the **lumbar plexus**, the roots of which lie within the psoas muscle. The **obturator nerve** (**A1**) and the **femoral nerve** (**A2**) originate from here, in addition to several short muscular branches. The remainder of the fourth lumbar nerve and the L5 nerve join to form the **lumbosacral trunk** (**A3**), which then unites in the small pelvis with sacral branches 1–3 to form the **sacral plexus**. The sacral branches emerge from the anterior sacral foramina of the sacrum and form together with the lumbosacral trunk the sacral plexus; the main nerves originating from here are the **sciatic nerve** (**A4**) (**common peroneal nerve** [**A5**] and **tibial nerve** [**A6**]).

Lumbar Plexus

The lumbar plexus gives off direct short *muscular branches* to the hip muscles, namely, to the greater and lesser psoas muscles (L1–L5), the lumbar quadrate muscle (T12–L3), and the lumbar intercostal muscles. The upper nerves of the plexus are still roughly organized in the same way as the intercostal nerves. Together with the *subcostal nerve* (**A7**), they represent transitional nerves between intercostal nerves and the lumbar nerves.

Iliohypogastric Nerve (T12, L1)

The *iliohypogastric nerve* (**A8**) initially runs on the inside of the lumbar quadrate muscle along the dorsal aspect of the kidney and then between the abdominal transverse muscle and the internal oblique muscle of the abdomen. It participates in the innervation of the broad abdominal muscles. It gives off two main branches, namely, the lateral cutaneous branch which supplies the lateral hip region, and the anterior cutaneous branch which penetrates the aponeurosis of the external oblique muscle of the abdomen cranially to the outer inguinal ring and supplies the skin of this region as well as the pubic region (p. 84, D15; p. 96, C16).

Ilioinguinal Nerve (L1)

The *ilioinguinal nerve* (**A9**) runs along the inguinal ligament and inguinal canal with the spermatic cord up to the scrotum, or with the round ligament of the uterus up to the greater lips in the female, respectively. It participates in the innervation of the broad abdominal muscles and supplies sensory fibers to the skin of the mons pubis and the upper part of the scrotum, or labia majora, respectively (p. 84, D16).

Genitofemoral Nerve (L1, L2)

The *genitofemoral nerve* (**A10**) divides already in, or on, the psoas muscle into two branches, the **genital branch** and the **femoral branch**. The *genital branch* runs in the abdominal wall along the inguinal ligament through the inguinal canal and reaches the scrotum with the spermatic cord or, in the female, the labia majora with the round ligament of the uterus. It innervates the cremaster muscle and supplies sensory fibers to the skin of the scrotum, or the labia majora, respectively, and the adjacent skin area of the thigh (p. 84, D17; p. 96, C15). The *femoral branch* continues to below the inguinal ligament and becomes subcutaneous in the saphenous hiatus. It supplies the skin of the thigh lateral to the region of the genital branch (p. 84, D18).

A11 Lateral cutaneous nerve of femur (p. 88, A).
A12 Posterior cutaneous nerve of femur (p. 90, D).
A13 Pudendal nerve (p. 96, AB1).
A14 Superior gluteal nerve (p. 90, E).

Spinal Cord

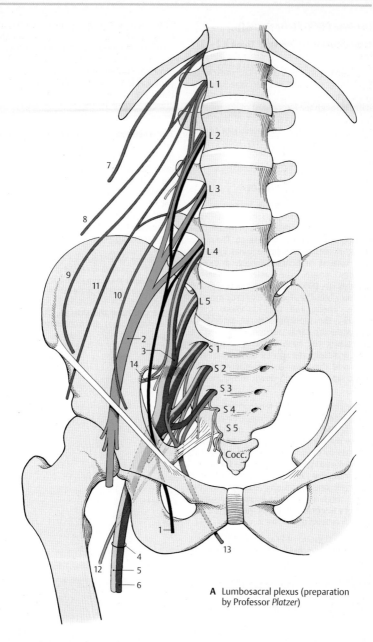

A Lumbosacral plexus (preparation by Professor *Platzer*)

Spinal Cord

Lumbar Plexus (continued)

Lateral Cutaneous Nerve of Thigh (L2–L3) (A)

The nerve runs over the iliac muscle to below the superior anterior iliac spine. It then extends underneath the inguinal ligament through the lateral part of the muscular lacuna to the outer aspect of the thigh and passes through the fascia lata to the skin. The nerve is exclusively sensory and supplies the skin of the lateral aspect of the thigh down to the level of the knee.

Femoral Nerve (L1–L4) (B–D)

The nerve runs along the margin of the greater psoas muscle up to the inguinal ligament and underneath it through the muscular lacuna to the front of the thigh. The nerve trunk divides below the inguinal ligament into several branches, namely, a mostly sensory group, the **anterior cutaneous branches** (**B–D1**), a lateral and medial group of motor branches for the extensor muscles of the thigh, and the *saphenous nerve* (**B–D2**). The saphenous nerve extends to the adductor canal and enters into it. It penetrates the vastoadductor membrane and runs along the medial side of the knee joint and the lower leg together with the great saphenous vein down to the medial ankle.

In the small pelvis, the femoral nerve gives off fine branches (**D3**) to the greater psoas muscle (**B4**) and to the iliac muscle (**B5**). Below the inguinal ligament, a branch (**D6**) extends to the pectineal muscle (**B7**). The *anterior cutaneous branches* (**B–D1**) originate slightly more distally, with the strongest one continuing along the middle of the thigh down to the knee. They supply sensory fibers to the skin of the anterior and medial aspects of the thigh.

The lateral group of branches (**D8**) consists of **muscular branches** for the sartorius muscle (**B9**), the rectus femoris muscle (**B10**), the lateral vastus muscle (**B11**), and the intermediate vastus muscle (**B12**). The muscular branch (**D13**) for the medial vastus muscle (**B14**) runs along the medial margin of the

sartorius muscle. The muscular branches always ramify into several branches for the proximal and distal portions of the muscles. The muscular branches also give off fine sensory branches to the capsule of the knee joint and the periosteum of the tibia. Fibers from the branch for the medial vastus muscle extend to the femoral artery and femoral vein.

The **saphenous nerve** (**CD2**) is exclusively sensory. Below the knee joint, it gives off the *infrapatellar branch* (**B–D15**) which supplies the skin below the patella. The remaining branches, the *medial crural cutaneous branches*, supply the skin of the anterior and medial aspects of the lower leg. The supplied area extends on the anterior side over the edge of the tibia and may reach to the great toe along the medial aspect of the foot.

Clinical Note: Injury to the femoral nerve makes it impossible to extend the leg in the knee joint. Flexion in the hip joint is reduced, and the patellar tendon reflex is absent.

Innervation of the skin (A, C). Autonomic zone (dark blue) and maximum zone (light blue).

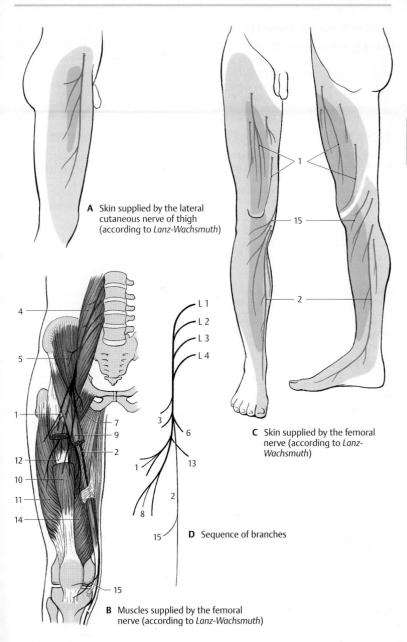

A Skin supplied by the lateral cutaneous nerve of thigh (according to *Lanz-Wachsmuth*)

C Skin supplied by the femoral nerve (according to *Lanz-Wachsmuth*)

B Muscles supplied by the femoral nerve (according to *Lanz-Wachsmuth*)

D Sequence of branches

Spinal Cord

Lumbar Plexus (continued) (A – C)

Obturator Nerve (L2 – L4)

The nerve provides motor innervation to the **adductor muscles** of the thigh. Medial to the greater psoas muscle, it extends along the lateral wall of the small pelvis down to the obturator canal through which it passes to reach the thigh. It gives off a muscular branch to the external obturator muscle (**AB1**) and then divides into a superficial branch and a deep branch. The **superficial branch** (**AB2**) runs between the long adductor muscle (**A3**) and short adductor muscle (**A4**) and innervates both. The nerve also gives off branches to the pectineal muscle and the gracilis muscle (**A5**) and finally terminates in a *cutaneous branch* (**A – C6**) to the distal region of the medial aspect of the thigh. The **deep branch** (**AB7**) runs along the external obturator muscle and then down to the great adductor muscle (**A8**).

▓▓ **Clinical Note:** Paralysis of the obturator nerve (for example, as a result of pelvic fracture) causes loss of adductor muscle function. This restricts standing and walking, and the affected leg can no longer be crossed over the other leg.

Sacral Plexus (D – F)

The *lumbosacral trunk* (parts of L4 and L5) and the anterior branches of S1 – S3 join on the anterior surface of the piriform muscle to form the sacral plexus. Direct branches extend from the plexus to the muscles of the pelvic region, namely, to the piriform muscle, the gemellus muscles (**F9**), the internal obturator muscle, and the quadrate muscle of thigh (**F10**).

Superior Gluteus Nerve (L4 – S1) (E)

The nerve extends across the upper margin of the piriform muscle in dorsal direction through the suprapiriform foramen to the gluteus medius (**E11**) and minimus (**E12**) muscles and supplies both with motor fibers. The nerve continues between the two muscles to the tensor muscle of the fascia lata (**E13**).

▓▓ **Clinical Note:** Paralysis of the nerve weakens abduction of the leg. Standing on the affected leg and lifting the healthy leg makes the pelvis of the other side drop (*Trendelenburg's symptom*).

Inferior Gluteus Nerve (L5 – S2) (F)

The nerve leaves the pelvis through the infrapiriform foramen and gives off several branches to the gluteus maximus muscle (**F14**).

▓▓ **Clinical Note:** Paralysis of the nerve weakens extension of the hip joint (for example, when standing up or climbing stairs).

Posterior Cutaneous Nerve of Thigh (S1 – S3) (D)

The nerve leaves the pelvis together with sciatic nerve and inferior gluteus nerve through the infrapiriform foramen and reaches below the gluteus maximus muscle to the posterior aspect of the thigh. Located directly beneath the fascia lata, it extends along the middle of the thigh into the popliteal fossa. This exclusively sensory nerve gives off branches to the lower part of the buttock, the **inferior cluneal nerves**, and to the perineal region, the **perineal branches**. It provides sensory innervation to the posterior aspect of the thigh from the lower buttock region into the popliteal fossa and reaches to the proximal aspect of the lower leg.

Innervation of the skin (C, D). Autonomic zone (dark blue) and maximum zone (light blue).

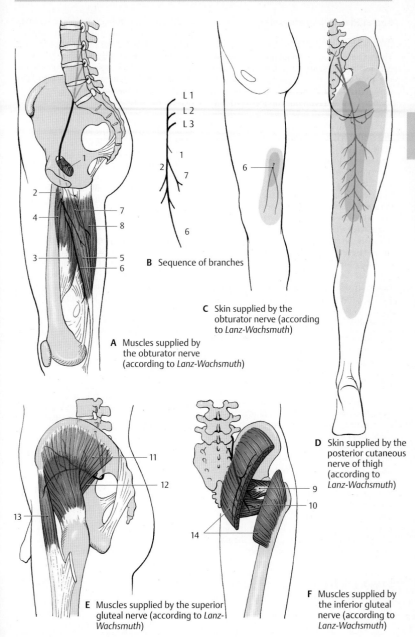

Spinal Cord

B Sequence of branches

L 1
L 2
L 3

C Skin supplied by the obturator nerve (according to *Lanz-Wachsmuth*)

A Muscles supplied by the obturator nerve (according to *Lanz-Wachsmuth*)

D Skin supplied by the posterior cutaneous nerve of thigh (according to *Lanz-Wachsmuth*)

E Muscles supplied by the superior gluteal nerve (according to *Lanz-Wachsmuth*)

F Muscles supplied by the inferior gluteal nerve (according to *Lanz-Wachsmuth*)

Spinal Cord

Sacral Plexus (continued)

Sciatic Nerve (L4–S3) (A–C)

The nerve has two components, the **common peroneal nerve** (*common fibular nerve*) and the **tibial nerve**; they appear as a uniform nerve trunk (**AC1**) because they are surrounded by a common connective-tissue sheath in the small pelvis and in the thigh. The sciatic nerve leaves the pelvis through the infrapiriform foramen and extends beneath the gluteus maximus muscle and biceps muscle along the posterior aspects of the internal obturator muscle, the quadrate muscle of femur, and the great adductor muscle in the direction of the knee joint. Peroneal nerve and tibial nerve separate above the knee joint. In the pelvis within the connective tissue sheath, the peroneal nerve is on the top and the tibial nerve below. In the thigh, the peroneal nerve lies laterally and the tibial nerve medially. However, both may run completely separately, in which case only the tibial nerve passes through the infrapiriform foramen, while the peroneal nerve penetrates the piriform muscle.

Common peroneal nerve (common fibular nerve) (L4 – S2). In the thigh, the peroneal part (**AC2**) of the sciatic nerve gives off a muscular branch to the short head of the biceps muscle of the thigh (**A3**).

After division of the sciatic nerve, the common peroneal nerve extends along the biceps muscle at the lateral edge of the popliteal fossa to the head of the fibula. It then winds around the neck of the fibula to the anterior aspect of the lower leg and enters into the long peroneal (fibular) muscle. The common peroneal nerve divides within this muscle into the *superficial peroneal nerve* (**AC4**) and the *deep peroneal nerve* (**AC5**). The superficial peroneal nerve is predominantly sensory and runs between the long peroneal muscle and the fibula to the back of the foot. The deep peroneal nerve is predominantly a motor nerve; it turns toward the front to the extensor muscles of the lower leg and extends on the lateral surface of the anterior tibial muscle to the back of the foot.

At the lateral margin of the popliteal fossa, the common peroneal nerve gives off two main branches for the skin, the **lateral sural cutaneous nerve** (**A–C6**), which supplies the skin at the lateral aspect of the lower leg, and the **fibular communicating branch** (**C7**), which joins the *medial sural cutaneous nerve* to form the *sural nerve*.

The **superficial peroneal nerve** gives off *muscular branches* (**AC8**) to the long (**A9**) and short (**A10**) peroneal muscles. The rest of the nerve is exclusively sensory; it ramifies into terminal branches, the **medial dorsal cutaneous nerve** (**BC11**) and the **intermediate dorsal cutaneous nerve** (**BC12**), which supply the skin of the back of the foot except for the interdigital space between great toe and second toe.

The **deep peroneal nerve** gives off several muscular branches (**AC13**) to the extensor muscles of the lower leg and the foot, namely, to the anterior tibial muscle (**A14**), the long (**A15**) and short (**A16**) extensor muscles of toes, and the long (**A17**) and short (**A18**) extensor muscles of the great toe. The terminal branch is sensory and supplies the apposing skin surfaces of the interdigital space between great toe and second toe (**B19**).

▬ **Clinical Note:** Injury to the nerve affects the extensor muscles of the foot. The foot can no longer be lifted in the ankle joint. When walking, the foot hangs down and the toes drag along the floor. The leg must be lifted higher than normal, resulting in the so-called *steppage gait*.

Innervation of the skin (B). Autonomic zone (dark blue) and maximum zone (light blue).

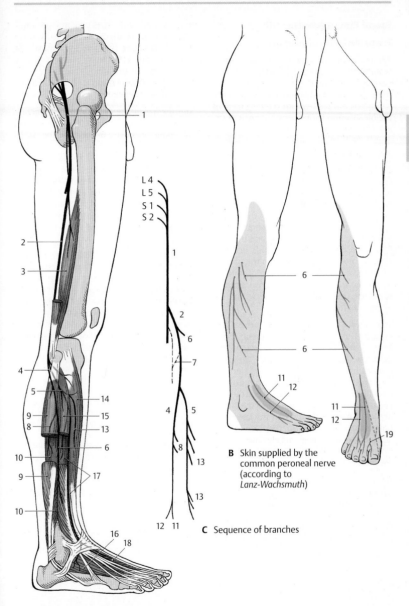

A Muscles supplied by the common peroneal nerve (according to *Lanz-Wachsmuth*)

B Skin supplied by the common peroneal nerve (according to *Lanz-Wachsmuth*)

C Sequence of branches

Sacral Plexus (continued)

Sciatic Nerve (continued) (A–D)

Tibial nerve (L4–S3). Several *motor branches* (**AC1**) originate from the tibial portion of the sciatic nerve, namely, those for the proximal and distal parts of the semitendinous muscle (**A2**), for the long head of the biceps muscle (**A3**), and a branch dividing further for the semimembraneous muscle (**A4**) and the medial part of the great adductor muscle (**A5**).

After the division of the sciatic nerve, the tibial nerve descends vertically through the middle of the popliteal fossa and underneath the gastrocnemius muscle. It then lies under the tendinous arch of the soleus muscle and, further distal, between the long flexor muscle of the great toe and the long flexor muscle of toes. It extends between the tendons of both muscles to the back of the medial ankle and winds around it. Below the ankle, it divides into two terminal branches, the *medial plantar nerve* and the *lateral plantar nerve*.

The **medial sural cutaneous nerve** (**C6**) branches off in the popliteal fossa; it descends between the two heads of the gastrocnemius muscle and joins the communicating branch of the peroneal nerve to form the **sural nerve** (**BC7**). The latter extends laterally from the Achilles tendon behind the lateral ankle and around it to the lateral aspect of the foot. It gives off the **lateral calcaneal branches** (**BC8**) to the skin of the lateral side of the heel and the **lateral dorsal cutaneous nerve** (**BC9**) to the lateral aspect of the foot.

Also in the popliteal fossa, *motor branches* (**AC10**) go off to the flexor muscles of the lower leg, namely, to the two heads of the gastrocnemius muscle (**A11**), to the soleus muscle (**A12**), to the plantar muscle and the popliteal muscle (**A13**). The *popliteal branch* gives rise to the *interosseous nerve of the leg* (**C14**), which runs along the dorsal surface of the interosseous membrane and provides sensory innervation to the periosteum of the tibia, the upper ankle joint, and the tibiofibular joint. The tibial nerve gives off *muscular branches* (**C15**) to the posterior tibial muscle (**A16**), the long flexor muscle of toes (**A17**), and the long flexor muscle of the great toe (**A18**). Before the nerve trunk ramifies into terminal branches, it sends off the **medial calcaneal branches** (**B19**) to the medial skin area of the heel.

The medial of the two terminal branches, the **medial plantar nerve** (**CD20**), innervates the abductor muscle of the great toe (**D21**), the short flexor muscle of toes (**D22**), and the short flexor muscle of the great toe (**D23**). Finally, it divides into the three **common plantar digital nerves** (**BC24**), which supply lumbrical muscles 1 and 2 (**D25**) and divide further into the proper plantar digital nerves (**BC26**) for the skin of the interdigital spaces from the great toe up to the fourth toe.

The second terminal branch, **lateral plantar nerve** (**CD27**), divides into a **superficial branch** (**C28**) with the *common plantar digital nerves* (**C29**) and *proper plantar digital nerves* (**BC30**) for the skin of the little toe area and into a **deep branch** (**CD31**) with the *muscular branches* for the interosseous muscles (**D32**), the adductor muscle of great toe (**D33**) and the three lateral lumbrical muscles. **D34**, short flexor muscle of the little toe.

▬ **Clinical Note:** Injury of the tibial nerve leads to paralysis of the flexor muscles of toes and foot. The foot can no longer be moved in plantar direction: tiptoeing becomes impossible.

Innervation of the skin (B). Autonomic zone (dark blue) and maximum zone (light blue).

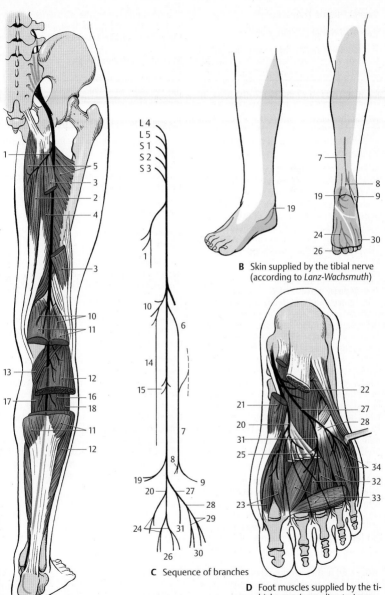

L 4
L 5
S 1
S 2
S 3

B Skin supplied by the tibial nerve (according to *Lanz-Wachsmuth*)

C Sequence of branches

D Foot muscles supplied by the tibial nerve (according to *Lanz-Wachsmuth*)

A Muscles supplied by the tibial nerve (according to *Lanz-Wachsmuth*)

Spinal Cord

Sacral Plexus (continued)

Pudendal Nerve (S2 – S4) (A, B)

The pudendal nerve (**AB1**) leaves the pelvis through the infrapiriform foramen (**AB2**), extends dorsally around the sciatic spine (**AB3**) and passes through the lesser sciatic foramen (**AB4**) into the ischioanal fossa. It then runs along the lateral wall of the fossa within the *pudendal canal (Alcock's canal)* to below the symphysis, sending its terminal branch to the dorsal side of the penis or clitoris, respectively.

Numerous branches are given off in the pudendal canal; the **inferior rectal nerves** (**A – C5**), which may also originate directly from the second to fourth sacral nerves, penetrate through the wall of the canal to the perineum and supply motor fibers to the external sphincter muscle of anus (**AB6**) and sensory fibers to the skin around the anus as well as the lower two-thirds of the anal canal.

The **perineal nerves** (**AB7**) subdivide into deep and superficial branches. The deep branches participate in the innervation of the external sphincter muscle of the anus. More superficially, they supply the bulbocavernous muscle, the ischiocavernous muscle, and the superficial transverse perineal muscle. The superficial branches supply sensory fibers to the posterior part of the scrotum (posterior scrotal nerves) (**AC8**) in males and to the labia majora (posterior labial nerves) (**BC9**) in females. They also supply the mucosa of the urethra and the bulb of penis in males, and the external urethral opening and the vestibule of vagina in females.

The terminal branch, the **dorsal nerve of the penis** (**A10**) or **dorsal nerve of the clitoris** (**B11**), respectively, sends motor branches to the deep transverse perineal muscle, the deep sphincter muscle, and the sphincter muscle of urethra (**B12**). After passing through the urogenital diaphragm (**AB13**), it gives off a branch to the cavernous body of the penis in males, and to the cavernous body of the clitoris in females. In males, the nerve runs along the dorsum of the penis and gives off

sensory branches to the skin of the penis and the glans. In females, it supplies sensory fibers to the clitoris including the glans.

Muscular Branches (S3, S4)

The levator ani muscle and the coccygeal muscle are supplied directly by nerve branches from the sacral plexus.

Coccygeal Plexus (S4 – Co) (A–C)

The anterior branches of the fourth and fifth sacral nerves form a fine plexus on the coccygeal muscle, the coccygeal plexus (**AB14**). The **anococcygeal nerves** originate from here; they supply sensory fibers to the skin over the coccyx and between coccyx and anus (**C14**).

Sensory Innervation of Pelvis and Perineum (C)

In addition to the sacral and coccygeal nerves, the following nerves participate: the *ilioinguinal nerve* and the *genitofemoral nerve* (**C15**), the *iliohypogastric nerve* (**C16**), the *obturator nerve* (**C17**), the *posterior cutaneous nerve of the thigh* (**C18**), the *inferior cluneal nerves* (**C19**), and the *medial cluneal nerves* (**C20**).

External openings of genitals, bladder, and rectum are border areas between involuntary smooth intestinal muscles and voluntary striated muscles. Accordingly, autonomic and somatomotor fibers are here intertwined. The pudendal nerve contains, apart from sensory, somatomotor, and sympathetic fibers, also parasympathetic fibers from the sacral spinal cord. Sympathetic fibers also originate as pelvic splanchnic nerves from the second to fourth sacral nerves.

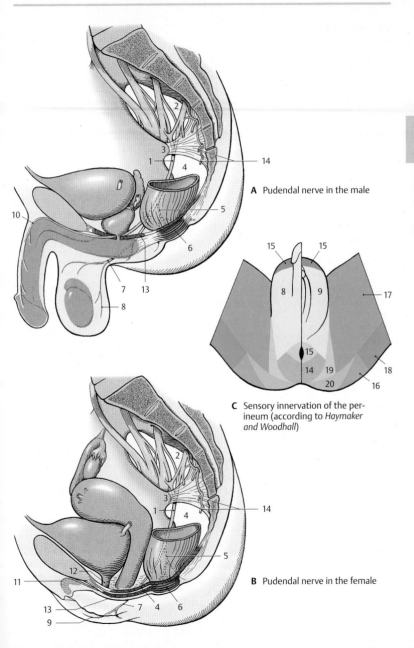

A Pudendal nerve in the male

C Sensory innervation of the perineum (according to *Haymaker and Woodhall*)

B Pudendal nerve in the female

Brain Stem and Cranial Nerves

Overview

The **brain stem**, or **encephalic trunk**, is sub-divided into three sections: the *medulla oblongata* (elongated spinal cord) (**C1**), the *pons* (bridge) (**C2**), and the *mesencephalon* (midbrain) (**C3**).

It is this part of the brain that is underlain by the chorda dorsalis (notochord) during embryonic development and from where ten pairs of genuine peripheral nerves (cranial nerves III – XII) emerge. The *cerebellum*, which in ontogenetic terms also belongs to it, will be discussed separately because of its special structure (p. 152).

The **medulla oblongata** between the pyramidal decussation and the lower border of the pons represents the transition from the spinal cord to the brain. The *anterior median fissure*, which is interrupted by the *pyramidal decussation* (**A4**), and the *anterolateral sulcus* (**AD5**) on each side extend up to the pons. The anterior funiculi thicken below the pons to form the *pyramids* (**A6**). Lateral to them on each side bulge the *olives* (**AD7**).

The **pons** forms a broad arching bulge with prominent transverse fibers. Here, descending pathways from the brain are relayed to neurons extending to the cerebellum.

The posterior surface of the brain stem is covered by the **cerebellum** (**C8**). Upon its removal, the *cerebellar peduncles* are cut through on both sides, namely, the *inferior cerebellar peduncle* (or *restiform body*) (**BD9**), the *middle cerebellar peduncle* (or *brachium pontis*) (**BD10**), and the *superior cerebellar peduncle* (or *brachium conjunctivum*) (**BD11**). Removal of the cerebellum opens the *fourth ventricle* (**C12**), the tent-shaped roof of which is formed by the *superior medullary velum* (**C13**) and the *inferior medullary velum* (**C14**). The floor of the fourth ventricle, the *rhomboid fossa*, thus becomes exposed (**B**). Medulla oblongata and pons together form the hindbrain, also known as **rhombencephalon**, named after this fossa. The posterior funiculi (see p. 56)

thicken on both sides to form the *tubercle of the cuneate nucleus* (**B15**) and the *tubercle of the gracile nucleus* (**B16**); they are bordered by the *posterior median sulcus* (**B17**) and by the *posterolateral sulcus* on each side (**B18**).

The **fourth ventricle** forms on each side the *lateral recess* (**B19**) which opens to the subarachnoid space by the *lateral aperture* (*foramen of Luschka*) (**B20**). An unpaired opening lies below the inferior medullary velum, the *median aperture* (*foramen of Magendie*) (p. 282, D14). The floor of the rhomboid fossa shows bulges near the *median sulcus* (**B21**); they are caused by cranial nerve nuclei, namely, the *medial eminence* (**B22**), the *facial colliculus* (**B23**), the *trigon of the hypoglossal nerve* (**B24**), the *trigon of the vagus nerve* (**B25**), and the *vestibular area* (**B26**). The rhomboid fossa is crossed by myelinated nerve fibers, the *medullary striae* (**B27**). The pigmented nerve cells of the *locus ceruleus* (**B28**) shine blueish through the floor of the rhomboid fossa. They are mostly noradrenergic and project into the hypothalamus, the limbic system, and the neocortex. The locus ceruleus also contains peptidergic neurons (enkephalin, neurotensin).

The anterior surface of the **midbrain**, or **mesencephalon**, is formed by the *cerebral peduncles* (**A29**) (descending cerebral pathways). Between them lies the *interpeduncular fossa* (**A30**); its floor is perforated by numerous vessels and is known as the *posterior perforated substance*. At the posterior surface of the midbrain lies the *tectal plate* (or *quadrigeminal plate*) (**BD31**) with two upper elevations, *superior colliculi* (**D32**), the relay station of the optic system, and two lower elevations, the *inferior colliculi* (**D33**), the relay station of the acoustic system.

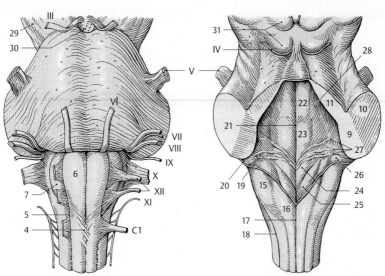

A Basal view of the brain stem

B Dorsal view of the brain stem, rhomboid fossa

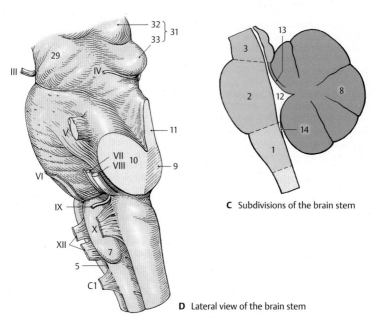

C Subdivisions of the brain stem

D Lateral view of the brain stem

Longitudinal Organization (A)

The longitudinal organization of the *neural tube* (**A1**) can still be recognized in the brain stem. However, it is modified through the widening of the central canal into the fourth ventricle (**A2**, **A3**).

The dorsoventral arrangement of motor **basal plate** (**A4**), visceromotor (**A5**) and viscerosensory (**A6**) regions, and sensory **alar plate** (**A7**) changes to a mediolateral arrangement on the floor of the rhomboid fossa when the neural tube unfolds (**A2**). The *somatomotor zone* lies medially, next comes the *visceromotor zone*; the *viscerosensory* and *somatosensory zones* are transposed laterally. The cranial nerve nuclei in the medulla oblongata are arranged according to this blueprint (**A3**) (p. 106, p. 108).

Cranial Nerves (B)

According to classical anatomical nomenclature, there are *12 pairs of cranial nerves*, although the first two pairs are not really peripheral nerves. The **olfactory nerve** (I) consists of the olfactory fibers, the **bundled processes of sensory cells** in **the olfactory epithelium** which enter the olfactory bulb (**B8**) (p. 228, A). The **optical nerve** (II) is a **cerebral pathway**; the origin of the optical fibers, the retina, together with the pigmented epithelium of the eyeball represents an evagination of the diencephalon (p. 342, A). *Optic chiasm* (**B9**), *optical tract* (**B10**).

The **eye-muscle nerves** (p. 138) are somatomotor nerves. The **oculomotor nerve** (III) leaves the brain on the floor of the *interpeduncular fossa* (**B11**); the **trochlear nerve** (IV) emerges at the dorsal surface of the midbrain and extends around the cerebral peduncles to the basal surface (p. 101, BD IV); the **abducens nerve** (VI) emerges from the lower border of the pons.

Five nerves have developed from the **branchial arch nerves** of lower vertebrates: the **trigeminal nerve** (V) (p. 124), the **facial nerve** (VII) (p. 122), the **glossopharyngeal**

nerve (IX) (p. 118), the **vagus nerve** (X) (p. 114), and the **accessory nerve** (XI) (p. 112, CD). The muscles supplied by these nerves are derived from the branchial arch muscles of the foregut. Hence, these nerves have originally been visceromotor nerves. In mammals, the branchial arch muscles have changed into the striated muscles of pharynx, oral cavity, and face. Unlike genuine striated muscles, they are not completely voluntary (e.g., facial expression in response to emotion).

The **vestibulocochlear nerve** (VIII) (p. 120) with its vestibular part represents a phylogenetically old **connection to the organ of balance** already present in lower vertebrates.

The trigeminal nerve (V) emerges from the lateral part of the pons. Its *sensory root* extends to the *trigeminal ganglion* (*semilunar ganglion, Gasser's ganglion*) (**B12**); its *motor root* (**B13**) bypasses the ganglion. The facial nerve (VII) and the vestibulocochlear nerve (VIII) leave the medulla oblongata at the cerebellopontine angle. The taste fibers of the facial nerve emerge as an independent nerve, the *intermediate nerve* (**B14**). The glossopharyngeal nerve (IX) and the vagus nerve (X) emerge dorsal to the olive. *Superior ganglion of the vagus nerve* (**B15**). The cervical roots of the accessory nerve (XI) unite to form the *spinal root* (**B16**). The upper fibers originating from the medulla oblongata form the *cranial root*; they run a short course in the nerve and change over to the vagus nerve as *internal branch* (**B17**).

The **hypoglossal nerve** (XII) (p. 112, AB) is a somatomotor nerve; in ontogenetic terms, it represents the remnants of several cervical nerves that have become included in the brain region secondarily and now have only rudimentary sensory roots.

B18 Olfactory tract.
B19 Lateral olfactory stria.
B20 Anterior perforated substance.
B21 Hypophyseal stalk.
B22 Choroid plexus (flower spray of Bochdalek) (p. 282, D15).

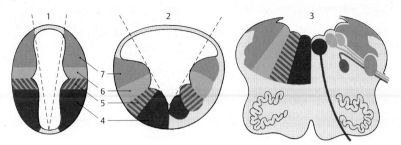

A Longitudinal organization of the medulla oblongata (according to *Herrick*)

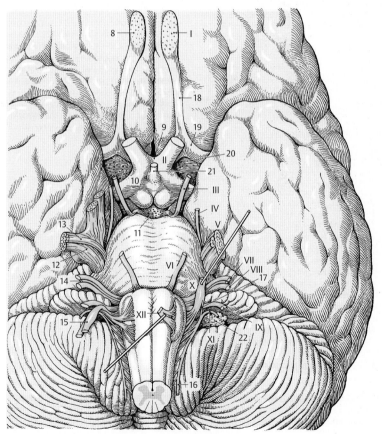

B Cranial nerves, base of the brain

Base of the Skull (A)

The base of the skull holds the brain. Three bony depressions on each side correspond to the basal aspects of the brain; the basal aspect of the frontal lobe lies in the **anterior cranial fossa** (**A1**), that of the temporal lobe in the **middle cranial fossa** (**A2**), and the basal aspect of the cerebellum in the **posterior cranial fossa** (**A3**). (For the participation of bones in and the boundaries of the cranial fossae, see vol. 1.) The cranial cavity is lined by a hard meninx, the *dura mater* (p. 288); its two layers form both a cover for the brain and the periosteum. Embedded between these two layers are large venous sinuses (p. 288). Nerves and blood vessels pass through numerous foramina in the base of the skull (see vol. 1).

On the floor of the anterior cranial fossa close to the midline, the *olfactory nerves* pass through the openings of the thin *lamina cribrosa* to the *olfactory bulb* (**A4**). The *sella turcica* rises between the two middle cranial fossae; its depression contains the *hypophysis* (**A5**), which is attached to the floor of the diencephalon. Lateral to the sella turcica, the *internal carotid artery* (**A6**) passes through the *carotid canal* into the cranial cavity. Its S-shaped course bypasses the *cavernous sinus* (**A7**). The **optic nerve** (**A8**) enters the cranial cavity through the optic canal in the medial area of the fossa, while the eye-muscle nerves leave the cavity through the superior orbital fissure (see vol. 1). The paths of the **abducens nerve** (**A9**) and the **trochlear nerve** (**A10**) are characterized by their intradural position. The abducens nerve enters the dura at the middle level of the clivus, and the trochlear nerve enters at the edge of the clivus at the attachment of the tentorium. The **oculomotor nerve** (**A11**) and the trochlear nerve run through the lateral wall of the cavernous sinus, and the abducens nerve through the laterobasal sinus of the internal carotid artery (see vol. 2). The **trigeminal nerve** (**A12**) reaches below a dural bridge into the middle cranial fossa where the **trigeminal ganglion** (**A13**) lies in a pocket formed by the two dural layers, the *trigeminal cavity*. The three trigeminal branches leave the cranial cavity through different openings; after passing through the wall of the cavernous sinus, the **ophthalmic nerve** (**A14**) extends with its branches through the *orbital fissure*, the **maxillary nerve** (**A15**) through the *round foramen*, and the **mandibular nerve** (**A16**) through the *oval foramen*.

The two posterior cranial fossae surround the *foramen magnum* (**A17**) to which the *clivus* (**A18**) descends steeply from the sella turcica. The brain stem rests on the clivus, and the cerebellar hemispheres fit into the two basal fossae. From the *confluence of sinuses* (**A19**), the *transverse sinus* (**A20**) embraces the posterior cranial fossa and opens into the *internal jugular vein* (**A21**). The **facial nerve** (**A22**) and the **vestibulocochlear nerve** (**A23**) enter the auditory canal, the *internal acoustic meatus*. Basal to the meatus, the **glossopharyngeal nerve** (**A24**), **vagus nerve** (**A25**), and **accessory nerve** (**A26**) pass through the anterior part of the *jugular fossa*. The fiber bundles of the **hypoglossal nerve** (**A27**) pass as a single nerve through the *hypoglossal canal*.

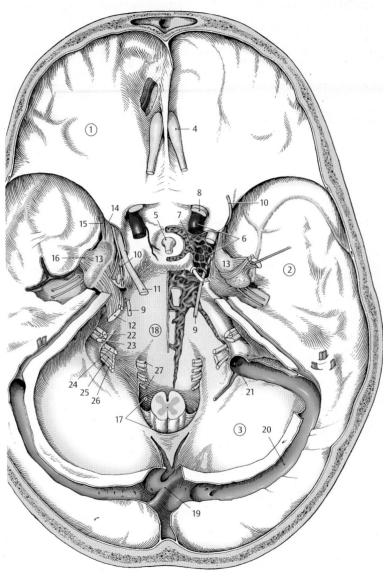

A Base of the skull, viewed from above (preparation by Professor *Platzer*)

Brain Stem

Cranial Nerve Nuclei

As in the spinal cord, where the anterior horn represents the area of origin of motor fibers and the posterior horn the area of termination of sensory fibers, the medulla oblongata contains the **nuclei of origin** (with the cell bodies of efferent fibers) and the **nuclei of termination** (for the axon terminals of afferent fibers), the pseudounipolar cells of which lie in sensory ganglia outside the brain stem.

The **somatomotor nuclei** lie close to the midline:

* The **nucleus of the hypoglossal nerve (AB1)** (tongue muscles)
* The **nucleus of the abducens nerve (AB2)**
* The **nucleus of the trochlear nerve (AB3)**
* The **nucleus of the oculomotor nerve (AB4)** (eye muscles)

The *visceromotor nuclei* follow laterally, namely, the genuine visceromotor nuclei belonging to the parasympathetic nervous system and the originally visceromotor nuclei of the transformed branchial arch muscles. The **parasympathetic nuclei** include:

* The **dorsal nucleus of vagus nerve (AB5)** (viscera)
* The **inferior salivatory nucleus (AB6)** (preganglionic fibers for the parotid gland)
* The **superior salivatory nucleus (AB7)** (preganglionic fibers for the submandibular and sublingual glands)
* The **Edinger–Westphal nucleus (accessory nucleus of oculomotor nerve) (AB8)** (preganglionic fibers for the sphincter muscle of pupil and the ciliary muscle)

The series of **motor nuclei of the branchial arch nerves** begins caudally with the **spinal nucleus of the accessory nerve (AB9)** (shoulder muscles), which extends into the cervical spinal cord. The series continues cranially with the **ambiguous nucleus (AB10)**, which is the motor nucleus of the vagus nerve and the glossopharyngeal nerve (muscles of pharynx and larynx), and the **nucleus of the facial nerve (AB11)** (facial muscles). The facial nucleus lies deep, as do all motor nuclei of

the branchial arch nerves. Its fibers run in a curve directed dorsally, extend on the floor of the rhomboid fossa (facial colliculus) around the abducens nucleus (internal genu of facial nerve) (A12), and then descend again to the lower border of the pons where they emerge from the medulla oblongata. The most cranial nucleus of the branchial arch nerves is the **motor nucleus of the trigeminal nerve (AB13)** (masticatory muscles).

The **sensory nuclei** are located laterally; most medially lies the viscerosensory **solitary nucleus (AB14)**, in which the sensory fibers of the vagus nerve and the glossopharyngeal nerve, as well as all taste fibers, terminate. Further laterally extends the nuclear area of the trigeminal nerve, which has the largest expanse of all cranial nerves and consists of:

* The **pontine nucleus of the trigeminal nerve** (principal sensory nucleus) **(AB15)**
* The **mesencephalic nucleus of the trigeminal nerve (AB16)**
* The **spinal nucleus of the trigeminal nerve (AB17)**

All fibers of the exteroceptive sensibility of face, mouth, and maxillary sinuses terminate in this area.

Finally, most laterally lies the area of the **vestibular nucleus (B18)** and the **cochlear nucleus (B19)**, in which the fibers of the vestibular root (organ of balance) and the cochlear root (organ of hearing) of the vestibulocochlear nerve terminate.

A20 Red nucleus.
A21 Olive.

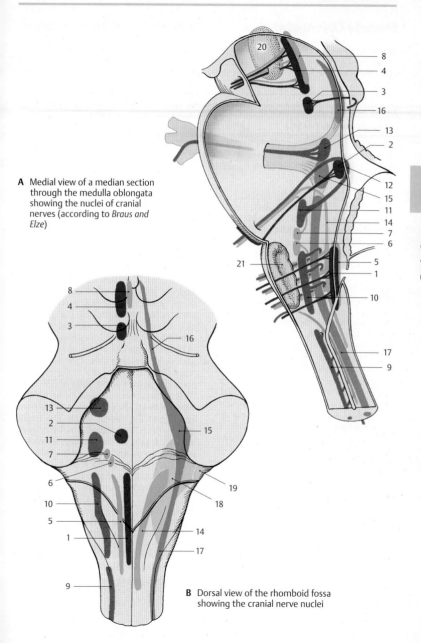

A Medial view of a median section through the medulla oblongata showing the nuclei of cranial nerves (according to *Braus and Elze*)

B Dorsal view of the rhomboid fossa showing the cranial nerve nuclei

Brain Stem

Medulla Oblongata

The semi-schematic cross sections show the cellular stain (Nissl) on the left and the corresponding fiber stain (myelin) on the right.

Cross Section at the Level of the Hypoglossal Nerve (A)

The dorsal part, the *tegmentum*, shows the cranial nerve nuclei, and the ventral part shows the *olive* (**AB1**) and the *pyramidal tract* (**AB2**).

In the tegmentum, the magnocellular *nucleus of the hypoglossal nerve* (**AB3**) lies medially, and dorsally to it lie the *posterior nucleus of the vagus nerve* (**AB4**) and the *solitary nucleus* (**AB5**); the latter nucleus contains a large number of peptidergic neurons. The posterior funiculi of the spinal cord terminate dorsolaterally in the *gracilis nucleus* (**A6**) and in the *cuneate nucleus* (**AB7**) where the secondary sensory pathway, the *medial lemniscus*, originates. Ventrally to the cuneate nucleus lies the *spinal nucleus of the trigeminal nerve* (**AB8**). The large cells of the *ambiguous nucleus* (**AB9**) stand out in the center of the field; they lie in the region of the *reticular formation*, of which only the slightly denser *lateral reticular nucleus* (**AB10**) can be delimited. The *olive* (**AB1**), the fibers of which extend to the cerebellum (p. 164, A11), is accompanied by two nuclei, the *posterior accessory olivary nucleus* (**AB11**) and the *medial accessory olivary nucleus* (**AB12**). Along the ventral aspect of the pyramid stretches the *arcuate nucleus* (**AB13**) where collaterals of the pyramidal tract synapse (*arcuatocerebellar tract*) (p. 164, C18).

The fibers of the *hypoglossal nerve* (**A14**) cross the medulla oblongata to reach their exit point between pyramid and olive. The *posterior longitudinal fasciculus* (*Schütz's bundle*) (**AB15**) (p. 144, B) lies dorsally to the hypoglossal nucleus; laterally lies the *solitary tract* (**AB16**) (p. 114, B12; p. 118, B10) and ventrally the *medial longitudinal fasciculus* (**AB17**) (p. 142, A). From the nuclei

of the posterior funiculus, the *internal arcuate fibers* (**AB18**) radiate broadly into the *medial lemniscus* (**AB19**) (p. 140, B). The *spinal tract of the trigeminal nerve* (**AB20**) (p. 124, B5) runs laterally, and the *central tegmental tract* (**A21**) (extrapyramidal motor tract, p. 144, A) descends dorsally to the main olivary nucleus. The fibers of the *olivocerebellar tract* (**AB22**) run through the hilum of the olive, while the *superficial arcuate fibers* (**AB23**) (arcuate nucleus, cerebellum) run along the lateral part of the olive. The ventral area is occupied by the *pyramidal tract* (**AB2**) (p. 140, A).

Cross Section at the Level of the Vagus Nerve (B)

The fourth ventricle has become wider. The same nuclear columns as in **A** are found in its floor. Ventrally to the *hypoglossal nucleus* (**AB3**) appears *Roller's nucleus* (**B24**) and dorsally the *intercalate nucleus* (*Staderini's nucleus*) (**B25**); the fiber connections of these two nuclei are not known. In the lateral field, the nuclei of the posterior funiculus disappear and make room for the *vestibular nuclei* (*medial vestibular nucleus*) (**B26**). In the middle of the medulla oblongata, the decussating fibers form a seam, the *raphe* (**B27**). On both sides of the raphe lie small groups of cells, the *raphe nuclei* (**B28**); their serotoninergic neurons project to the hypothalamus, the olfactory epithelium, and the limbic system. Along the lateral aspect, spinal fibers running to the cerebellum aggregate in the *inferior cerebellar peduncle* (*restiform body*) (**B29**). Afferent and efferent fibers of the *vagus nerve* (**B30**) cross the medulla oblongata. Ventrally to them, the *spinothalamic tract* (**B31**) (p. 140, B8) and the *spinocerebellar tract* (**B32**) (p. 164, A1; p. 166, B14) ascend along the lateral aspect. The *olivocerebellar fibers* (**B33**) (p. 144, A12), which run to the inferior cerebellar peduncle, aggregate dorsally to the olive.

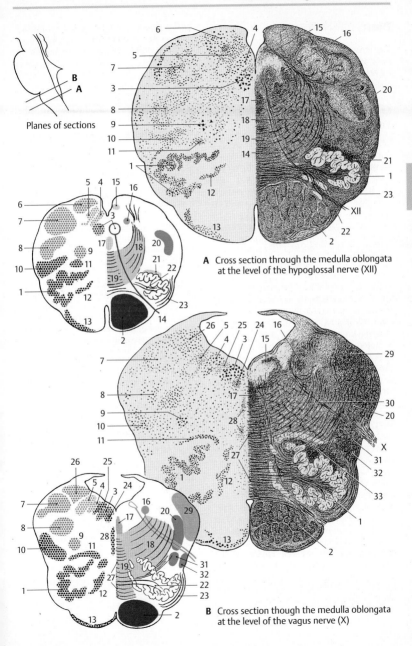

A Cross section through the medulla oblongata
at the level of the hypoglossal nerve (XII)

B Cross section though the medulla oblongata
at the level of the vagus nerve (X)

Planes of sections

Brain Stem

Pons

The semi-schematic cross sections show the cellular stain (Nissl) on the left and the corresponding fiber stain (myelin) on the right.

Cross Section at the Level of the Genu of the Facial Nerve (A)

Beneath the floor of the rhomboid fossa lies the magnocellular *nucleus of the abducens nerve* (**A1**) and, ventrolaterally to it, the *nucleus of the facial nerve* (**A2**). The visceroefferent *superior salivatory nucleus* (**A3**) is seen between the abducens and facial nuclei. The lateral field is occupied by the sensory terminal nuclei of the *vestibular nerve* and the *trigeminal nerve*, namely, the *medial vestibular nucleus* (*Schwalbe's nucleus*) (**A4**), the *lateral vestibular nucleus* (*Deiters' nucleus*) (**A5**), and the *spinal nucleus of the trigeminal nerve* (**A6**).

The fibers of the facial nerve bend around the abducens nucleus (**A1**) and form the *facial colliculus* (**A7**). We distinguish an ascending limb (**A8**) and, cranial to the illustrated section, a descending limb. The apex is the *internal genu of the facial nerve* (**A9**). The fibers of the *abducens nerve* (**A10**) descend through the medial field of the tegmentum. The *medial longitudinal fasciculus* (**AB11**) is seen medially and the *posterior longitudinal nucleus* (*Schütz's bundle*) (**AB12**) dorsally to the abducens nucleus. Deep in the pontine tegmentum run the *central tegmental tract* (**AB13**) and the *spinothalamic tract* (**A14**). Secondary fibers of the auditory pathway collect from the anterior cochlear nucleus as a wide fiber bundle, the *trapezoid body* (**AB15**); they cross ventral to the *medial lemniscus* (**A16**) to the opposite side where they ascend in the *lateral lemniscus* (**B17**). They synapse in part in the adjacent nuclei of the trapezoid body, namely, the *anterior nucleus of the trapezoid body* (**A18**) and the *posterior nucleus of the trapezoid body* (*superior olive*) (**AB19**). The *spinal tract of the trigeminal nerve* (**A20**) lies in the lateral field.

The pontine bulb is formed by the *transverse pontine fibers* (**A21**). They are corticopontine fibers, which synapse in the *pontine nuclei* (**A22**), and pontocerebellar fibers, which are postsynaptic and extend to the cerebellum in the *medial cerebellar peduncle* (*brachium pontis*) (**A23**). Embedded in the middle of the longitudinally cut fiber bundles is the transversely cut fiber bundle of the *pyramidal tract* (**AB24**).

Cross Section at the Level of the Trigeminal Nerve (B)

The medial field of the pontine tegmentum is occupied by the nuclei of the tegmentum. These nuclei, of which only the *inferior central tegmental nucleus* (**B25**) is well defined, belong to the reticular formation. In the lateral field, the trigeminal complex has reached its widest expansion; lateral is the *pontine nucleus of the trigeminal nerve* (principal nucleus) (**B26**), medial to it the *motor nucleus of the trigeminal nerve* (**B27**), and dorsal the nucleus of the mesencephalic trigeminal root (**B28**). Afferent and efferent fibers unite to form the strong trunk emerging at the anterior aspect of the pons.

The *lateral lemniscus* (**AB17**), the *trapezoid body* (**AB15**), and the adjacent *posterior nucleus of the trapezoid body* (**AB19**) lie ventrally to the trigeminal nuclei. The following ascending and descending pathways can be recognized: the *posterior longitudinal fasciculus* (**AB12**), the *medial longitudinal fasciculus* (**AB11**), and the *central tegmental tract* (**AB13**).

AB29 Pontine tegmentum.
AB30 Pontine bulb.

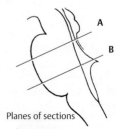

Planes of sections

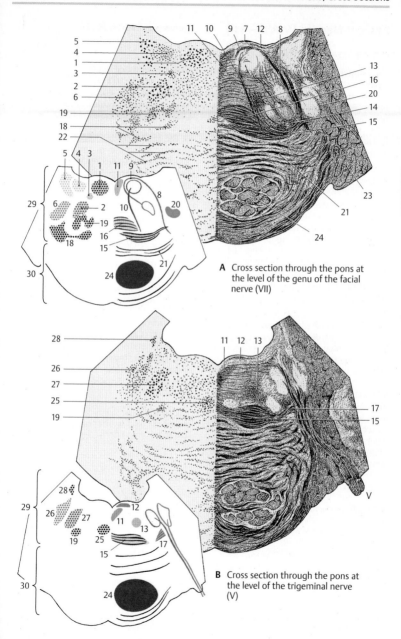

A Cross section through the pons at the level of the genu of the facial nerve (VII)

B Cross section through the pons at the level of the trigeminal nerve (V)

Cranial Nerves (V, VII – XII)

Hypoglossal Nerve (A, B)

The **twelfth cranial nerve** is an **exclusively somatomotor nerve** for the tongue muscles. Its nucleus, the **nucleus of the hypoglossal nerve** (**B1**), forms a column of large multipolar neurons in the floor of the rhomboid fossa (**trigon of hypoglossal nerve**). It consists of a number of cell groups, each of which innervates a particular muscle of the tongue. The nerve fibers emerge between pyramid and olive and form two bundles that then combine into a nerve trunk.

The nerve leaves the skull through the *canal of the hypoglossal nerve* (**B2**) and descends laterally to the vagus nerve and the internal carotid artery. It forms a loop, the **arch of the hypoglossal nerve** (**A3**), and reaches the root of the tongue slightly above the hyoid bone between the *hypoglossal muscle* and the *mylohyoid muscle*, where it ramifies into terminal branches.

Fiber bundles of the first and second cervical nerves adhere to the hypoglossal nerve. They form the **deep cervical ansa** (branches for the lower hyoid bone muscles) by branching off again as **superior root** (**A4**) and combining with the **inferior root** (**A5**) (second and third cervical nerve). The cortical fibers for the *geniohyoid muscle* (**A6**) and the *thyrohyoid muscle* (**A7**) continue to run in the hypoglossal nerve. The hypoglossal nerve gives off the **lingual branches** to the *hypoglossal muscle* (**A8**), the *genioglossal muscle* (**A9**), the *styloglossal muscle* (**A10**), and to the intrinsic muscles of the body of the tongue (**A11**). Innervation of the tongue muscles is strictly ipsilateral.

▓▓▓ **Clinical Note:** Injury to the hypoglossal nerve causes hemilateral shrinkage of the tongue (*hemiatrophy*). When the tongue is stuck out, it turns to the affected side because the genioglossal muscle, which moves the tongue to the front, dominates on the healthy side.

Accessory Nerve (C, D)

The **eleventh cranial nerve** is **exclusively a motor nerve**; its external branch supplies the **sternocleidomastoid muscle** (**D12**) and the **trapezius muscle** (**D13**). Its nucleus, the **spinal nucleus of the accessory nerve** (**C14**), forms a narrow cell column from C1 to C5 or C6. The large multipolar neurons lie at the lateral aspect of the anterior horn. The cells of the caudal section supply the trapezius muscle, and those of the cranial section supply the sternocleidomastoid muscle. The nerve fibers emerge from the lateral aspect of the cervical spinal cord between posterior root and anterior root and combine to form a bundle that enters the skull as the **spinal root** (**C15**) alongside the spinal cord through the foramen magnum. Fiber bundles from the caudal part of the *ambiguous nucleus* join the nerve here as **cranial roots** (**C16**). Both components pass through the *jugular foramen* (**C17**). Immediately after passing, the fibers change from the ambiguous nucleus as **internal branches** (**C18**) over to the *vagus nerve* (**C19**). The fibers from the cervical spinal cord form the **external branch** (**C20**), which supplies as *accessory nerve* the sternocleidomastoid muscle and the trapezius muscle. It passes through the sternocleidomastoid muscle and reaches the trapezius muscle with its terminal branches.

▓▓▓ **Clinical Note:** Injury to the accessory nerve causes the head to tilt (plagiocephaly). The arm can no longer be lifted above the horizontal.

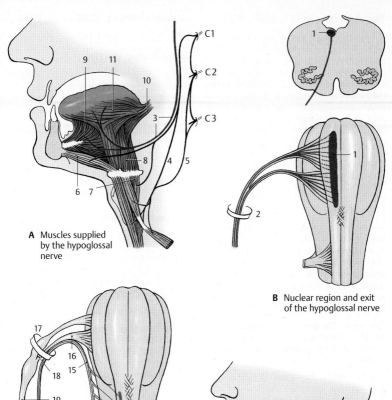

A Muscles supplied by the hypoglossal nerve

B Nuclear region and exit of the hypoglossal nerve

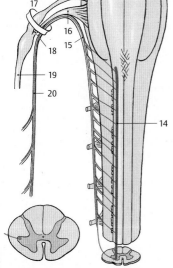

C Nuclear region and exit of the accessory nerve

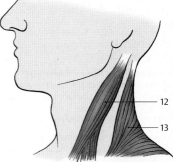

D Muscles supplied by the accessory nerve

Brain Stem

Vagus Nerve (A – F)

The **tenth cranial nerve** not only supplies areas in the head region like the other cranial nerves, it also descends into thorax and abdomen where it ramifies in the viscera like a plexus. It is the **strongest para-sympathetic nerve** of the autonomic nervous system and, hence, the **most important antagonist of the sympathetic nervous system** (p. 292).

It has the following components:

- Motor fibers (branchial arch muscles)
- Exteroceptive sensory fibers
- Visceromotor fibers
- Viscerosensory fibers
- Taste fibers

The fibers emerge directly from behind the olive, unite to form the nerve trunk, and leave the skull through the *jugular foramen* (**B1**). In the foramen, the nerve forms the **superior ganglion of the vagus nerve** (*jugular ganglion*) (**B2**) and, after passing through it, the much larger **inferior ganglion of the vagus nerve** (*nodose ganglion*) (**B3**).

The **motor fibers** for the branchial arch muscles (**AB4**) originate from large multipolar neurons in the **ambiguous nucleus** (**AB5**).

The **visceromotor fibers** (**AB6**) originate in the parvocellular **posterior nucleus of the vagus nerve** (**AB7**), which lies laterally to the nucleus of the hypoglossal nerve in the floor of the rhomboid fossa.

The **exteroceptive sensory fibers** (**AB8**) originate from neurons in the *superior ganglion*. They descend with the *terminal trigeminal root* (**B9**) and terminate in the **spinal nucleus of the trigeminal nerve** (**AB10**).

The cells of the **viscerosensory fibers** (**AB11**) lie in the *inferior ganglion* (*nodose ganglion*). The fibers run as part of the *solitary tract* (**B12**) in caudal direction and terminate at various levels of the **solitary nucleus** (**AB13**). This nucleus is rich in peptidergic neurons (VIP, corticoliberin, dynorphin).

The **taste fibers** (**AB14**) also originate from cells of the *inferior ganglion* and terminate

in the cranial part of the *solitary nucleus* (p. 328, B7).

Head Region (B – D)

In addition to a *meningeal branch* (sensory supply to the dura mater in the posterior cranial fossa), the vagus nerve gives off the **auricular branch** (**B15**). The latter branches off at the superior ganglion, passes through the *mastoid canaliculus*, and reaches the external meatus through the *tympanomastoid fissure*. It supplies the skin of the meatus in the dorsal and caudal region (**D**) and a small area of the auricula (**C**) (exteroceptive sensory component of the nerve).

Cervical Region (B, E, F)

Inside a common connective-tissue sheath, the nerve descends in the neck together with the internal carotid artery, the common carotid artery, and the internal jugular vein; it emerges with them through the upper thoracic aperture.

It gives off four branches:

1 The **pharyngeal branches** (**B16**) at the level of the inferior ganglion. They combine in the pharynx with fibers of the *glossopharyngeal nerve* and the *sympathetic nervous system* to form the **pharyngeal plexus**. At the outer surface of the muscles and in the submucosa of the pharynx, the latter forms a meshwork of fine fibers with groups of neurons. The vagal fibers provide sensory innervation to the muscosa of the trachea and mucosa of the esophagus including the epiglottis (**E, F**). The taste buds (**E**) on the epiglottis are also supplied by the vagus nerve. (*Continued p. 116.*)

B17 Superior laryngeal nerve (p. 116, A2).

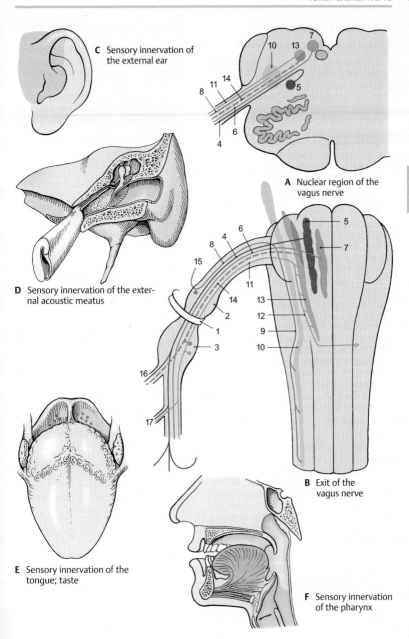

C Sensory innervation of the external ear

A Nuclear region of the vagus nerve

D Sensory innervation of the external acoustic meatus

B Exit of the vagus nerve

E Sensory innervation of the tongue; taste

F Sensory innervation of the pharynx

Brain Stem

Vagus Nerve (continued)

Cervical Region (continued) (A – CEF)

1 The **pharyngeal branches** (*continued*). The motor fibers of the vagus nerve innervate muscles of the soft palate and the pharynx; these are the muscles of the tonsillar sinus, the *levator muscle of the velum palatinum*, and the *constrictor muscles of the pharynx* (**B1**).

2 The **superior laryngeal nerve** (**A2**) originates below the inferior ganglion (nodose ganglion) and divides at the level of the hyoid bone into an *external branch* (motor branch for the cricothyroid muscle) and an *internal branch* (sensory branch for the mucosa of the larynx as far down as the vocal cords).

3 The **recurrent laryngeal nerve** (**A3**) branches off in the thorax after the vagus nerve has extended on the left over the *arch of the aorta* (**A4**) and on the right across the *subclavian artery* (**A5**). It passes on the left around the aorta and the ligamentum arteriosum and on the right around the subclavian artery and then ascends behind the artery. Between trachea and esophagus, to which it gives off the **tracheal branches** (**A6**) and **esophageal branches**, it extends up to the larynx. Its terminal branch, the **inferior laryngeal nerve** (**A7**), supplies motor fibers to all laryngeal muscles except for the cricothyroid muscle, and sensory fibers to the laryngeal mucosa below the vocal cords.

The *motor fibers* originate from the **ambiguous nucleus**, the cell groups of which show a somatotopic organization: The fibers of the *glossopharyngeal nerve* arise in the cranial part, those of the *superior laryngeal nerve* further below, and those of the *inferior laryngeal nerve* caudally, whereby the neurons for abduction and adduction are arranged one below the other (**C**).

4 The **cervical cardiac branches** (*preganglionic parasympathetic fibers*). The *superior branches* (**A8**) depart at various levels and run with the large vessels to the heart, where they terminate in the para-sympathetic ganglia of the *cardiac plexus*. One of the branches carries viscerosensory fibers that transmit information about aortic wall tension. Stimulation of these fibers causes a fall in blood pressure (*depressor nerve*). The *inferior cervical cardiac branches* (**A9**) depart from the recurrent laryngeal nerve or from the main trunk and terminate in the ganglia of the *cardiac plexus*.

▬ **Clinical Note:** Injury to the vagus nerve leads to deficiencies in both pharynx and larynx (**F**) (see vol. 2). In the case of unilateral paralysis of the *levator muscle of the velum palatinum* (**F18**), the soft palate and uvula are pulled toward the healthy side. The vocal cord of the affected side (**F19**) remains immobilized in the *cadaver position* owing to paralysis of the internal laryngeal muscles (*recurrent laryngeal paralysis*). **E** shows the normal positions.

Thoracic and Abdominal Part (A, D)

The vagus nerve loses its identity as a single nerve; as a visceral nerve, it spreads out like a network. It forms the *pulmonary plexus* (**A10**) at the hilum of the lung, which it crosses dorsally, and the *esophageal plexus* (**A11**), from which the *anterior vagal trunk* (**A12**) and the *posterior vagal trunk* (**A13**) extend to the anterior and posterior aspects of the stomach, forming the *anterior* (**A14**) and *posterior gastric branches*. The *hepatic branches* (**A15**) run to the *hepatic plexus*, the *celiac branches* (**A16**) to the *celiac plexus*, and the *renal branches* (**A17**) to the *renal plexus*.

The preganglionic *visceromotor* (parasympathetic) fibers originate from the **posterior nucleus of the vagus nerve**, in which a somatotopic organization of the visceral supply can be recognized (**D**).

Brain Stem

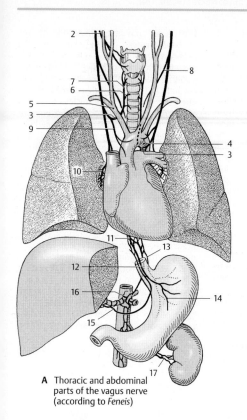

A Thoracic and abdominal parts of the vagus nerve (according to *Feneis*)

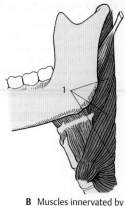

B Muscles innervated by the vagus nerve

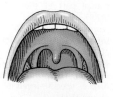

E Velum palatinum and vocal cords (normal)

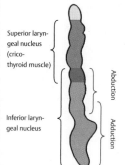

Superior laryngeal nucleus (crico- thyroid muscle)

Inferior laryngeal nucleus

Abduction

Adduction

C Somatotopic organization of the ambiguous nucleus (according to *Crosby, Humphrey, and Lauer*)

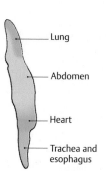

Lung

Abdomen

Heart

Trachea and esophagus

D Somatotopic organization of the posterior nucleus of the vagus nerve (according to *Getz and Sienes*)

F Velum palatinum and vocal cords in case of paralysis of the left vagus nerve

Glossopharyngeal Nerve (A – E)

The **ninth cranial nerve** supplies sensory fibers for the *middle ear*, areas of the *tongue* and of the *pharynx*, and motor fibers for the muscles of the *pharynx*. It contains motor, visceromotor (parasympathetic), viscerosensory, and taste fibers. It emerges from the medulla oblongata behind the olive right above the vagus nerve and leaves the skull together with the vagus nerve through the *jugular foramen* (**B1**). In the foramen, it forms the **superior ganglion** (**B2**) and, after passing through it, the larger **inferior ganglion** (*petrosal ganglion*) (**B3**). Laterally to the internal carotid artery and the pharynx, it forms an arch to the root of the tongue where it ramifies into several terminal branches.

The **motor fibers** (**AB4**) originate from the cranial part of the **ambiguous nucleus** (**AB5**), while the **visceroefferent fibers** (secretory fibers) (**AB6**) originate from the **inferior salivatory nucleus** (**AB7**). The cells of the **viscerosensory fibers** (**AB8**) and the **taste fibers** (**AB9**) lie in the *inferior ganglion* and descend in the **solitary tract** (**B10**) to terminate at specific levels of the **solitary nucleus** (**AB11**).

The first branch, the **tympanic nerve** (**B12**), originates from the inferior ganglion with viscerosensory and preganglionic secretory fibers in the petrosal fossula. It runs through the *tympanic canaliculus* into the *tympanic cavity*, where it receives fibers from the plexus of the internal carotid artery via the *caroticotympanic nerve* and forms the **tympanic plexus**. It supplies sensory fibers to the mucosa of tympanic cavity and *auditory (eustachian) tube* (**C**). The secretory fibers run as lesser petrosal nerve to the otic ganglion (p. 130).

Apart from connections with the vagus nerve, facial nerve, and sympathetic nervous system, the inferior ganglion gives off the (viscerosensory) **branch of the carotid sinus** (**B13**), which descends to the bifurcation of the common carotid artery and terminates in the wall of the *carotid sinus* (**B14**) and in the *carotid glomus* (**B15**) (see vol. 2).

The nerve transmits impulses of the *mechanoreceptors* of the sinus and the *chemoreceptors* of the glomus to the medulla oblongata and via collaterals to the posterior nucleus of the vagus nerve (afferent limb of the sinus reflex). Preganglionic fibers run from the vagal nucleus to groups of neurons in the cardiac atria, the axons of which (postganglionic parasympathetic fibers) terminate at the sinoatrial node and the atrioventricular node (efferent limb of the sinus reflex). This system registers and regulates the blood pressure and heart rate.

The nerve also gives off the **pharyngeal branches** (**B16**); together with the part of the vagus nerve, they form the pharyngeal plexus and participate in the sensory (**E**) and motor supply of the pharynx. A motor branch, the **stylopharyngeal branch** (**B17**), innervates the *stylopharyngeal muscle*, while some sensory **tonsillar branches** (**D18**) extend to the tonsils and the soft palate. The nerve divides below the tonsils into the **lingual branches** (**D19**), which supply the posterior third of the tongue including the vallate papillae with sensory as well as taste fibers (**D20**).

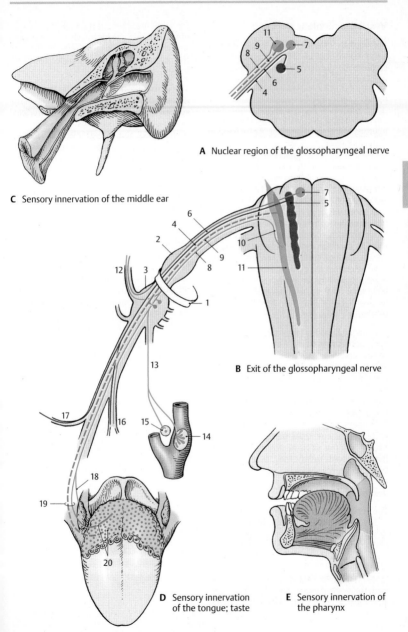

A Nuclear region of the glossopharyngeal nerve

C Sensory innervation of the middle ear

B Exit of the glossopharyngeal nerve

D Sensory innervation of the tongue; taste

E Sensory innervation of the pharynx

Vestibulocochlear Nerve

The **eighth cranial nerve** is an *afferent nerve* consisting of two components, the *cochlear root* for the **organ of hearing** and the *vestibular root* for the **organ of balance**.

Cochlear Root (A)

The nerve fibers originate from the bipolar neurons of the **spiral ganglion** (**A1**), a band of cells following the spiral course of the cochlea. The peripheral processes of the cells terminate at the hair cells of *Corti's organ*; the central processes form small bundles that organize into the **foraminous spiral tract** (**A2**) and combine in the floor of the inner auditory canal, the *internal acoustic meatus*, to form the **cochlear root** (**A3**). The latter extends, together with the *vestibular root* (**B**) inside a common connective-tissue sheath, through the *internal acoustic meatus* into the cranial cavity. At the entrance of the eighth cranial nerve into the medulla oblongata at the cerebellopontine angle, the cochlear component lies dorsally and the vestibular component ventrally.

The cochlear fibers terminate in the **anterior cochlear nucleus** (**A4**) and in the **posterior cochlear nucleus** (**A5**). From the anterior nucleus, the fibers cross over to the opposite side (**trapezoid body**) (**A6**) (p. 110, AB15); after partly synapsing in the *trapezoid nuclei* (**B7**), they ascend as **lateral lemniscus** (**A8**) (central auditory tract, p. 378). The fibers originating from the posterior cochlear nucleus cross partly as **medullary striae** (*posterior acoustic striae*) just below the rhomboid fossa; they ascend in the lateral lemniscus as well.

Vestibular Root (B)

The nerve fibers originate from bipolar neurons of the **vestibular ganglion** (**B9**) which lies in the *internal acoustic meatus*. The peripheral processes of these cells terminate at the sensory epithelia of the *semicircular ducts* (**B10**), the *saccule* (**B11**), and the *utricle* (**B12**) (p. 377, D). Their central processes unite to form the **vestibular root** (**B13**) and terminate, after bifurcation into the ascending and descending branches, in the vestibular nuclei of the medulla oblongata. Only a small portion reaches directly into the cerebellum via the *inferior cerebellar peduncle* (restiform body).

The vestibular nuclei lie in the floor of the rhomboid fossa below the lateral recess: the **superior nucleus** (*Bechterew's nucleus*) (**B14**), the **medial nucleus** (*Schwalbe's nucleus*) (**B15**), the **lateral nucleus** (*Deiters' nucleus*) (**B16**), and the **inferior nucleus** (**B17**). The primary vestibular fibers terminate mostly in the medial nucleus. Secondary fibers run from the vestibular nuclei to the cerebellum and into the spinal cord (**vestibulospinal tract**) (**B18**).

The **function of the vestibular apparatus** plays an important role for *balance* and *upright posture*. The tracts to the cerebellum and the spinal cord serve this purpose. The vestibulospinal tract has an effect on *muscular tension* in various parts of the body. The vestibular apparatus controls especially *movements of the head* and *fixation of vision* during movement (tracts to the eye-muscle nuclei) (p. 383, C).

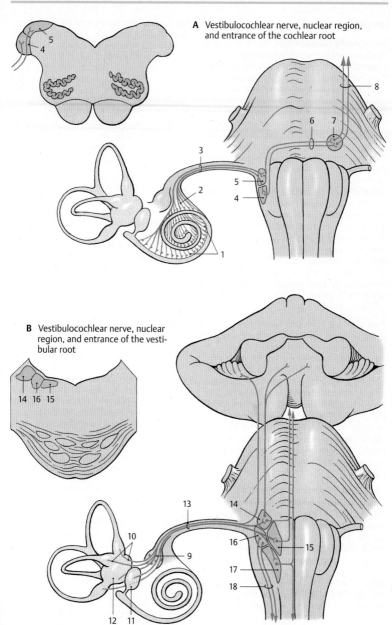

A Vestibulocochlear nerve, nuclear region, and entrance of the cochlear root

B Vestibulocochlear nerve, nuclear region, and entrance of the vestibular root

Facial Nerve (A – F)

The **seventh cranial nerve** supplies motor fibers to the muscles of facial expression; in a nerve bundle emerging separately from the brain stem, called the intermediate nerve, it carries taste fibers and viscero-efferent secretory (parasympathetic) fibers. The **motor fibers** (**AB1**) originate from the large, multipolar neurons in the **nucleus of the facial nerve** (**AB2**). They arch around the *abducens nucleus* (**AB3**) (*internal genu of the facial nerve*) and emerge on the lateral aspect of the medulla oblongata from the lower border of the pons. The cells of the **preganglionic secretory fibers** (**AB4**) form the **superior salivatory nucleus** (**AB5**). The **taste fibers** (**AB6**) originate from the pseudo-unipolar cells in the **geniculate ganglion** (**BC7**) and terminate in the cranial section of the **solitary nucleus** (**AB8**). The visceroefferent and taste fibers do not arch around the abducens nucleus but join the ascending limb of the nerve and emerge as **intermediate nerve** (**B9**) between the facial nerve and the vestibulocochlear nerve.

Both parts of the nerve pass through the inner auditory canal, the *internal acoustic meatus* (petrous part of temporal bone, internal acoustic pore, see vol. 1), and enter the *facial canal* as a nerve trunk. At the bend of the nerve in the petrous bone (*external genu of the facial nerve*) lies the *geniculate ganglion* (**BC7**). The canal continues above the tympanic cavity (p. 367, A10) and turns caudally toward the *stylomastoid foramen* (**BC10**), through which the nerve leaves the skull. The nerve ramifies into terminal branches (**parotid plexus**) (**E11**) in the *parotid gland*.

The greater petrosal nerve (**BC12**), the stapedius nerve (**BC13**), and the chorda tympani (**BC14**) branch off inside the *facial canal*. The **greater petrosal nerve** (preganglionic secretory fibers for the lacrimal gland, nasal glands, and palatal glands) originates from the geniculate ganglion, extends through the *hiatus for the lesser petrosal nerve* into the cranial cavity and over the anterior aspect of the petrous bone through the *foramen lacerum* and finally

through the *pterygoid canal* to the *pterygo-palatine ganglion* (**C15**). The **stapedius nerve** supplies the stapedius muscle in the middle ear. The **chorda tympani** (**BC14**) branches off above the stylomastoid foramen, runs beneath the mucosa through the tympanic cavity (p. 365, A22) and further to the petrotympanic fissure, and finally joins the *lingual nerve* (**C16**). It contains taste fibers for the anterior two-thirds of the tongue (**D**) and preganglionic fibers for the submandibular and sublingual glands as well as various lingual glands.

Before it enters the parotid gland, the facial nerve gives off the **posterior auricular nerve** (**E17**) as well as branches to the posterior belly of the *digastric muscle* (**CE18**) and to the *stylohyoid muscle* (**C19**). The parotid plexus gives off the *temporal branches* (**E20**), the *zygomatic branches* (**E21**), the *buccal branches* (**E22**), the *marginal mandibular branch* (**E23**), and the *cervical branch* (**E24**) for the *platysma* (see vol. 1). The branches provide innervation to all the muscles of facial expression.

Ramifications of the cervical branch lying beneath the platysma form the *superficial cervical ansa* by anastomosing with branches of the sensory transverse cervical nerve (p. 72, BC17). The small branches departing from the ansa are mixed sensorimotor nerves. The terminal ramifications of temporal branches, buccal branches, and marginal mandibular branch form similar plexuses with branches of the trigeminal nerve.

▬ **Clinical Note:** Injury to the nerve results in atony of all muscles of the affected half of the face. The mouth region drops, and the eye can no longer close (**F**). There is increased sensitivity to sound, *hyperacusis* (p. 366).

C25 Trigeminal ganglion.

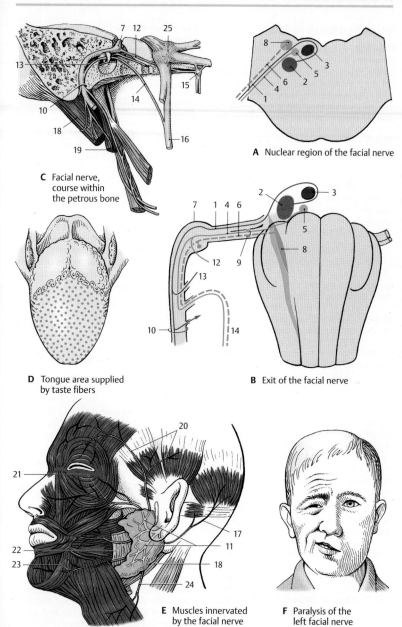

A Nuclear region of the facial nerve

C Facial nerve, course within the petrous bone

D Tongue area supplied by taste fibers

B Exit of the facial nerve

E Muscles innervated by the facial nerve

F Paralysis of the left facial nerve

Brain Stem

Trigeminal Nerve (A-F)

The **fifth cranial nerve** carries sensory fibers for the facial skin and mucosa and motor fibers for the masticatory muscles, for the *mylohyoid muscle*, the *anterior belly of the digastric muscle*, and probably also for the *tensor muscle of the velum palatinum* and the *tensor muscle of the tympanic membrane*. It emerges from the pons with a thick **sensory root** (*greater portion*) and a thinner **motor root** (*lesser portion*) and then passes to the front over the petrous bone. The trigeminal ganglion (semilunar ganglion, Gasser's ganglion) lies in a dural pocket, the trigeminal cave, and gives off three main branches, namely, the *ophthalmic nerve*, the *maxillary nerve*, and the *mandibular nerve* (see p. 104, A14 – A16).

The **sensory fibers** (**B1**) originate from the pseudounipolar cells in the **trigeminal ganglion** (semilunar ganglion, Gasser's ganglion) (**BE2**); the central processes of these cells terminate in the sensory trigeminal nuclei. Most of the fibers of the **epicritic sensibility** (p. 322) terminate in the **pontine nucleus of the trigeminal nerve** (*principal nucleus*) (**AB3**), while those of the **protopathic sensibility** (p. 324) terminate in the **spinal nucleus of the trigeminal nerve** (**BC4**). Fibers descend as **spinal tract** (**B5**) down to the upper cervical spinal cord and terminate in a somatotopic arrangement (**C**): fibers for the perioral region terminate cranially, those for the adjacent skin areas more caudally. The fibers for the outermost semicircle terminate furthest caudally (onion skin arrangement of central sensory innervation). The **mesencephalic tract** (**B6**) carries *proprioceptive* impulses from the masticatory muscles.

The **mesencephalic nucleus of the trigeminal nerve** (**AB7**) consists of pseudounipolar neurons, the processes of which run through the trigeminal ganglion without interruption. These are the only sensory fibers for which the cells of origin do not lie in a ganglion outside the CNS but in a nucleus of the brain stem, so to speak representing a sensory ganglion located inside the brain.

The **motor fibers** originate from large, multipolar neurons in the **motor nucleus of the trigeminal nerve** (**AB8**).

Innervation of mucosa (D)

The *ophthalmic nerve* supplies the frontal and sphenoid sinuses as well as the nasal septum (**D9**); the *maxillary nerve* supplies the maxillary sinuses, nasal conchae, palate, and gingiva (**D10**); and the *mandibular nerve* supplies the lower region of the oral cavity (**D11**) and the cheeks.

Ophthalmic Nerve (E)

The ophthalmic nerve (**E12**) gives off a recurrent *tentorial branch* and divides into the *lacrimal nerve* (**E13**), the *frontal nerve* (**E14**), and the *nasociliary nerve* (**E15**). These branches pass through the superior orbital fissure into the orbit; the nasociliary nerve enters through the medial section of the fissure, the two other branches enter through the lateral section.

The **lacrimal nerve** runs to the lacrimal gland (**E16**) and innervates the skin of the lateral corner of the eye. Via a communicating branch, it receives postganglionic secretory (parasympathetic) fibers from the *zygomatic nerve* for innervation of the lacrimal gland.

The **frontal nerve** divides into the **supratrochlear nerve** (**E17**) (medial corner of the eye) and the **supraorbital nerve** (**E18**), which passes through the *supraorbital notch* (conjunctiva, upper eyelid, and the skin of the forehead).

The **nasociliary nerve** runs to the medial corner of the eye, which it supplies with its terminal branch, the **infratrochlear nerve** (**E19**). The nasociliary nerve gives off the following branches: a communicating branch to the *ciliary ganglion* (**E20**), the **long ciliary nerves** (**E21**) to the eyeball, the **posterior ethmoidal nerve** (**E22**) to the sphenoidal and ethmoidal sinuses, and the **anterior ethmoidal nerve** (**E23**); the latter runs through the *anterior ethmoidal foramen* to the ethmoidal plate and through the plate into the nasal cavity. Its terminal branch, the *external nasal branch*, supplies the skin of the dorsum and the tip of the nose.

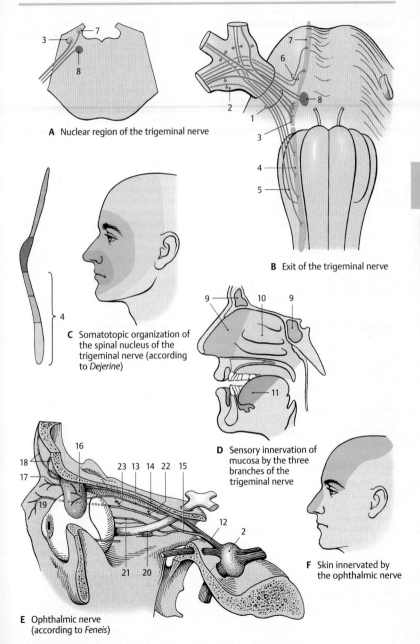

A Nuclear region of the trigeminal nerve

B Exit of the trigeminal nerve

C Somatotopic organization of the spinal nucleus of the trigeminal nerve (according to *Dejerine*)

D Sensory innervation of mucosa by the three branches of the trigeminal nerve

E Ophthalmic nerve (according to *Feneis*)

F Skin innervated by the ophthalmic nerve

Brain Stem

Trigeminal Nerve (continued)

Maxillary Nerve (A, B)

The maxillary nerve (**A1**) gives off a *meningeal branch* and then passes through the *round foramen* (**A2**) into the *pterygopalatine fossa*, where it divides into the *zygomatic nerve*, the *ganglionic branches* (*pterygopalatine nerves*), and the *infraorbital nerve*.

The **zygomatic nerve** (**A3**) reaches through the inferior orbital fissure to the lateral wall of the orbit. It gives off a communicating branch, which contains postganglionic secretory (parasympathetic) fibers from the pterygopalatine ganglion for the lacrimal gland, to the lacrimal nerve and divides into the **zygomaticotemporal branch** (**A4**) (temple) and the **zygomaticofacial branch** (**A5**) (skin over the zygomatic arch).

The **ganglionic branches** (**A6**) are two or three fine filaments running to the pterygopalatine ganglion (p. 128, A10). The fibers provide sensory innervation to the upper pharynx, nasal cavity, and hard and soft palates.

The **infraorbital nerve** (**A7**) reaches through the inferior orbital fissure into the orbit and through the infraorbital canal (**A8**) to the cheek, where it supplies the skin between lower eyelid and upper lip (**B**). It gives off the **posterior superior alveolar nerves** (**A9**) (molar teeth), the **middle superior alveolar nerve** (**A10**) (premolar teeth), and the **anterior superior alveolar nerves** (**A11**) (incisors). The nerves form the **superior dental plexus** above the alveoli.

Mandibular Nerve (C – F)

After passing through the *oval foramen* and giving off a *meningeal branch* (**C12**) in the *infratemporal fossa*, the nerve divides into the *auriculotemporal nerve*, the *lingual nerve*, the *inferior alveolar nerve*, the *buccal nerve*, and the pure *motor branches*.

The **pure motor branches** leave the mandibular nerve shortly after its passage through the foramen: the **masseteric nerve** (**C13**) for the *masseter muscle* (**F14**), the **deep temporal nerves** (**C15**) for the *temporal muscle* (**F16**),

and the **pterygoid nerves** (**C17**) for the *pterygoid muscles* (**F18**). Motor fibers for the *tensor tympani muscle* and for the *tensor muscle of the velum palatinum* run to the otic ganglion (p. 131, AB1) and emerge from it as **nerve of the tensor tympani muscle** and as **nerve of the tensor veli palatini muscle**.

The **auriculotemporal nerve** (**C19**) (temporal skin, external acoustic meatus, and tympanic membrane) usually originates with two roots that embrace the middle meningeal artery and then unite to form the nerve (p. 131, A15). The **lingual nerve** (**C20**) descends in an arch to the base of the tongue and supplies sensory fibers to the anterior two-thirds of the tongue (**D**). It receives its taste fibers from the chorda tympani (facial nerve). The **inferior alveolar nerve** (**C21**) contains motor fibers for the *mylohyoid muscle* and the anterior belly of the *digastric muscle*; furthermore, it contains sensory fibers, which enter the *mandibular canal* and give off numerous **inferior dental branches** (**C22**) for the teeth of the lower jaw. The main branch of the nerve, the **mental nerve** (**C23**), passes through the *mental foramen* and supplies sensory fibers to the chin, the lower lip, and the skin over the body of the mandible (**E**). The **buccal nerve** (**C24**) passes through the buccinator muscle (**C25**) and supplies the mucosa of the cheek.

BC26 trigeminal ganglion.

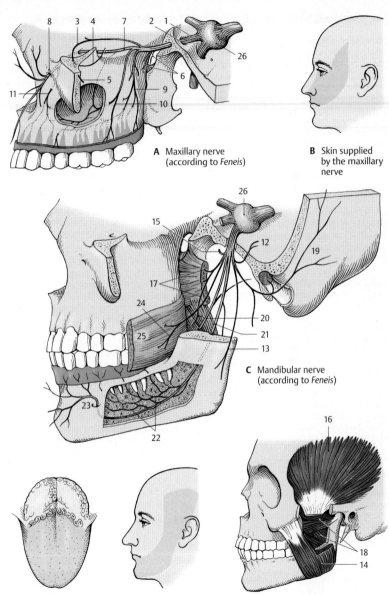

A Maxillary nerve (according to *Feneis*)

B Skin supplied by the maxillary nerve

C Mandibular nerve (according to *Feneis*)

D Sensory innervation of the tongue

E Skin supplied by the mandibular nerve

F Innervation of muscle

Brain Stem

Parasympathetic Ganglia

The fibers of the visceroefferent nuclei (visceromotor and secretory) synapse in parasympathetic ganglia to form postganglionic fibers. Apart from the *parasympathetic root* (preganglionic fibers), each ganglion has a *sympathetic root* (synapsing in the ganglia of the sympathetic chain, p. 296) and a *sensory root*, the fibers of which pass the ganglion without interruption. Thus, the branches leaving the ganglion contain sympathetic, parasympathetic, and sensory fibers.

Ciliary Ganglion (A, B)

The ciliary ganglion (**AB1**) is a small, flat body lying laterally to the optical nerve in the orbit. Its parasympathetic fibers from the Edinger–Westphal nucleus run in the *oculomotor nerve* (**AB2**) and cross over to the ganglion as oculomotor root (**AB3**) (*parasympathetic root*). The preganglionic sympathetic fibers originate from the lateral horn of the spinal cord C8–T2 (**ciliospinal center**) (**B4**) and synapse in the *superior cervical ganglion* (**B5**). The postganglionic fibers ascend in the carotid plexus (**B6**) as *sympathetic root* (**B7**) to the ciliary ganglion. Sensory fibers originate from the *nasociliary nerve* (nasociliary root) (**AB8**).

The **short ciliary nerves** (**AB9**) extend from the ganglion to the eyeball and penetrate the sclera to enter the interior of the eyeball. Their parasympathetic fibers innervate the *ciliary muscle* (accommodation) and the *sphincter pupillae muscle*; the sympathetic fibers innervate the *dilator pupillae muscle* (p. 358).

▬▬▬ **Clinical Note:** The pupil is antagonistically innervated by parasympathetic fibers (constriction of pupil) and sympathetic fibers (dilatation of pupil). Injury to the ciliospinal center or the spinal roots C8, T1 (paralysis of the lower brachial plexus, p. 74) results in *ipsilateral constriction of the pupil*.

Pterygopalatine Ganglion (A, B)

The pterygopalatine ganglion (**AB10**) lies at the anterior wall of the pterygopalatine fossa below the maxillary nerve (**AB11**), which gives off **ganglionic branches** (*pterygopalatine nerves*) (**AB12**) to the ganglion (*sensory root*). The parasympathetic secretory fibers from the superior salivatory nucleus extend in the facial nerve (intermediate nerve) (**AB13**) up to the genu of the facial nerve where they branch off as the **greater petrosal nerve** (**AB14**). The nerve passes through the *foramen lacerum* to the base of the skull and through the *pterygoid canal* to the ganglion (*parasympathetic root*). Sympathetic fibers from the carotid plexus form the **deep petrosal nerve** (**AB15**) (*sympathetic root*) and join the greater petrosal nerve to form the **nerve of the pterygoid canal** (**AB16**).

The branches leaving the ganglion carry secretory fibers for the lacrimal gland and for the glands of the nasal cavity. The parasympathetic fibers (**B17**) for the lacrimal gland (**AB18**) synapse in the ganglion. The postganglionic fibers run in the ganglionic branches (**AB12**) to the maxillary nerve (**AB11**) and reach the lacrimal gland via the zygomatic nerve (**AB19**) and its anastomosis (**A20**) with the lacrimal nerve (**A21**).

The remaining parasympathetic secretory fibers run in the **orbital branches** (**B22**) to the posterior ethmoidal cells, in the **lateral posterior nasal branches** (**B23**) to the nasal conchae, in the **nasopalatine nerve** across the nasal septum and through the *incisive canal*, and in the **palatine nerve** (**AB24**) to the hard and soft palates.

The *taste fibers* (**B25**) for the soft palate run in the palatine nerves and in the greater petrosal nerve.

A26 Trigeminal ganglion.

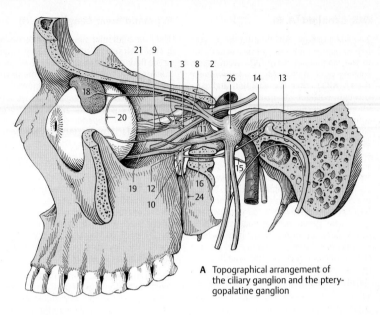

A Topographical arrangement of the ciliary ganglion and the ptery-gopalatine ganglion

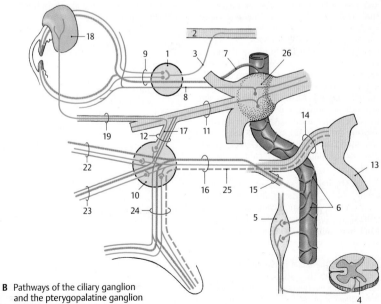

B Pathways of the ciliary ganglion and the pterygopalatine ganglion

Otic Ganglion (A, B)

The otic ganglion (**AB1**) is a flat body lying below the oval foramen on the medial side of the *mandibular nerve* (**A2**), from where sensory and motor fibers (*sensorimotor roots*) (**AB3**) enter the ganglion and pass through without synapsing. The preganglionic parasympathetic fibers originate from the *inferior salivatory nucleus*. They run in the glossopharyngeal nerve and branch off, together with the tympanic nerve, from the inferior ganglion of the glossopharyngeal nerve in the petrous fossula to the tympanic cavity. The fibers leave the tympanic cavity through the hiatus for the lesser petrosal nerve as a fine branch, the **lesser petrosal nerve** (**AB4**) (*parasympathetic root*). The nerve runs beneath the dura mater along the surface of the petrous bone and reaches the otic ganglion after passing through the foramen lacerum. The fibers of the *sympathetic root* (**AB5**) originate from the plexus of the middle meningeal artery.

The motor fibers from the motor root of the trigeminal nerve pass through the ganglion and leave it in the **nerve to tensor veli palatini** (**B6**) (soft palate) and in the **nerve to tensor tympani** (**B7**) (for the muscle that tightens the tympanic membrane). *Motor fibers* (**B8**) for the *levator veli palatini* from the facial nerve (VII) are thought to run in the chorda tympani (**AB9**) and cross over into the ganglion via the *communicating branch with chorda tympani* (**AB10**). They pass through without synapsing and enter via a communicating branch (**A11**) the *greater petrosal nerve* (**A12**), in which they reach the *pterygopalatine ganglion* (**A13**). They pass to the palate in the *palatine nerves* (**A14**).

The postganglionic secretory (parasympathetic) fibers together with sympathetic fibers enter the *auriculotemporal nerve* (**AB15**) via a communicating branch and from here into the *facial nerve* (**AB16**) via another anastomosis. The fibers then ramify in the *parotid gland* (**AB17**) together with branches of the facial nerve. Apart from the parotid gland, they supply the buccal and labial glands via the *buccal nerve* and the *inferior alveolar nerve*.

Submandibular Ganglion (A, B)

The submandibular ganglion (**AB18**) lies together with several small secondary ganglia in the floor of the mouth above the *submandibular gland* (**AB19**) and below the *lingual nerve* (**AB20**), to which it connects via several ganglionic branches. Its preganglionic *parasympathetic fibers* (**B21**) originate from the *superior salivatory nucleus*, run in the facial nerve (intermediate nerve), and leave the nerve together with the taste fibers (**B22**) in the *chorda tympani* (**AB9**). In the latter, the fibers reach the lingual nerve (**AB20**) and extend in it to the floor of the mouth where they cross over into the ganglion (**AB18**). Postganglionic *sympathetic fibers* from the plexus of the external carotid artery reach the ganglion via the *sympathetic branch* (**B23**) given off by the plexus of the facial artery; they pass through the ganglion without synapsing.

The postganglionic parasympathetic and sympathetic fibers pass partly in the *glandular branches* to the *submandibular gland*, partly in the lingual nerve to the *sublingual gland* (**AB24**) and to the glands in the distal two-thirds of the tongue.

A25 Ciliary ganglion.
A26 Trigeminal ganglion.

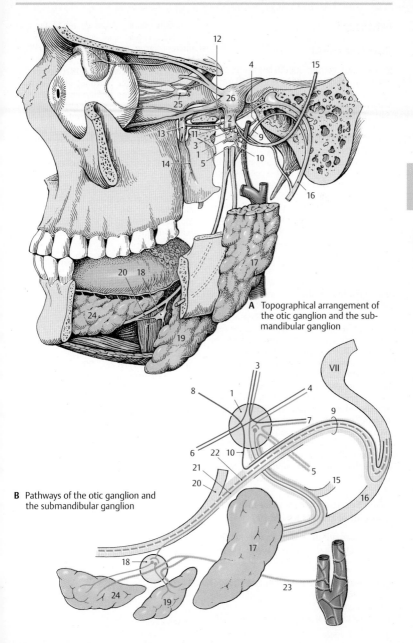

A Topographical arrangement of the otic ganglion and the sub-mandibular ganglion

B Pathways of the otic ganglion and the submandibular ganglion

Brain Stem

Midbrain

Structure (A–C)

Apart from certain modifications in the *medulla oblongata* (**A1**), *pons* (**A2**), and *mesencephalon* (**A3**), the brain stem has a uniform structure. The phylogenetically old part of the brain stem, which is common to all three parts and contains the cranial nerve nuclei, is the **tegmentum** (**A4**). At the level of the medulla oblongata and the pons, it is overlain by the cerebellum and in the mesencephalon by the **tectum** (*quadrigeminal plate*) (**A5**). The ventral part of the brain stem mainly contains the large tracts descending from the telencephalon; they form the *pyramids* (**A6**) in the medulla oblongata, the *pontine bulb* (**A7**) in the pons, and the *cerebral peduncles* (**A8**) in the mesencephalon.

The ventricular system undergoes considerable narrowing in the midbrain, the **aqueduct of the mesencephalon** (*cerebral aqueduct, aqueduct of Sylvius*) (**A–D9**). During development, the lumen of the neural tube becomes increasingly narrowed as the tegmentum of the midbrain increases in volume (**B**), while the blueprint of the neural tube survives. The motor derivatives of the **basal plate** lie ventrally: the *nucleus of the oculomotor nerve* (**BC10**), the *trochlear nucleus* (eye muscles), the *red nucleus* (**C11**), and the *substantia nigra* (**C12**) (consisting of an outer reticular part and an inner compact part). The sensory derivatives of the alar plate lie dorsally: the *tectum of the mesencephalon* (quadrigeminal plate) (**C13**) (synaptic relay station for auditory and visual pathways).

Cross Section Through the Inferior Colliculi of the Midbrain (D)

The *inferior colliculus* with its nucleus (nucleus of inferior colliculus) (**D14**) (synaptic relay station of the central auditory pathway) is seen dorsally. The transitional region between pons and cerebral peduncles and the most caudal cell groups of the *substantia nigra* (**D15**) lie ventrally. The magnocellular *nucleus of the trochlear nerve* (**D16**)

is clearly visible in the center of the tegmentum below the aqueduct, and the *lateroposterior tegmental nucleus* (**D17**) is situated dorsal to it. Further lateral are the cells of the *locus ceruleus* (**D18**) (the pontine respiratory center reaching into the midbrain; it contains noradrenergic neurons, p. 100, B28). The relatively large, scattered cells dorsal to the locus ceruleus form the *mesencephalic nucleus of the trigeminal nerve* (**D19**). The lateral field is occupied by the *pedunculopontine tegmental nucleus* (**D20**). At the ventral margin of the tegmentum lies the *interpeduncular nucleus* (**D21**), which is rich in peptidergic neurons (predominantly enkephalin). The *habenulo interpeduncular tract* (fasciculus retroflexus, Meynert's bundle) (p. 176, A11), which descends from the *habenular nucleus*, terminates here.

The *lateral lemniscus* (**D22**) radiates ventrally into the nucleus of the inferior colliculus (**D14**) (p. 378, A5). The fibers of the *peduncle of the inferior colliculus* (**D23**) aggregate at the lateral aspect and run to the *medial geniculate body* (central auditory pathway, p. 378). In the medial field lie the *medial longitudinal fasciculus* (**D24**) (p. 142) and the *decussation of the superior cerebellar peduncle* (**D25**) (p. 166, B5). The fiber plate of the *medial lemniscus* (**D26**) (p. 140, B) lies laterally. The fibers of the *cerebral peduncle* (**D27**) are cut transversely and are interspersed with a few pontine fibers running across.

D28 Periaqueductal gray substance, *central gray.*
C29 Edinger–Westphal nucleus (accessory nucleus of the oculomotor nerve).

Plane of section

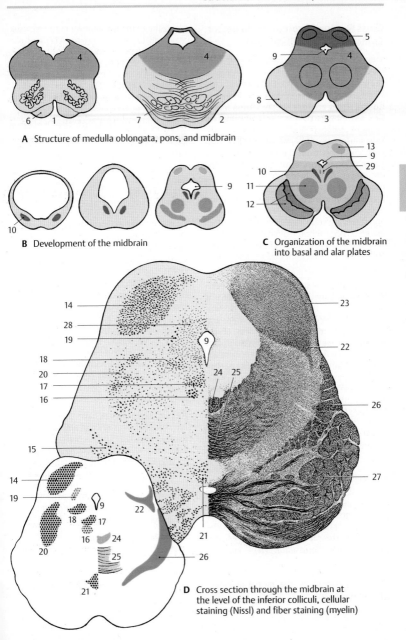

A Structure of medulla oblongata, pons, and midbrain

B Development of the midbrain

C Organization of the midbrain into basal and alar plates

D Cross section through the midbrain at the level of the inferior colliculi, cellular staining (Nissl) and fiber staining (myelin)

Brain Stem

Cross Section Through the Superior Colliculi of the Midbrain (A)

The two *superior colliculi* (**A1**) are seen dorsally. In lower vertebrates, they represent the most important visual center and consist of several layers of cells and fibers. In humans, they are only a relay station for reflex movements of the eyes and pupillary reflexes, and their stratification is rudimentary. In the *superficial gray layer* (**A2**) terminate the fibers from the occipital fields of the cortex (*corticotectal tract*) (**A3**). The *optic layer* (**A4**), which in lower vertebrates consists of fibers of the optical tract, is formed in humans by fibers from the lateral genicular body. The deeper layers of cells and fibers are collectively known as *stratum lemnisci* (**A5**). Here terminate the spinotectal tract (p. 56, A5), fibers of the medial and lateral lemnisci, and fiber bundles of the inferior colliculi.

The aqueduct is surrounded by the periaqueductal gray, or *central gray* (**AB6**). It contains a large number of peptidergic neurons (VIP, enkephalin, cholecystokinin, and others). The **mesencephalic nucleus of the trigeminal nerve** (**A7**) lies laterally to it, and ventrally to it lie the **nucleus of the oculomotor nerve** (**A8**) and the **Edinger-Westphal nucleus** (accessory oculomotor nucleus) (**A9**) (p. 138, AD19). Dorsally to both nuclei runs the *posterior longitudinal fasciculus* (Schütz's bundle) (p. 144, B) and ventrally to them the *medial longitudinal fasciculus* (**A10**) (p. 142). The main nucleus of the tegmentum is the **red nucleus** (**AB11**) (p. 136; p. 148, A2); it is delimited by a capsule consisting of afferent and efferent fibers (among others, the *dentatorubral fasciculus*) (**A12**). At its medial margin descend *fiber bundles of the oculomotor nerve* (**A13**) in ventral direction. Tectospinal fibers (pupillary reflex) and tectorubral fibers cross the midline in the *superior tegmental decussation* (*Meynert's decussation*) (**A14**) and tegmentospinal fibers in the *inferior decussation* (*Forel's decussation*) (**A15**). The lateral field is occupied by the **medial lemniscus** (**AB16**) (p. 140, B).

Ventrally to the tegmentum border the **substantia nigra** (*pars compacta* [**A17**] and *pars reticulata* [**A18**], p.136; p. 148, A1). The ventral aspect on both sides is formed by the corticofugal fiber masses of the **cerebral peduncles** (**AB19**). The dorsal aspect is formed by the *medial genicular body* (**AB20**).

Cross Section Through the Pretectal Region of the Midbrain (B)

The **pretectal region** (**B21**) situated orally to the superior colliculi represents the transition from the midbrain to the diencephalon. Hence, the cross section already contains structures of the diencephalon: dorsally on each side lies the *pulvinar* (**B22**), in the middle the *epithalamic commissure* (**B23**), and ventrally the *mamillary bodies* (**B24**). The pretectal region extends dorsolaterally with the **principal pretectal nucleus** (**B25**). The latter is an important relay station for the pupillary reflex (p. 358, A2). The fibers of the optical tract and the fibers of the occipital cortical fields terminate here. An efferent pathway of the nucleus extends across the epithalamic commissure to the Edinger–Westphal nucleus (accessory oculomotor nucleus). Ventral to the aqueduct are the **Darkshevich's nucleus** (**B26**) and the **interstitial nucleus** (*of Cajal*) (**B27**), the relay stations in the system of the medial longitudinal fascicle (p. 142, A8, A9).

Animal experiments have shown that the *interstitial nucleus of Cajal* and the *prestitial nucleus* situated further orally are important relay stations for automatic motions (p. 192, B) within the extrapyramidal motor system (p. 310). The essential synapses for the rotation of the body around its longitudinal axis lie in the interstitial nucleus, those for raising the head and upper body in the prestitial nucleus.

B28 Supramamillary commissure.

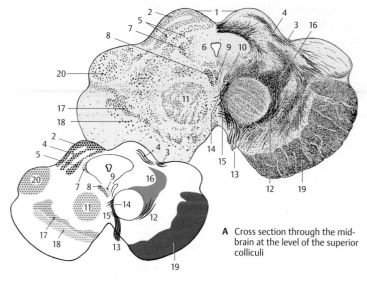

A Cross section through the mid-brain at the level of the superior colliculi

Planes of sections

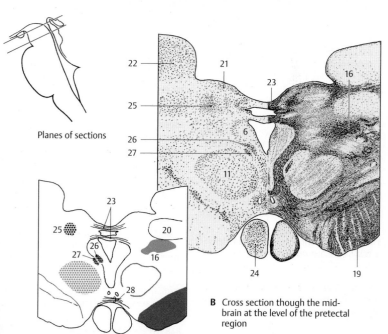

B Cross section though the mid-brain at the level of the pretectal region

Red Nucleus and Substantia Nigra

Lateral View of the Brain Stem (A)

The two large nuclei extend far toward the diencephalon. The **substantia nigra** (**AB1**) reaches from the oral part of the *pons* (**A2**) to the *pallidum* (**AB3**) in the diencephalon. Both nuclei are important relay stations of the extrapyramidal system (p. 310).

Red Nucleus (A, B)

The nucleus (**AB4**) appears reddish in a fresh brain section (high iron content, p. 148, A). It consists of the parvocellular *neorubrum* and the magnocellular *paleorubrum* situated ventrocaudally.

Afferent connections

- The *dentatorubral fasciculus* (**B5**) of the *dentate nucleus* (**B6**) of the cerebellum runs in the *superior cerebellar peduncle* and terminates in the contralateral red nucleus.
- The *tectorubral tract* (**B7**) of the superior colliculus terminates in the ipsilateral and contralateral paleorubrum.
- The *pallidorubral tract* (**B8**) consists of pallidotegmental bundles from the inner segment of the pallidum.
- The *corticorubral tract* (**B9**) from the frontal and precentral cortex terminates in the ipsilateral red nucleus.

Efferent connections

- The *rubroreticular* and *rubro-olivary fibers* (**B10**) run in the *central tegmental tract* (p. 144, A) and terminate primarily in the *olive* (neuronal circuit: dentate nucleus – red nucleus – olive – cerebellum).
- The *rubrospinal tract* (**B11**) (poorly developed in humans) crosses in Forel's tegmental decussation and terminates in the cervical spinal cord.

Functional significance. The red nucleus is a relay and control station for cerebellar, pallidal, and corticomotor impulses that are important for *muscle tone, posture,* and *locomotion.* Injury to this nucleus causes passive tremor (shaking), changes in muscle tone, and choreic-athetoid hyperactivity.

Substantia Nigra (A–C)

This consists of the dark **compact part** (nerve cells with black melanin pigment) (**C**) and the **reticular part** (of reddish color and rich in iron). The tracts of the substantia nigra form only loose pathways of fine fibers rather than compact bundles.

Afferent connections terminating in the anterior part

- Fibers of the caudate nucleus, *strionigral fasciculus* (**B12**)
- Fibers of the frontal cortex (areas 9 to 12), *corticonigral fibers* (**B13**)

Afferent connections terminating in the caudal part

- Fibers of the putamen (**B14**)
- Fibers of the precentral cortex (areas 4 and 6) (**B15**)

Efferent connections

- **Nigrostriatal fibers** (**B16**), running from the compact part to the striatum
- Fibers of the reticular part, running to the thalamus

The majority of efferent fibers ascend to the striatum, to which the substantia nigra is closely connected functionally by the **nigrostriatal system**. In the axons of the **dopaminergic nigral neurons** (compact part), dopamine is transported to the striatum, where it is released by the axon terminals. There is a topographical relationship between the substantia nigra and the striatum (caudate nucleus and putamen); cranial and caudal segments of the substantia nigra are connected with the corresponding segments of caudate nucleus and putamen. Caudate nucleus and putamen are under the control of massive input from totally different neocortical zones (**B17**).

Functional significance. The substantia nigra is of special importance for the control of involuntary coordinated movement and the rapid onset of movement (*starter function*). Injury causes muscle stiffness, passive tremor, and loss of coordinated movement and facial expression (*masklike expression of the face*).

B18 Posterior thalamus.

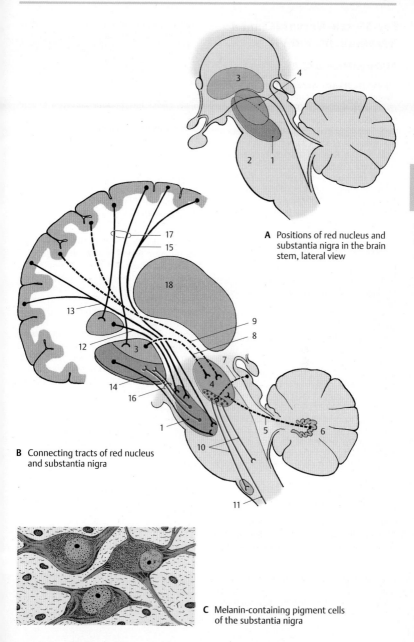

A Positions of red nucleus and substantia nigra in the brain stem, lateral view

B Connecting tracts of red nucleus and substantia nigra

C Melanin-containing pigment cells of the substantia nigra

Eye-Muscle Nerves (Cranial Nerves III, IV, and VI)

Abducens Nerve (C, E)

The **sixth cranial nerve** (**C1**) is an exclusively *somatomotor nerve*, which innervates the **lateral rectus muscle** (**E2**) of the *extra-ocular muscles*. Its fibers originate from the large, multipolar neurons of the **nucleus of the abducens nerve** (**C3**), which lies in the pons in the floor of the rhomboid fossa (p. 110, A1). The fibers exit at the basal margin of the pons above the pyramid. After taking a long intradural course, the nerve passes through the cavernous sinus and leaves the cranial cavity through the *superior orbital fissure*.

Trochlear Nerve (B, C, E)

The **fourth cranial nerve** (**BC4**) is an exclusively *somatomotor nerve* and innervates the **superior oblique muscle** (**E5**) of the *extra-ocular muscles*. Its fibers originate from the large, multipolar neurons of the **nucleus of the trochlear nerve** (**BC6**) (p. 132, D16), which lies in the midbrain below the aqueduct at the level of the inferior colliculi. The fibers ascend dorsally in an arch, cross above the aqueduct, and leave the midbrain at the lower margin of the inferior colliculi. The nerve is the only cranial nerve leaving the brain stem at its dorsal aspect. It descends in the subarachnoid space (p. 289, A13) to the base of the skull, where it enters the dura mater at the margin of the tentorium and continues through the lateral wall of the cavernous sinus. It enters the orbit through the *superior orbital fissure*.

Oculomotor Nerve (A, C–F)

The **third cranial nerve** (**AC7**) contains *somatomotor* and *visceromotor* (parasympathetic) (**A8**) fibers. It innervates the *remaining outer eye muscles* (**E**) and, with its visceromotor portion, the *intra-ocular muscles*. The fibers exit from the floor of the interpeduncular fossa at the medial margin of the cerebral peduncle in the oculomotor

sulcus. Laterally to the sella turcica, they penetrate the dura mater, run through the roof and then through the lateral wall of the cavernous sinus, and enter the orbit through the *superior orbital fissure*. Here, the nerve divides into a *superior branch*, which supplies the *levator muscle of the upper eyelid* and the *superior rectus muscle* (**E9**), and into an *inferior branch*, which supplies the *inferior rectus muscle* (**E10**), the *medial rectus muscle* (**E11**), and the *inferior oblique muscle* (**E12**).

The somatomotor fibers originate from large multipolar neurons of the **nucleus of the oculomotor nerve** (**AC13**) (p. 134, A8), which lies in the midbrain below the aqueduct at the level of the superior colliculi.

The longitudinally arranged cell groups innervate specific muscles. The neurons for the inferior rectus muscle (**D14**) lie dorsolaterally, those for the superior rectus muscle (**D15**) dorsomedially; below them lie the neurons for the inferior oblique muscle (**D16**), those for the medial rectus muscle (**D17**) ventrally, and those for the levator muscle of the upper eyelids (caudal central oculomotor nucleus) (**D18**) dorsocaudally. In the middle third between the two paired main nuclei there usually lies an unpaired cell group, *Perlia's nucleus*, which is thought to be associated with ocular convergence (p. 358, C).

The preganglionic visceromotor (parasympathetic) fibers originate from the parvocellular **Edinger–Westphal nucleus**, the *accessory oculomotor nucleus* (**ACD19**). They run from the oculomotor nucleus to the ciliary ganglion where they synapse. The postganglionic fibers enter through the sclera into the eyeball and innervate the *ciliary muscle* (**F20**) and the *sphincter pupillae muscle* (**F21**) (p. 358, A, B).

(For extra-ocular muscles, see p. 340.)

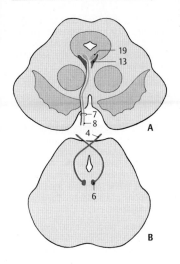

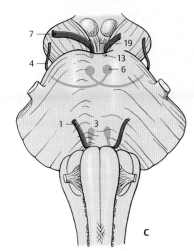

A–C Nuclear regions and exits of abducens nerve, trochlear nerve, and oculomotor nerve

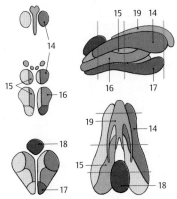

D Somatotopic arrangement of neurons in the oculomotor nucleus (according to *Warwick*)

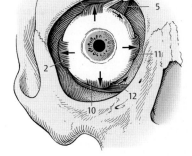

E Extra-ocular muscles

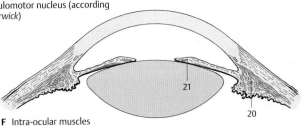

F Intra-ocular muscles

Brain Stem

Long Pathways

Corticospinal Tract and Corticonuclear Fibers (A)

The **pyramidal tract**, or **corticospinal tract** (p. 58, A; p. 308), runs through the basal part of the brain stem and forms the pyramids in the medulla oblongata (p. 100, A6).

Some of the pyramidal tract fibers terminate in the motor nuclei of cranial nerves (**corticonuclear fibers**):

- *Bilaterally* in the oculomotor nucleus (III), in the motor nucleus of the trigeminal nerve (V), in the caudal part of the facial nucleus (VII) (forehead muscles), and in the ambiguous nucleus (X)
- *After crossing* to the contralateral nucleus: in the abducens nucleus (VI), in the rostral part of the facial nucleus (VII) (facial muscles with the exception of the forehead muscles), and in the hypoglossal nucleus (XII)
- *Uncrossed* in the ipsilateral trochlear nucleus (IV)

▣ **Clinical Note:** In *central facial paralysis*, paralysis of facial muscles is caused by injury to the corticobulbar fibers, yet the mobility of the bilaterally innervated forehead muscles is retained.

Aberrant fibers (*Déjérine*) (**A1**). At various levels of the midbrain and the pons, fine fiber bundles branch off from the corticonuclear fibers and unite to form the *mesencephalic aberrant tract* and the *pontine aberrant tract*. Both descend in the *medial lemniscus* (**A2**) and terminate in the contralateral abducens nucleus (VI) and hypoglossal nucleus (XII), in the two ambiguous nuclei (X), and in the spinal accessory nucleus (XI).

Medial Lemniscus (B)

This fiber system includes the *most important ascending pathways of the exteroceptive sensibility* from the spinal cord and the brain stem. It is subdivided into the *spinal lemniscus* and the *trigeminal lemniscus*. The spinal lemniscus contains the sensory pathways for trunk and limbs (bulbothalamic tract, spinothalamic tract, spinotectal tract), while the trigeminal lemniscus contains the sensory pathways for the face (anterior tegmental fasciculus).

1. Bulbothalamic tract (B3). The fibers represent the extension of the posterior funiculi of the spinal cord (**B4**) (*epicritic sensibility*). They originate in the gracile nucleus (**B5**) and the cuneate nucleus (**B6**), cross as arcuate fibers (decussation of lemnisci) (**B7**), and form the medial lemniscus in the narrower sense (p. 108, A19). The cuneate fibers initially lie dorsally to the gracile fibers, while they lie medially to them in the pons and midbrain. They terminate in the thalamus.

2. Spinothalamic tract (*lateral and anterior*) (**B8**). The fibers (*protopathic sensibility, pain, temperature, coarse tactile sensation*) have already crossed to the contralateral side at various levels of the spinal cord and form slightly spread, loose bundles (spinal lemniscus) in the medulla oblongata. They join the medial lemniscus (p. 133, D26; p. 135, A16) in the midbrain.

3. Spinotectal tract (**B9**). The fibers run together with those of the lateral spinothalamic tract. They form the lateral tip of the lemniscus in the midbrain and terminate in the superior colliculi (*pupillary reflex on sensation of pain*).

4. Anterior tegmental fasciculus (*Spitzer*) (**B10**). The fibers (*protopathic and epicritic sensibilities of the face*) cross in small bundles from the spinal nucleus of the trigeminal nerve (principle nucleus) to the contralateral side (trigeminal lemniscus) and join the medial lemniscus at the level of the pons. They terminate in the thalamus.

5. Secondary taste fibers (**B11**). These originate from the rostral part of the solitary nucleus (**B12**), probably cross to the contralateral side, and occupy the medial margin of the lemniscus. They terminate in the thalamus.

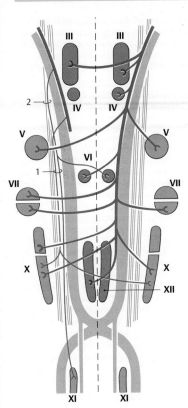

A Pyramidal system: corticospinal tract and corticonuclear fibers

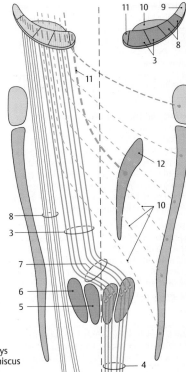

B Ascending pathways of the medial lemniscus

Medial Longitudinal Fasciculus (A)

The **medial longitudinal fasciculus** is not a uniform fiber tract but contains different fiber systems that enter and exit at various levels. It reaches from the rostral midbrain into the spinal cord and interconnects numerous nuclei of the brain stem. On cross sections through the brain stem, it is found in the middle of the tegmentum, ventrally from the central gray (p. 109, AB17; p. 111, A11; p. 133, D24).

Vestibular part. Crossed and uncrossed fibers run in the longitudinal fasciculus from the lateral (**A1**), medial (**A2**), and inferior (**A3**) vestibular nuclei to the abducens nucleus (**A4**) and to the motor cells of the anterior horn of the cervical spinal cord. From the superior vestibular nucleus (**A5**), fibers ascend to the ipsilateral trochlear nucleus (**A6**) and oculomotor nucleus (**A7**). The vestibular fibers finally terminate in the ipsilateral or contralateral interstitial nucleus of Cajal (**A8**) and in Darkshevich's nucleus (**A9**) (decussation of the epithalamic commissure [**A10**]). The longitudinal fasciculus connects the vestibular apparatus with the eye and neck muscles and with the extrapyramidal system (p. 382).

Extrapyramidal part. The interstitial nucleus of Cajal and Darkshevich's nucleus are intercalated in the course of the longitudinal fasciculus. They receive fibers from the striatum and pallidum and crossed fibers from the cerebellum. They send a fiber tract, the *interstitiospinal fasciculus* (**A11**), in the longitudinal fasciculus to the caudal brain stem and into the spinal cord.

Internuclear part. This consists of connecting fibers between motor nuclei of cranial nerves, namely, between abducens nucleus (**A4**) and oculomotor nucleus (**A7**), facial nucleus (**A12**) and oculomotor nucleus, facial nucleus and motor nucleus of the trigeminal nerve (**A13**), hypoglossal nucleus (**A14**) and ambiguous nucleus (**A15**).

The interconnections of motor nuclei of cranial nerves allow certain muscle groups to interact functionally, for example, during the coordination of eye muscles with the movements of the eyeball, coordination of eyelid muscles during opening and closing of the eyelids, and coordination of masticatory muscles and muscles of tongue and pharynx during swallowing and speaking.

Internuclear Connections of the Trigeminal Nuclei

Only a few secondary trigeminal fibers enter the medial longitudinal fasciculus. The majority of fibers run primarily uncrossed in the dorsolateral region of the tegmentum to the motor nuclei of cranial nerves; they form the basis of numerous important reflexes. Crossed and uncrossed fibers run to the facial nucleus as the basis of the *corneal reflex* (eyelids close upon touching the cornea). There are connections to the superior salivatory nucleus for the *lacrimal reflex.* Fibers to the hypoglossal nucleus, to the ambiguous nucleus, and to the anterior horn cells of the cervical spinal cord (cells of origin of the phrenic nerve) are the basis of the *sneezing reflex.* The *pharyngeal reflex* is based on fiber connections to the ambiguous nucleus, the posterior vagus nucleus, and the motor nucleus of the trigeminal nerve. Connections with the posterior vagus nucleus are the basis of the *oculocardial reflex* (slow heart rate upon pressure on the eyeballs).

Brain Stem

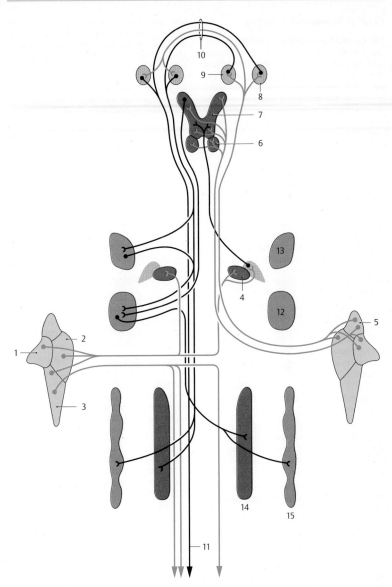

A Medial longitudinal fasciculus (according to *Crosby, Humphrey, and Lauer*)

Central Tegmental Tract (A)

The **central tegmental tract** is the most important efferent pathway of the extrapyramidal motor system (p. 310). It runs from the midbrain to the lower portion of the olive (**A1**) where the majority of its fibers terminate. The remaining fibers are thought to continue into the spinal cord via short neurons that synapse in series (*reticuloreticular fibers*) (**A2**). In the caudal midbrain, the tract lies dorsolaterally from the decussation of the superior cerebellar peduncles; it forms a large, not clearly demarcated fiber plate (p. 111, AB13) in the pons.

The tract consists of three components:

- The **pallido-olivary fibers** (**A3**) from the *striatum* (**A4**) and the *pallidum* (**A5**), which extend in the pallidotegmental bundle (**A6**) to the capsule of the red nucleus (**A7**) and further to the olive. Fibers from the *zona incerta* (**A8**) join them.
- The **rubro-olivary fibers** (**A9**) from the parvocellular part (neorubrum) of the red nucleus. In humans, they form a strong fiber tract, the *rubro-olivary fasciculus*, representing the most important descending pathway of the red nucleus.
- The **reticulo-olivary fibers** (**A10**) join the tegmental tract from various levels, namely, from the red nucleus, the central gray of the aqueduct (**A11**), and the reticular formation of pons and medulla oblongata.

Impulses received by the olive from the extrapyramidal motor centers, and probably also from the motor cortex, are relayed to the cerebellar cortex via the *olivocerebellar fibers* (**A12**).

Posterior Longitudinal Fasciculus (B)

The **posterior longitudinal fasciculus** (*Schütz's bundle*) (p. 197, B11) contains ascending and descending fiber systems that connect the hypothalamus with various nuclei of the brain stem and provide connections between the visceroefferent parasympathetic nuclei. A large portion of the fibers are peptidergic (somatostatin, among others). They originate or terminate, respectively, in the *septum*, the *oral hypothalamus*, the *gray tubercle* (**B13**), and the *mamillary bodies* (**B14**). They aggregate in the midbrain below the ependyma (p. 284) of the aqueduct and form the longitudinal fasciculus, which runs beneath the ependyma on the floor of the fourth ventricle to the lower portion of the medulla oblongata (p. 107, AB15; p. 111, AB12).

Fibers branch off to the *superior colliculi* (**B15**) and to the parasympathetic nuclei, namely, the *Edinger–Westphal nucleus* (accessory nucleus of oculomotor nerve) (**B16**), the *superior* (**B17**) and *inferior* (**B18**) *salivatory nuclei*, and the *posterior vagus nucleus* (**B19**). Other fibers terminate in the cranial nerve nuclei, namely, in the *motor nucleus of the trigeminal nerve* (**B20**), the *facial nucleus* (**B21**), and the *hypoglossal nucleus* (**B22**). Fibers are also exchanged with the nuclei of the reticular formation.

The posterior longitudinal fascicle receives olfactory impulses via the *lateroposterior tegmental nucleus* (habenular nucleus – interpeduncular nucleus – lateroposterior tegmental nucleus).

Long ascending pathways. Fibers, probably taste fibers, ascend from the *solitary nucleus* (**B23**) to the hypothalamus. The fibers of serotoninergic neurons can be traced by fluorescence microscopy from the *posterior raphe nucleus* (**B24**) into the region of the septum.

The posterior longitudinal fasciculus receives hypothalamic, olfactory, and gustatory impulses that are relayed to the motor and sensory nuclei of the brain stem (reflex movement of the tongue, secretion of saliva).

A25 Subthalamic nucleus.

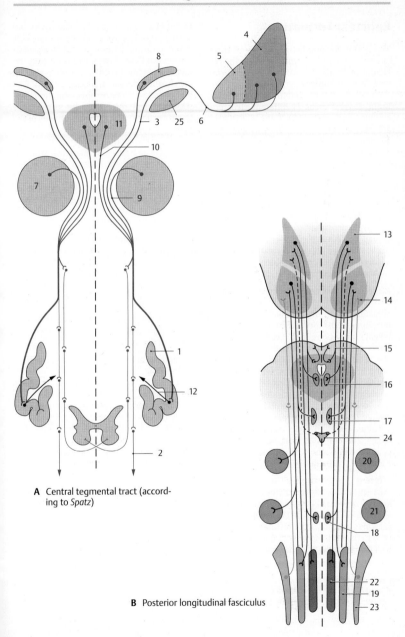

A Central tegmental tract (according to *Spatz*)

B Posterior longitudinal fasciculus

Brain Stem

Reticular Formation

The scattered neurons of the tegmentum and their network of processes form the **reticular formation**. This occupies the central area of the tegmentum and expands from the medulla oblongata into the rostral midbrain. Several areas of different structure can be distinguished (**A**). In the *medial part* are magnocellular nuclei from where long **ascending and descending fiber tracts** originate. The parvocellular *lateral part* is regarded as an **association area**.

Many of the neurons have long ascending or descending axons, or axons bifurcating into an ascending and a descending branch. As shown by Golgi impregnation, such a neuron (**B1**) can simultaneously reach *caudal cranial nerve nuclei* (**B2**) and *diencephalic nuclei* (**B3**). The reticular formation contains a large number of peptidergic neurons (enkephalin, neurotensin, and others).

Afferent connections. The reticular formation is reached by impulses of *all sensory modalities*. Sensory spinoreticular fibers terminate in the medial field of medulla oblongata and pons, and so do secondary fibers of the trigeminal and vestibular nuclei. Collaterals of the lateral lemniscus bring in acoustic impulses, while fibers of the tectoreticular fasciculus bring in optic impulses. Experimental studies on stimulation have shown that reticular neurons are excited more by sensory (pain), acoustic and vestibular stimuli than by optic stimuli. Other afferent fibers originate from the cerebral cortex, the cerebellum, the red nucleus, and the pallidum.

Efferent connections. The **reticulospinal tract** (p. 58, B5, B6) runs from the medial field of medulla oblongata and pons into the spinal cord. Bundles of the **reticulothalamic fasciculus** ascend to the intralaminar nuclei of the thalamus (truncothalamus) (p. 180, B). Fiber bundles from the midbrain terminate in the oral hypothalamus and in the septum.

Respiratory and cardiovascular control centers. Groups of neurons regulate respiration (**C**), heart rate, and blood pressure (changes upon physical activity or emotion). The neurons for *inspiration* are localized in the central field of the lower portion of the medulla oblongata (**C4**), those for *expiration* are further dorsal and lateral (**C5**). The higher relay stations for inhibition and stimulation of respiration lie in the pons (*locus ceruleus*). The autonomic nuclei of the glossopharyngeal nerve and the vagus nerve are involved in regulating heart rate and blood pressure (**D**). Electrical stimulation in the caudal central field of the medulla oblongata causes a drop in blood pressure (*depressor center*) (**D6**), while electrical stimulation of the remaining reticular formation in the medulla oblongata (**D7**) leads to an increase in blood pressure.

Effect on the motor system. The reticular formation has a differential effect on the spinal motor system. In the medial field of the medulla oblongata lies an *inhibition center*; upon stimulation, the muscle tone drops, reflexes fail, and the electric stimulation of the motor cortex no longer triggers a reaction. By contrast, the reticular formation in pons and midbrain has an *enhancing effect* on the motor system.

Ascending activation systems. The reticular formation has an effect on consciousness via connections to the intralaminar nuclei of the thalamus. When strongly stimulated by sensory or cortical input, the organism suddenly becomes fully alert, a prerequisite for attention and perception. Upon electrical stimulation of the reticular formation, this **wake-up function** can be objectively assessed by electroencephalography (EEG).

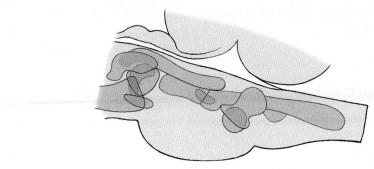

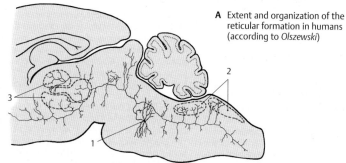

A Extent and organization of the reticular formation in humans (according to *Olszewski*)

B Neuron with branching dendrites; reticular formation in the rat (according to *Scheibel and Scheibel*)

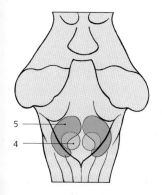

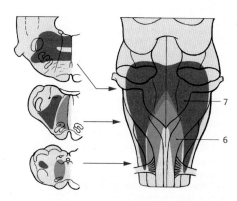

C Respiratory center in the brain stem of the monkey (according to *Beaton and Magoun*)

D Cardiovascular center in the brain stem of the cat (according to *Alexander*)

Histochemistry of the Brain Stem

Different regions of the brain stem are characterized by different contents in chemical substances. The delimitation of areas according to their chemical composition is called **chemoarchitectonics**. Substances can be demonstrated by quantitative chemical analysis after homogenization of the brain tissue, or by treating histological sections with certain chemicals that make it possible to show the exact localization of a substance in the tissue. The methods complement each other.

Iron was one of the first substances for which different distributions were demonstrated. By means of the Berlin blue reaction, a high iron content can be demonstrated in the substantia nigra (**A1**) and in the pallidum, while a lower iron content is found in the red nucleus (**A2**), in the dentate nucleus of the cerebellum, and in the striatum. The iron is contained in neurons and glial cells in the form of small particles. This high iron content is a characteristic of the nuclei that make up the extrapyramidal system (p. 310).

Neurotransmitter substances and the enzymes required for their synthesis and degradation show marked regional variations. While *catecholaminergic* and *serotoninergic neurons* form specific nuclei in the tegmentum (p. 33), the motor nuclei of cranial nerves are characterized by a high content in *acetylcholine* and *acetylcholine esterase*. Quantitative chemical analysis of brain tissue yields a relatively high content of *norepinephrine* in the tegmentum of the midbrain (**B3**), but a considerably lower content in the tectum (**B4**) and in the tegmentum of the medulla oblongata (**B5**). The content of *dopamine* is particularly high in the substantia nigra (**B1**) and very low in the rest of the brain stem.

Metabolic enzymes (**C**) also show regional variations in their distribution. Activity of *oxidative enzymes* is generally higher in gray matter than in white matter. In the brain stem, activity is particularly high in the cranial nerve nuclei, the lower portion of the olive, and the pontine nuclei. The differences refer not only to the individual areas but also to the localization of enzyme activity within the cell bodies (*somatic type*) or in the neuropil (*dendritic* type).

Neuropil. The substance between the cell bodies, which appears amorphous in Nissl-stained material, is called the neuropil. It consists mainly of dendrites and also of axons and glial processes. The majority of all synaptic contacts are found in the neuropil.

The distribution in the medulla oblongata of *succinate dehydrogenase* (an enzyme of the citric acid cycle) serves as an example for different localizations of an oxidative metabolic enzyme within the tissue: in the *oculomotor nucleus* (**C6**), its activity in the perikarya and in the neuropil is high, while it is low at both locations in the *solitary nucleus* (**C7**). In the *posterior nucleus of the vagus nerve* (**C8**), the cell bodies contrast with the neuropil owing to their high activity. By comparison, the highly active neuropil in the *gracile nucleus* (**C9**) lets the poorly reacting perikarya appear as light spots. Fiber tracts (for example, the *solitary tract*) (**C10**) show very low activity. The distribution of enzymes is characteristic for each nuclear area and is referred to as the **enzyme pattern**.

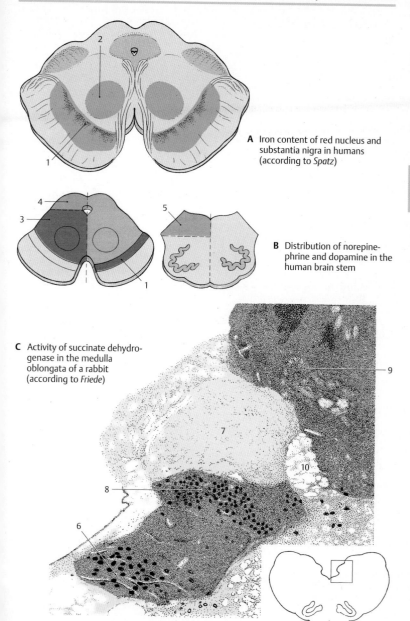

A Iron content of red nucleus and substantia nigra in humans (according to *Spatz*)

B Distribution of norepinephrine and dopamine in the human brain stem

C Activity of succinate dehydrogenase in the medulla oblongata of a rabbit (according to *Friede*)

Cerebellum

Structure

Subdivision (A–D)

The **cerebellum** is the integrative organ for the *coordination and fine-tuning of movement* and for the *regulation of muscle tone*. It develops from the alar plate of the brain stem and forms the roof of the fourth ventricle. The **superior surface** (C) is covered by the cerebrum. Embedded into the **inferior surface** (D) is the medulla oblongata (see p. 101, C). There is an unpaired central part, the **vermis of the cerebellum** (**ACD1**, **B**), and the two **cerebellar hemispheres**. This tripartition is only visible at the inferior surface, where the vermis forms the floor of a fossa, the *vallecula of the cerebellum* (**D2**). The surface of the cerebellum exhibits numerous narrow, almost parallel convolutions, the *folia of the cerebellum*.

Phylogenetic studies indicate that the cerebellum consists of old portions (developed early in evolution, present in all vertebrates) and new portions (developed late, present only in mammals). Accordingly, the cerebellum is subdivided into two parts, the **flocculonodular lobe** and the **cerebellar body** (**A3**). The two are separated by the *posterolateral fissure* (**A4**). The cerebellar body is further subdivided by the *primary fossa* (**AC5**) into **anterior lobe** and **posterior lobe**.

Flocculonodular Lobe (A6)

Together with the *lingula* (**AB7**), this is the oldest portion (**archicerebellum**). Functionally, it is connected to the vestibular nuclei through its fiber tracts (**vestibulocerebellum**) (p. 164, B).

Anterior Lobe of the Cerebellar Body (A8)

This is a relatively old component; together with its central sections, which belong to the vermis (*central lobule* [**A–C9**], *culmen* [**A–C10**]) and other sections of the vermis (*uvula* [**ABD11**], *pyramid* [**ABD12**]), it forms the **paleocerebellum**. It receives the spinocerebellar tracts for proprioceptive sensibility from the muscles (**spinocerebellum**) (p. 164, A).

Posterior Lobe of the Cerebellar Body (A13)

This is the new portion (**neocerebellum**); its enormous enlargement in primates contributes significantly to the formation of the cerebellar hemispheres. It receives the large corticocerebellar tracts from the cerebral cortex via the pontine nuclei (**pontocerebellum**) and represents the apparatus for fine-tuning of voluntary movements.

Traditional Nomenclature

The individual sections of the cerebellum have traditional names unrelated to their development or function. According to this classification, most sections of the vermis are associated with a pair of hemispheric lobes: the *central lobule* (**A–C9**) with the *wing of the central lobule* (**A14**) on each side, the *culmen* (**A–C10**) with the *quadrangular lobule* (**AC15**), the *declive* (**A–C16**) with the *simple lobule* (**AC17**), the *folium* (**A–C18**) with the *superior semilunar lobule* (**ACD19**), the *tuber* (**ABD20**) with the *inferior semilunar lobule* (**AD21**) and part of the *gracile lobule* (**AD22**), the *pyramid* (**ABD12**) with part of the *gracile lobule* and the *biventral lobule* (**AD23**), the *uvula* (**AB11**) with the *tonsilla* (**A24**) and the *paraflocculus* (**A25**), and the *nodulus* (**AB26**) with the *flocculus* (**AD27**). Only the *lingula* (**AB7**) is not associated with any lateral lobe.

The pale red arrow A in diagram **B** refers to the direction of viewing the anterior surface of the cerebellum as illustrated on p. 155, A.

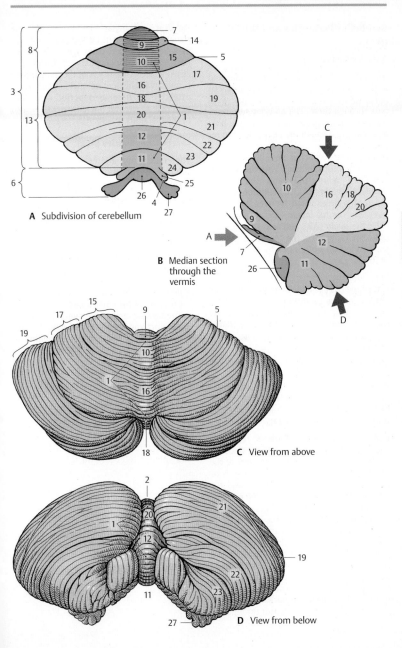

A Subdivision of cerebellum

B Median section through the vermis

C View from above

D View from below

Cerebellar Peduncles and Nuclei (A–C)

Anterior Surface (A)

On both sides, the cerebellum is connected to the brain stem by the **cerebellar peduncles** (**A1**). All afferent and efferent pathways pass through them. The anterior surface of the cerebellum becomes fully visible only after cutting through the peduncles and removing the pons and medulla oblongata. Between the cerebellar peduncles lies the roof of the fourth ventricle with the *superior medullary velum* (**A2**) and the *inferior medullary velum* (**A3**). The anterior parts of the vermis are exposed, namely, the *lingula* (**A4**), the *central lobule* (**A5**), the *nodulus* (**A6**), the *uvula* (**A7**), and also the *flocculus* (**A8**). The vallecula of the cerebellum (**A9**) is surrounded on both sides by the *tonsillae* (**A10**).

The following parts are also visible: biventral lobule (**A11**), superior semilunar lobule (**A12**), inferior semilunar lobule (**A13**), simple lobule (**A14**), quadrangular lobule (**A15**), and wing of the central lobule (**A16**).

Nuclei (B)

The cross section shows cortex and nuclei of the cerebellum. The sulci are heavily branched, resulting in a leaflike configuration of the cross sectioned folia. The sagittal section thus shows a tree-like image, the *arbor vitae* (tree of life) (**C17**).

Deep in the white matter are the cerebellar nuclei. The **fastigial nucleus** (**B18**) lies close to the median line in the white matter of the vermis. It receives fibers from the cortex of the vermis, the vestibular nuclei, and the olive. It sends fibers to the vestibular nuclei and other nuclei of the medulla oblongata. The **globose nucleus** (**B19**), too, is thought to receive fibers from the cortex of the vermis and to send fibers to the nuclei of the medulla oblongata. Fibers of the cerebellar cortex from the region between vermis and hemisphere (intermediate part) are thought to terminate at the hilum of the dentate nucleus in the **emboliform nucleus** (**B20**). The fibers of the latter nucleus run through the

superior cerebellar peduncle to the thalamus. The **dentate nucleus** (**B21**) appears as a heavily folded band with the medial part remaining open (*hilum of dentate nucleus*). The cortical fibers of the hemisphere terminate in the dentate nucleus, and fibers extend from here as superior cerebellar peduncle to the red nucleus (p. 137, B) and to the thalamus (p. 185, A).

Cerebellar Peduncles (A, C)

The efferent and afferent pathways of the cerebellum run through three cerebellar peduncles:

- The **inferior cerebellar peduncle** (restiform body) (**AC22**), which ascends from the lower medulla oblongata; it contains the spinocerebellar tracts and the connections to the vestibular nuclei
- The **medial cerebellar peduncle** (brachium pontis) (**AC23**) with the fiber masses from the pons, which originate from the pontine nuclei and represent the continuation of the corticopontine tracts
- The **superior cerebellar peduncle** (brachium conjunctivum) (**AC24**), which contains the efferent fiber systems extending to the red nucleus and the thalamus

C25 Tectal plate.
C26 Medial lemniscus.
C27 Lateral lemniscus.
C28 Trigeminal nerve.
C29 Facial nerve.
C30 Vestibulocochlear nerve.
C32 Olive.
C32 Central tegmental tract.
C33 Anterior cerebellar tract.

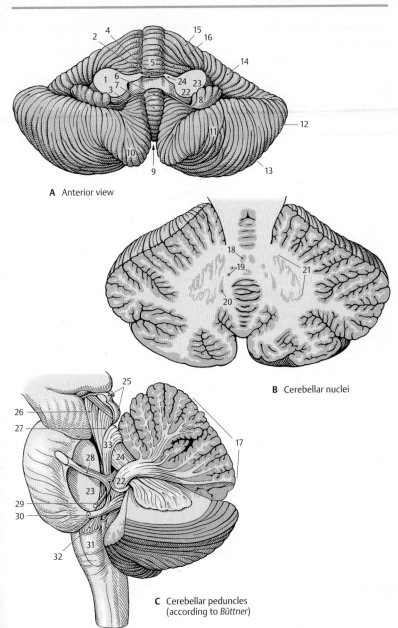

A Anterior view

B Cerebellar nuclei

C Cerebellar peduncles
(according to *Büttner*)

Cerebellar Cortex (A – D)

Overview (A)

The cortex lies immediately below the surface and follows the course of the sulci and folia. Projection of the convoluted relief of the human cerebellum onto a plane results in an expanse of 1 m in length in the oro-caudal dimension (from the lingula to the nodulus). The cortex is regularly structured throughout all regions of the cerebellum. It consists of three layers:

- The molecular layer
- The Purkinje cell layer
- The granular layer

The **molecular layer** (**A1**) lies beneath the surface; it contains few cells and consists mainly of unmyelinated fibers. Among its neurons we can distinguish the outer *stellate cells* (lying close to the surface) and the inner *basket cells*. The narrow **Purkinje cell layer** (ganglionic layer) (**A2**) is formed by the large neurons of the cerebellum, the *Purkinje cells*. Then follows the **granular layer** (**A3**). It is very rich in cells, consisting of densely packed, small neurons, the *granule cells*. There are also scattered larger cells, the *Golgi cells*.

Purkinje Cells (B – D)

The Purkinje cell represents the largest and most characteristic cell of the cerebellum. The Nissl stain shows the pear-shaped cell body (**B4**) filled with coarse Nissl bodies. Also visible are the basal portions of two or three dendrites (**B5**) at the upper pole of the cell. However, the cell's entire expanse with all its processes can only be visualized by Golgi impregnation or intracellular staining. The primary stems of the dendrites ramify into further branches, and these again into fine arborizations that form the **dendritic tree** (**B6**). The dendrites spread in a two-dimensional plane like the branches of an espalier tree. The Purkinje cells are arranged in a strictly geometric fashion; spaced at relatively regular intervals, they form a row between granular and molecular layers and send their dendritic trees into the molecular layer toward the surface of the folium.

Without exception, the flattened dendritic trees extend in a plane perpendicular to the longitudinal axis of the cerebellar folium (**D**).

The initial branches of the dendritic tree (primary and secondary dendrites) have a *smooth surface* (**C7**) and are covered with synapses. The fine terminal branches are dotted with short spines (**C8**). Each Purkinje cell carries approximately 60 000 *spinous synapses*. Different fiber systems terminate at the smooth and spiny sections of the cell: the *climbing fibers* end at the smooth section and the *parallel fibers* at the spiny section (p. 160).

The axon (**B9**) departs from the base of the Purkinje cell and extends through the granular layer into the white matter. The axons of Purkinje cells terminate at neurons of the cerebellar nuclei (p. 160, D). They give off recurrent collaterals. Purkinje cells use GABA as neurotransmitter.

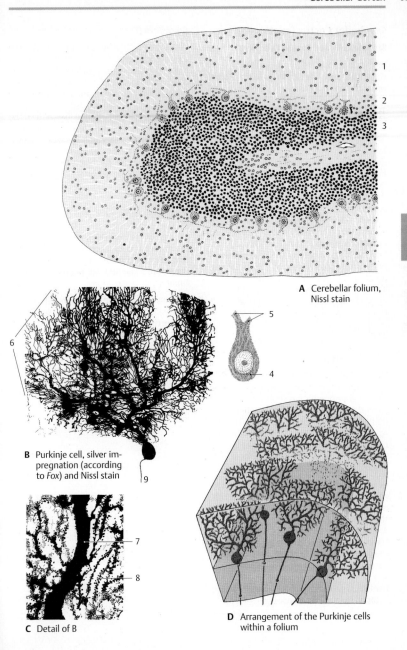

A Cerebellar folium, Nissl stain

B Purkinje cell, silver impregnation (according to *Fox*) and Nissl stain

C Detail of B

D Arrangement of the Purkinje cells within a folium

Cerebellum

Cerebellar Cortex (continued)

Stellate Cells and Basket Cells (A, B)

In the upper half of the molecular layer lie the **stellate cells**. The dendrites of these small neurons run in all directions and reach approximately 12 Purkinje dendritic trees. Their axons terminate either at the Purkinje cell bodies or run horizontally below the surface of the folium.

In the lower third of the molecular layer lie the slightly larger **basket cells** (**A1**). Their long axons run horizontally above the Purkinje cell bodies and give off collaterals, the terminal branches of which form networks (baskets) around the Purkinje cell bodies. The electron-microscopic image shows that the basket cell fibers form numerous synaptic contacts (**B2**) with the Purkinje cell, namely, at the base of the cell body (axon hillock) and at the initial segment of the axon up to where the myelin sheath begins. The rest of the Purkinje cell body is enveloped by *Bergmann's glial cells* (**B3**). The positioning of the synapses at the axon hillock indicates the *inhibitory character* of the basket cells.

Granule Cells (C)

These small, densely packed neurons form the granular layer. At high magnification, the Golgi impregnation shows three to five short dendrites which carry clawlike thickenings at their terminal branches. The thin axon (**C4**) of the granule cell ascends vertically through the Purkinje cell layer into the molecular layer, where it bifurcates at right angles into two *parallel fibers* (p. 161, C5).

Cerebellar Glomeruli. The granular layer contains small, cell-free islets (glomeruli), in which the clawlike dendritic endings of the granule cells form synaptic contacts with the axon terminals of afferent nerve fibers (*mossy fibers*, p. 161, B3). In addition, the short axons of the *Golgi cells* terminate here. The electron-microscopic image shows complex, large synapses (glomerulus-like synaptic complexes), which are enveloped by glial processes.

Golgi Cells (E)

These are much larger than the granule cells and are scattered throughout the granular layer, usually slightly below the Purkinje cells (p. 161, C9). Their dendritic trees, which ramify predominantly in the molecular layer and extend toward the surface of the folium, are not flattened like those of the Purkinje cells but are spread in all directions. The cells have short axons, which either terminate in a glomerulus or ramify into a fine, dense fiber network. The Golgi cells belong to the *inhibitory interneurons*.

Glia (D)

Apart from the regular glial cell types, such as the *oligodendrocytes* (**D5**) and protoplasmic *astrocytes* (**D6**) commonly found in the granular layer, there are also glial cells that are characteristic for the cerebellum: *Bergmann's glia* and the penniform *glia of Fañanás*.

The cell bodies of the **Bergmann's cells** (**D7**) lie between the Purkinje cells and send long supporting fibers vertically toward the surface, where their small end-feet form a limiting glial membrane against the meninges. The supporting fibers carry leaflike processes and form a dense scaffold. Bergmann's glia begins to proliferate where Purkinje cells go dead. The **Fañanás cells** (**D8**) have several short processes with a characteristic penniform structure.

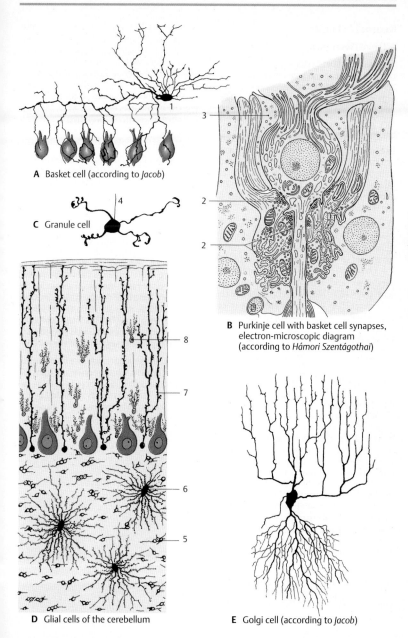

A Basket cell (according to *Jacob*)

C Granule cell

B Purkinje cell with basket cell synapses, electron-microscopic diagram (according to *Hámori Szentágothai*)

D Glial cells of the cerebellum

E Golgi cell (according to *Jacob*)

Cerebellum

Neuronal Circuits

Afferent Fibers (A, B)

The afferent fiber systems terminate in the cerebellar cortex and give off axon collaterals to the cerebellar nuclei. There are two different types of terminals: climbing fibers and mossy fibers.

The **climbing fibers** (**AC1**) terminate at the Purkinje cells by splitting up and attaching like tendrils to the ramifications of the dendritic tree. Each climbing fiber terminates at a single *Purkinje cell* and via axon collaterals also at some stellate and basket cells. The climbing fibers originate from neurons of the olive and its accessory nuclei.

The **mossy fibers** (**BC2**) divide into widely divergent branches and finally give off numerous lateral branches with small rosettes of spheroid terminals. These fit into the clawlike terminals of the *granule cell dendrites* and form synaptic complexes with them (**B3**). Mossy fibers terminate the spinocerebellar and pontocerebellar tracts and also fibers from nuclei of the medulla.

Cortex (C)

The structure of the cerebellar cortex is determined by the transverse orientation of the flattened dendritic trees of the Purkinje cells (**ACD4**) and the longitudinally extending **parallel fibers** of the granule cells (**B–D5**), which form synapses with the Purkinje cell dendrites. The Purkinje cells represent the efferent elements of the cortex. They receive **excitatory input** through direct contact with climbing fibers (**C1**) and indirectly through mossy fibers (**C2**) via interposed granule cells (**CD6**). The axons of granule cells bifurcate in the molecular layer into two parallel fibers each, which measure approximately 3 mm in total length and travel through approximately 350 dendritic trees. About 200 000 parallel fibers are thought to pass through each dendritic tree.

Stellate cells (**C7**), basket cells (**C8**), and Golgi cells (**C9**) are **inhibitory interneurons** which inhibit Purkinje cells. They

are coexcited by every incoming impulse, either via synapses in the glomeruli, via synapses with parallel fibers, via synapses of Golgi cells with mossy fibers, or via axon collaterals of afferent fibers. In this way, the excitation of one row of Purkinje cells inhibits all neighboring Purkinje cells; this sharpens the boundaries between active Purkinje cell clusters (contrast formation).

Functional Principle of the Cerebellum (D)

The axons of the Purkinje cells (**D4**) terminate at the neurons of the subcortical nuclei (**D10**) (cerebellar nuclei and vestibular nuclei). *Purkinje cells are inhibitory neurons* with a high GABA content. They have a strong inhibitory effect on the neurons of the cerebellar nuclei, which continuously receive excitatory input exclusively via axon collaterals (**D11**) of the afferent fibers (climbing fibers and mossy fibers) (**D12**). However, these impulses cannot be passed on further, because the nuclei are under the inhibitory control of the Purkinje cells. Only when Purkinje cells are inhibited by inhibitory interneurons (**D13**) is their inhibiting effect lifted so that excitation is passed on to the corresponding segments of the nuclei.

Hence, the cerebellar nuclei are *independent synaptic centers*, which receive and transmit impulses and in which there is continuous excitation. Transmission is regulated by the cerebellar cortex by means of fine-tuned inhibition and disinhibition.

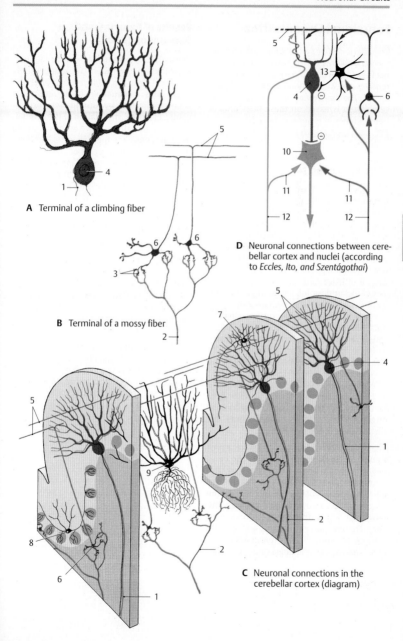

A Terminal of a climbing fiber

B Terminal of a mossy fiber

D Neuronal connections between cerebellar cortex and nuclei (according
to *Eccles, Ito,* and *Szentágothai*)

C Neuronal connections in the
cerebellar cortex (diagram)

Cerebellum

Functional Organization

The afferent fiber systems terminate in circumscribed areas of the cortex; in turn, corticofugal fibers of different cortical areas extend to specific portions of the cerebellar nuclei. The functional organization of the cerebellum is based on these fiber projections. Studies in experimental animals (rabbits, cats, monkeys) have cast light on these relationships.

Fiber Projection (A, B)

The way in which the **afferent fiber systems** terminate reveals a functional tripartition of the cerebellum into *vermis*, *hemisphere*, and the *intermediate zone* lying between them. The **spinocerebellar fibers**, namely, *posterior spinocerebellar tract*, *cuneocerebellar tract* (p. 164, A1 and A2), and *anterior spinocerebellar tract* (p. 166, B14), terminate as mossy fibers in the vermis of the anterior lobe, in the pyramid and uvula, and in the intermediate zone lying laterally to them (**A1**). The **corticopontocerebellar fibers**, which enter through the middle cerebellar peduncle, terminate as mossy fibers in the cerebellar hemisphere (**A2**). The **vestibulocerebellar fibers** terminate in the flocculonodular lobe and in the uvula (**A3**). Afferent pathways that synapse in the olive and its accessory nuclei and in the lateral reticular nucleus terminate according to their origin: the spinal ones in the vermis, the cortical ones in the hemisphere.

The tripartition is also obvious from the projection of **corticofugal axons** to the cerebellar nuclei. In the vestibular nuclei (**B4**) terminate fibers from the vermis (anterior lobe, pyramid, uvula, and nodulus) and the flocculus (**B5**). In the fastigial nucleus (**B6**) terminate fibers from the entire vermis (**B7**). The emboliform nucleus and globose nucleus (a complex in experimental animals) (**B8**) receive the fibers of the intermediate zone (**B9**) and, finally, in the dentate nucleus (**B10**) terminate the fiber masses of the hemisphere (**B11**).

Results of Experimental Stimulation (C, D)

The pathways terminate following a set pattern in which lower limb, trunk, upper limb, and head area are arranged one after the other. This *somatotopic organization* has been confirmed by stimulation experiments. Electrical stimulation of the cerebellar cortex in a decerebrate animal resulted in contractions and tonic changes in the extensor and flexor muscles of the limbs, in conjugate eye movements, and in contractions of the facial and cervical muscles (**C**) (the figure shows upper and lower surfaces of the cerebellum projected onto the same plane).

Similar results were obtained by tactile stimulation of various body parts and simultaneous electrical recording of the resulting potentials in the cerebellar cortex (*evoked potentials*) (**D**). In addition, localization of the potentials demonstrated the ipsilateral representation of the body half in the anterior lobe and simple lobule (**D12**) and the bilateral representation in the paramedian lobule (**D13**).

As this somatotopic organization has been demonstrated in rabbits, cats, dogs, and monkeys, we may assume that the same organization is present in all mammals. The somatotopic organization hypothesized for the human brain is illustrated in **E**.

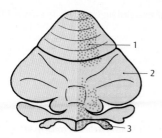

A Terminals of the vestibulocerebellar tract and the spinocerebellar tract (according to *Brodal*)

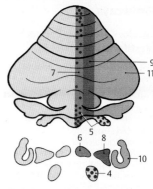

B Projection of the cerebellar cortex to the cerebellar nuclei and the vestibular nucleus (according to *Jansen and Brodal*)

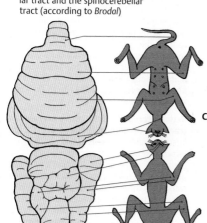

C Somatotopic organization of the cerebellar cortex, motor effects in the cat (according to *Hampson, Harrison, and Woolsey*)

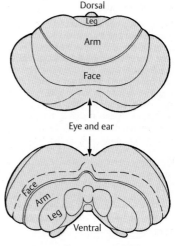

E Hypothetical somatotopic organization of the human cerebellar cortex (according to *Hampson, Harrison, and Woolsey*)

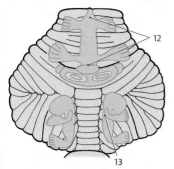

D Somatotopic organization of the cerebellar cortex, evoked potentials during sensory stimulation (according to *Snider*)

Cerebellum

Pathways

Inferior Cerebellar Peduncle (Restiform Body) (A – C)

The **inferior cerebellar peduncle** contains the following fiber systems:

Posterior spinocerebellar tract and cuneocerebellar tract (A). The fibers of the *posterior spinocerebellar tract* (Flechsig's tract) (**A1**) originate from cells of the *posterior thoracic nucleus* (*Clarke's column*) (**A3**), in which the afferent fibers of the proprioceptive sensibility terminate (tendon organs, muscle spindles, pp. 312, 314). The region supplied by this tract is restricted to the lower extremity and the lower trunk. The fibers of the posterior spinocerebellar tract terminate as mossy fibers in the vermis and intermediate zone of the anterior lobe and in the pyramid. The corresponding fibers for the upper extremity and the upper part of the trunk collect in the *lateral cuneate nucleus* (*Monakow's nucleus*) (**A4**) and extend as *cuneocerebellar tract* (**A2**) to the same areas. The anterior spinocerebellar tract reaches the cerebellum via the superior cerebellar peduncle (brachium conjunctivum) (p. 166, B).

Vestibulocerebellar tract (B). The cerebellum receives primary and secondary vestibular fibers. The primary fibers (**B5**) originate from the *vestibular ganglion* (**B6**) (predominantly from the semicircular ducts) and run to the cerebellum without synapsing. The secondary fibers (**B7**) synapse in the *vestibular nuclei* (**B8**). Nearly all fibers terminate in the nodulus, flocculus (**B9**), and fastigial nucleus (**B10**) but some terminate in the uvula. The connection with the vestibular nuclei also contains cerebellofugal fibers (*cerebellovestibular tract*), which originate from the terminal areas just mentioned and from the vermis of the anterior lobe. Some of them synapse in the lateral vestibular nucleus and extend in the *vestibulospinal tract* (p. 382, A9) to the spinal cord.

Olivocerebellar tract (A). The olive (**A11**), which may be regarded as a cerebellar nucleus transposed ventrally, sends all its fibers to the cerebellum. The olive and its accessory nuclei receive ascending fibers from the spinal cord (*spino-olivary tract*) (**A12**), fibers from the cerebral cortex and from extrapyramidal nuclei (*central tegmental tract*, p. 144, A). The fibers synapse in specific segments of the olive to form the *olivocerebellar tract* (**A13**), which crosses to the opposite side and extends to the contralateral half of the cerebellum. The fibers of the olivary complex terminate as climbing fibers in the cerebellar cortex: the fibers of the accessory nuclei (termination zone of the spino-olivary tract) run to the cortex of vermis and intermediate zone of the anterior lobe, while the fibers of the main nucleus (termination zone of cortical fibers and tegmental tract) run to the cerebellar hemispheres.

Reticulocerebellar tract, nucleocerebellar tract, and arcuatocerebellar tract (C). The *lateral reticular nucleus* (**C14**) receives exteroceptive sensory fibers which ascend together with the spinothalamic tracts. The postsynaptic fibers run as *reticulocerebellar tract* (**C15**) through the ipsilateral cerebellar peduncle to the vermis and the hemisphere. The *nucleocerebellar tract* (**C16**) transmits tactile impulses of the facial area primarily from the trigeminal nuclei (**C17**) to the cerebellum. The fibers of the *arcuatocerebellar tract* (**C18**) originate in the arcuate nucleus (**C19**) and run to the floor of the fourth ventricle, where they form the *medullary striae*. They run crossed and uncrossed and are thought to terminate in the flocculus.

The *uncinate fasciculus of cerebellum*, a cerebellospinal tract originating in the contralateral fastigial nucleus, has not been unequivocally demonstrated in the human brain.

Cerebellum

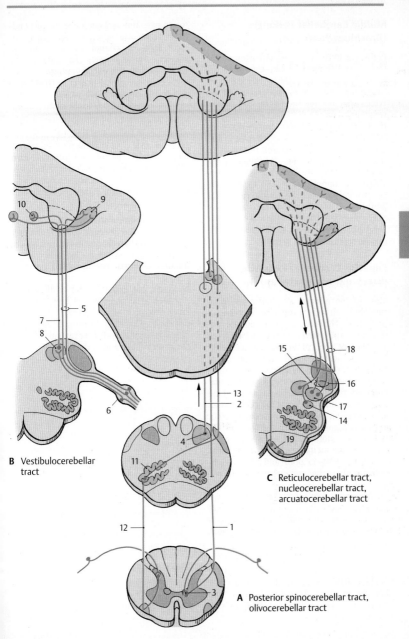

B Vestibulocerebellar tract

C Reticulocerebellar tract, nucleocerebellar tract, arcuatocerebellar tract

A Posterior spinocerebellar tract, olivocerebellar tract

Middle Cerebellar Peduncle (Brachium Pontis) (A)

Pathways from the cerebral cortex of the frontal and temporal lobes extend to the pons. Together with the pyramidal tract they form the *cerebral peduncles* (**A1**) in which they occupy the lateral and medial segments: laterally lies the *temporopontine tract* (Türck's bundle) (**A2**) and medially the *frontopontine tract* (Arnold's bundle) (**A3**). The fibers of the first neuron terminate in the pontine nuclei (**A4**). The fibers of the second neuron cross from the pontine nuclei to the opposite side and form as the *pontocerebellar tract* the **middle cerebellar peduncle**. The fibers terminate as mossy fibers mostly in the contralateral cerebellar hemisphere but some also bilaterally in the middle segment of the vermis.

Superior Cerebellar Peduncle (Brachium conjunctivum) (B)

Most of the efferent cerebellar pathways run through the **superior cerebellar peduncle**. The only afferent bundle entering it is the *anterior spinocerebellar tract*. The efferent fibers of the superior cerebellar peduncle enter the midbrain tegmentum at the level of the inferior colliculi and cross in the **decussation of the superior cerebellar peduncle** (**B5**) to the opposite side, where they divide into a descending (**B6**) and an ascending limb (**B7**). The descending fiber bundles originate from the fastigial nucleus (**B8**) and the globose nucleus (**B9**). They terminate in the medial nuclei of the reticular formation (**B10**) in pons and medulla oblongata, where they synapse to form the *reticulospinal tract*. Thus, cerebellar impulses are transmitted to the spinal cord via two pathways, namely, the *reticulospinal tract* and the *vestibulospinal tract*. From both tracts, interneurons affect the anterior horn cells.

The fibers of the stronger ascending limb originate predominantly from the dentate nucleus (**B11**), but partly also from the emboliform nucleus. They terminate in two areas: (1) in the red nucleus (**B12**), its sur-roundings, and various nuclei of the midbrain tegmentum (Edinger–Westphal nucleus, accessory oculomotor nucleus, Darkshevich's nucleus, etc.) that connect the cerebellum to the extrapyramidal system, and (2) in the dorsal thalamus (**B13**) from where the impulses are passed on to the cerebral cortex, mainly to the motor cortex.

These connecting tracts create a large neuronal circuit; cerebellar impulses affect the cerebral cortex via brachium conjunctivum and thalamus. The cerebral cortex, in turn, affects the cerebellum via the corticopontocerebellar and cortico-olivocerebellar systems. Thus, motor cortex and cerebellum are under mutual control.

Anterior spinocerebellar tract (Gowers' tract) (**B14**). The fibers originate in the posterior horn where fibers primarily from the tendon organs synapse. The postsynaptic bundles run crossed or uncrossed; however, they do not enter the inferior cerebellar peduncle but extend as far as the upper margin of the pons, where they turn and enter through the superior cerebellar peduncle (**C**). They terminate as mossy fibers in vermis and intermediate zone of the anterior lobe and in the uvula.

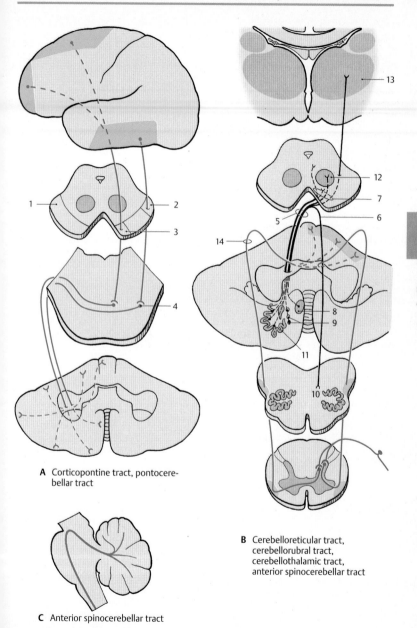

A Corticopontine tract, pontocerebellar tract

C Anterior spinocerebellar tract

B Cerebelloreticular tract, cerebellorubral tract, cerebellothalamic tract, anterior spinocerebellar tract

Cerebellum

Diencephalon

Development of the Prosencephalon

Brain and spinal cord develop from the neural tube which forms several brain vesicles at its anterior segment, namely, the *rhombencephalon* (**A1**), *mesencephalon* (**A2**), *diencephalon* (**A3**), and *telencephalon* (**A4**). The lateral walls of the vesicles thicken to become the brain substance proper in which the nerve cells and their processes differentiate. The developmental process begins in the rhombencephalon and spreads to the mesencephalon and diencephalon. Development of the telencephalon is greatly delayed. A thin-walled vesicle forms on each side so that the telencephalon becomes subdivided into three parts, namely, the two symmetrical *hemispheres* (**A5**) and the unpaired median portion (**A6**), which forms the anterior wall of the third ventricle (*terminal lamina*).

The vesicles of the telencephalon begin to cover the diencephalon. As they expand, particularly in caudal direction, the *telodiencephalic boundary* becomes displaced. Initially, it represents the frontal border line (**A7**) but then runs more and more obliquely (**A8**) until it finally becomes the lateral border (**A9**) of the diencephalon. The diencephalon thus comes to lie between the two hemispheres, having almost no outer surface. The three brain sections that are initially arranged one behind the other, namely, mesencephalon, diencephalon (red), and telencephalon (yellow), become largely positioned one within the other in the mature brain.

Telodiencephalic Boundary (B – E)

Only the floor of the diencephalon is visible at the surface of the brain; it forms the *optic chiasm*, the *tuber cinereum*, and the *mamillary bodies* at the base of the brain (p. 12, A). The roof of the diencephalon becomes visible only after cutting horizontally to remove the corpus callosum (**B**). This exposes the roof of the third ventricle and the two thalami. The entire region is covered by a vascularized connective tissue plate, the *tela choroidea* (**D10**), the removal of which opens the third ventricle (**C**). Over the third ventricle and in the medial wall of the hemisphere, the brain tissue is extremely thin and becomes invaginated into the ventricular cavity by protruding vascular loops (p. 282, A). The vascular convolutions lying in the ventricle form the *choroid plexus* (**D11**) (production of cerebrospinal fluid). Upon removal of tela choroidea and choroid plexus, the thinned wall of the cerebral hemisphere is torn away and only the separation line remains as the *choroid line* (**C12**). The surface of the *thalamus* (**C13**) becomes exposed up to this separation line, while it is still covered laterally by the thinned wall of the hemisphere. The segment of the thinned wall of the hemisphere between the attachment of the plexus and the *thalamostriate vein* (**C–E14**) is called the **lamina affixa** (**CD15**). It adheres to the dorsal surface of the thalamus and fuses with it in the mature brain (**E16**). The thalamostriate vein (**C–E14**), which runs between thalamus and caudate nucleus (**C–E17**), marks the boundary between diencephalon and telencephalon when viewed from above.

BCE18 Fornix.
BC19   Epiphysis.
C20   Quadrigeminal plate.
C21   Habenula.
D22   Telodiencephalic fissure.
BCE23 Corpus callosum.

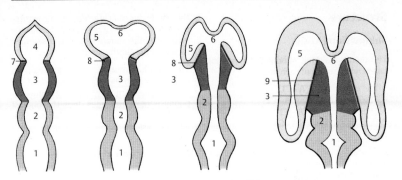

A Development of the prosencephalon (according to *Schwalbe*)

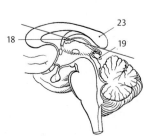

B Plane of section shown in **C**

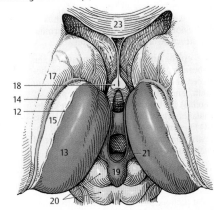

C Diencephalon viewed from above, horizontal section after removal of corpus callosum, fornix, and choroid plexus

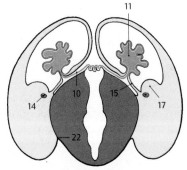

D Lamina affixa in the embryonic brain, frontal section

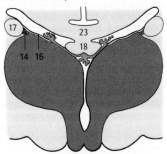

E Lamina affixa in the mature brain, frontal section

Diencephalon

Structure

Subdivision (A–C)

The diencephalon is subdivided into four layers lying on top of each other:

- The *epithalamus* (**A–C1**)
- The *dorsal thalamus* (**A–C2**)
- The *subthalamus* (**A–C3**)
- The *hypothalamus* (**A–C4**)

The simple arrangement of these layers is still clearly visible in the embryonic brain. However, it changes considerably during development owing to differences in regional growth. In particular, the extraordinary increase in mass of the dorsal thalamus and the expansion of the hypothalamus in the region of the tuber cinereum determine the structure of the diencephalon.

The **epithalamus** (p. 176) consists of the *habenulae*, a relay station for pathways between the olfactory centers and the brain stem, and of the *pineal gland* (epiphysis cerebri). Owing to the increasing size of the thalamus, the dorsally situated epithalamus (**B1**) becomes medially transposed and appears only as an appendage of the dorsal thalamus (**C1**).

The **dorsal thalamus** (p. 178) is the terminal station of sensory pathways (cutaneous sensibility; taste; visual, acoustic, and vestibular pathways). It is connected to the cerebral cortex by efferent and afferent fiber systems.

The **subthalamus** (p. 192) is the continuation of the midbrain tegmentum. It contains nuclei of the extrapyramidal motor system (zona incerta, subthalamic nucleus, globus pallidus) and may be regarded as the motor zone of the diencephalon.

The **globus pallidus**, or pallidum (**CD5**), is a derivative of the diencephalon. It becomes separated from the other gray regions of the diencephalon as a result of the ingrowing fiber masses of the internal capsule (**CD6**) during development and finally becomes displaced into the telencephalon. Only a small medial rest of the pallidum remains within the unit of the diencephalon; this is the entopeduncular nucleus. As a constituent of the extrapyramidal system, the globus pallidus should logically be regarded as part of the subthalamus.

The **hypothalamus** (p. 194) is derived from the lowest layer and forms the floor of the diencephalon from which the neurohypophysis (**A7**) protrudes. It is the highest regulatory center of the autonomous nervous system.

Frontal Section at the Level of the Optic Chasm (D)

A section through the anterior wall of the third ventricle shows parts of the diencephalon and telencephalon. Ventrally lies the fiber plate of the decussation of the optic nerve, the *optic chasm* (**D8**). A rostral excavation of the third ventricle, the *preoptic recess* (**D9**), is seen above it. The globus pallidus (**CD5**) appears laterally to the internal capsule. All other structures belong to the telencephalon: the two lateral ventricles (**D10**) and the *septum pellucidum* (**D11**) enclosing the *cave of the septum pellucidum* (**D12**), the *caudate nucleus* (**D13**), the *putamen* (**CD14**), and at the base, the *olfactory area* (**D15**) (anterior perforated substance). The *corpus callosum* (**D16**) and the *anterior commissure* (**D17**) connect the two hemispheres. Other fiber systems shown in the section are the *fornix* (**D18**) and the *lateral olfactory stria* (**D19**).

D

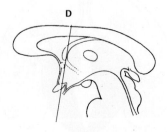

Plane of section

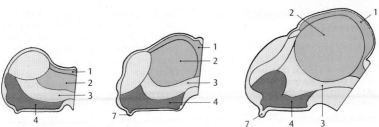

A Development of the layers of the diencephalon

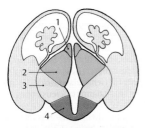 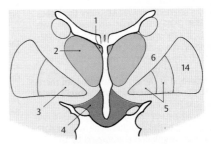

B Structure of the diencephalon in the embryonic brain

C Structure of the diencephalon in the adult brain

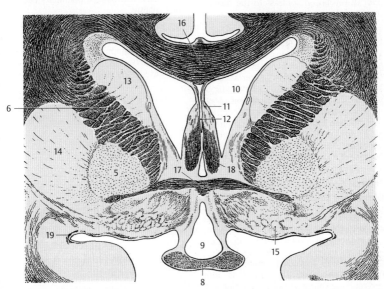

D Frontal section through the rostral wall of the third ventricle (according to *Villiger and Ludwig*)

Diencephalon

Frontal Section Through the Tuber Cinereum (A)

The plane of section lies just behind the interventricular foramen (foramen of Monro). Lateral ventricle and third ventricle are separated by the thin base of the *choroid plexus* (**A1**). From here to the *thalamostriate vein* (**AB2**) extends the *lamina affixa* (**AB3**). It covers the dorsal surface of the *thalamus* (**A4**), of which only the anterior nuclei are visible. Ventrolaterally to it and separated by the *internal capsule* (**AB5**) lies the **globus pallidus** (**AB6**), which is divided into two parts, the *internal* (**A7**) and the *external segment of the pallidum* (**A8**). It stands out against the adjacent *putamen* (**AB9**) because of its higher myelin content. At the basal margin and at the tip of the pallidum there exit the *lenticular fasciculus* (Forel's field H2) and the *lenticular ansa* (**A10**). The latter forms an arch in dorsal direction around the medial tip of the pallidum. The ventral part of the diencephalon is occupied by the hypothalamus (*tuber cinereum* [**A11**] and *infundibulum* [**A12**]), which appears markedly poor in myelin, in contrast to the heavily myelinated *optic tract* (**AB13**).

The diencephalon is enclosed on both sides by the telencephalon, but without a clearly visible boundary. The closest nuclei of the telencephalon are the *putamen* (**AB9**) and the *caudate nucleus* (**AB14**). Anterior to the globus pallidus lies a nucleus belonging to the telencephalon, the **basal nucleus** (*Meynert's nucleus*) (**A15**). It receives fibers from the midbrain tegmentum. Its large cholinergic neurons project diffusely into the entire neocortex. The *fornix* (**AB16**) is seen twice because of its arched course (p. 233, C).

Frontal Section at the Level of the Mamillary Bodies (B)

The section shows both thalami; their increase in volume has lead to secondary fusion in the median line, resulting in the *interthalamic adhesion* (**B17**). Myelinated fiber lamellae, the *medullary layers of the thalamus*, subdivide the thalamus into several large complexes of nuclei. Dorsally lies the *anterior nuclear group* (**B18**), ventrally to it the *medial nuclear group* (**B19**), which borders medially on several small paraventricular nuclei (**B20**) and is laterally separated by the *internal medullary layer* (**B21**) from the *lateral nuclear group* (**B22**). Subdivision of the latter into a dorsal and a ventral area is less distinct. The entire complex is enclosed by a narrow, shell-shaped nucleus, the *reticular nucleus of the thalamus* (**B23**), which is separated from the lateral nuclear group by the *external medullary layer* (**B24**).

Ventrally to the thalamus lies the **sub-thalamus** with the *zona incerta* (**B25**) and the subthalamic nucleus (Luys' body) (**B26**). The zona incerta is delimited by two myelinated fiber plates, dorsally by *Forel's field H1* (*thalamic fasciculus*) (**B27**) and ventrally by *Forel's field H2* (*lenticular fasciculus*) (**B28**). Below the subthalamic nucleus appears the rostral pole of the *substantia nigra* (**B29**). The floor of the diencephalon is formed by the two **mamillary bodies** (**B30**). The *mamillothalamic fasciculus* (*Vicq d'Azyr's bundle*) (**B31**) ascends from the mamillary body to the thalamus.

AB32 Corpus callosum.
A33 Amygdaloid body (amygdala).
B34 Hippocampus.
A35 Anterior commissure.
B36 Medullary stria (p. 176, A2).

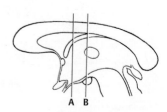

A B

Planes of sections

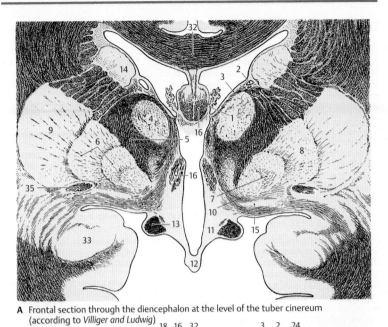

A Frontal section through the diencephalon at the level of the tuber cinereum (according to *Villiger and Ludwig*)

B Frontal section through the diencephalon at the level of the mamillary bodies (according to *Villiger and Ludwig*)

Diencephalon

Diencephalon *(side margin)*

Epithalamus

The epithalamus includes the *habenula* (with *habenular nuclei, habenular commissure,* and *medullary stria*), the *pineal gland*, and the *epithalamic commissure* (*posterior commissure*).

Habenula (A)

The **habenula** (**A1**) (p. 171, C21) with its afferent and efferent pathways forms a relay system in which olfactory impulses are transmitted to efferent (salivatory and motor) nuclei of the brain stem. In this way, olfactory sensation is thought to affect food intake. The habenular nucleus contains numerous peptidergic neurons.

The afferent pathways reach the habenular nuclei via the **medullary stria of the thalamus** (**A2**). It contains fibers from the *septal nuclei* (**A3**), the *anterior perforated substance* (olfactory area) (**A4**), and the *preoptic region* (**A5**). Furthermore, it receives fibers from the *amygdaloid body* (amygdala) (**A6**) crossing over from the *terminal stria* (**A7**).

The efferent pathways extend into the midbrain. The **habenulotectal tract** (**A8**) transmits olfactory impulses to the superior colliculi. The **habenulotegmental tract** (**A9**) terminates in the *dorsal tegmental nucleus* (**A10**), from where there is a link to the posterior longitudinal fasciculus (p. 144, B) with connections to the salivatory and motor nuclei of the masticatory and deglutitory muscles (olfactory stimuli leading to secretion of saliva and gastric juice). The **habenulo-interpeduncular tract**, *Meynert's bundle* (**A11**), terminates in the *interpeduncular nucleus* (**A12**) (p. 132, D21) which is connected to various nuclei of the reticular formation.

Pineal Gland (B – D)

The **pineal gland** (*pineal body, epiphysis cerebri*) (**A13, B**), is a small peg-shaped body at the posterior wall of the third ventricle above the quadrigeminal plate (p. 170 BC19). Its cells, the *pinealocytes*, are grouped into lobules by connective tissue septa. In silver-impregnated sections, they show long processes with club-shaped terminal swellings (**C**), which terminate predominantly at blood vessels (**D**). In adults, the pineal gland contains large foci of *calcification* (**B14**), which are visible on radiographs.

In lower vertebrates, the pineal gland is a photosensitive organ; it registers changes from light to dark either by a special parietal eye or just by the light penetrating through the thin roof of the skull. By doing so, it influences the day and night rhythm of the organism. For example, it regulates the color change in amphibians (dark pigmentation during the day, pale pigmentation at night) and the corresponding change in the animal's behavior. The pineal gland also registers the transition from bright summertime to dark wintertime and thus brings about seasonal changes in the gonads.

In higher vertebrates, the light does not penetrate the thick roof of the skull. The rhythm of day and night is transmitted to the pineal gland through the following route: via retinal fibers to the suprachiasmatic nucleus in the hypothalamus, then via efferent hypothalamic fibers to the intermediolateral nucleus. and finally via postganglionic fibers of the cervical sympathetic chain to the pineal gland.

In humans, the pineal gland is thought to inhibit maturation of the genitals until puberty. As in animals, it is supposed to have an antigonadotropic action. Hypergonadism has been observed in some cases of pineal gland destruction in children.

Epithalamic commissure (posterior commissure) (B). Not all fiber systems that pass through the epithalamic commissure (**B15**) are known. Habenulotectal fibers cross in it. From the various pretectal nuclei that send fibers through the commissure, the interstitial nucleus of Cajal and Darkshevich's nucleus are the most important ones. Vestibular fibers are also thought to cross in this commissure (p. 135, B23).

A16 Olfactory bulb.
A17 Chiasm.
A18 Hypophysis.
B19 Pineal recess.
B20 Habenular commissure.

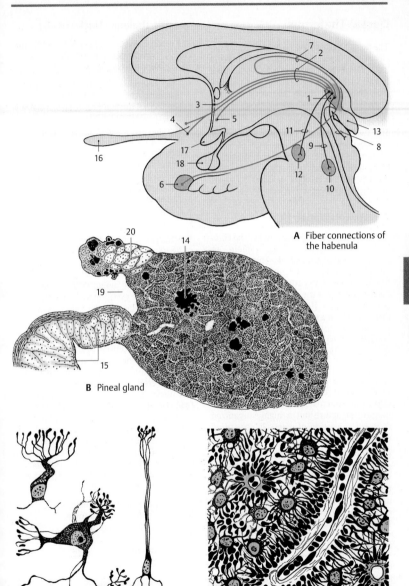

A Fiber connections of the habenula

B Pineal gland

C Pinealocytes, silver impregnation (according to *Hortega*)

D Histological appearance of the pineal gland, silver impregnation (according to *Hortega*)

Diencephalon

Dorsal Thalamus

The dorsal thalami are two large, ovoid nuclear complexes. Their medial surfaces form the wall of the third ventricle, while their lateral surfaces border on the internal capsule. They extend from the interventricular foramen (foramen of Monro) to the quadrigeminal plate of the midbrain.

The two thalami are the *relay stations for most sensory pathways*, almost all of which terminate in the contralateral thalamus. Fiber bundles connect the thalami also with the cerebellum, globus pallidus, striatum, and hypothalamus.

Thalamic Radiation (A)

The thalamus (**A1**) is connected to the cerebral cortex by the *corona radiata*, or **thalamic radiation** (**A2 – 4**). The fibers run obliquely through the internal capsule toward the cerebral cortex. The more prominent bundles are the **anterior thalamic radiation** (**A2**) (to the frontal lobe), the **superior thalamic radiation** (**A3**) (to the parietal lobe), the **posterior thalamic radiation** (**A4**) (to the occipital lobe), and the **inferior thalamic radiation** (to the temporal lobe).

The great variety of fiber connections indicates the central function of the thalamus, which is directly or indirectly integrated into most systems. Accordingly, it is not a uniform structure but a highly organized complex consisting of diversely structured nuclear groups.

Based on their fiber connections, two types of thalamic nuclei are distinguished. Nuclei with fiber connections to the cerebral cortex are collectively called **specific thalamic nuclei**; nuclei without any connection to the cortex but with connections to the brain stem are the **nonspecific thalamic nuclei**.

▇▇ **Clinical Note:** Knowledge of the basic structure of the thalamus is of practical importance because motor disturbances and conditions of pain can be treated by stereotactic surgery at the thalamus.

Specific Thalamic Nuclei (B – D)

The *cortex-dependent* **specific thalamic nuclei** are subdivided into the following nuclear groups (or complexes):

- The anterior nuclear group, **anterior thalamic nuclei** (green) (**B – D5**).
- The medial nuclear group, **medial thalamic nuclei** (red) (**BD6**).
- The lateral nuclear group, **ventrolateral thalamic nuclei** (blue) (**CD7**); this group is further divided into a lateral tier, the *lateral nuclei*, and a ventral tier, the *ventral nuclei*
- The **lateral geniculate nucleus** (**BC8**).
- The **medial geniculate nucleus** (**BC9**).
- The **pulvinar** (**BC10**).
- The **reticular nucleus of the thalamus** (**D11**).

The nuclear groups are separated by layers of fibers: the **internal medullary lamina** (**D12**) (between the medial nuclear group and the lateral and anterior nuclear groups) and the **external medullary lamina** (**D13**) (between the lateral nuclear group and the reticular nucleus which encloses the lateral surface of the thalamus).

The reticular nucleus and nonspecific nuclei, with the exception of the *centromedian nucleus* (**B14**), have been omitted from the reconstruction of nuclear groups (**B, C**). The most anterior nuclear groups are the anterior nuclei (**B5**) to which the medial nuclei (**B6**) border caudally. In the lateral complex, we distinguish between a nuclear group located dorsally (the lateral nuclei, *lateral dorsal nucleus* [**C15**] and *lateral posterior nucleus* [**C16**]), and a group located ventrally (the ventral nuclei, *ventral anterior nucleus* [**C17**], *ventral lateral nucleus* [**C18**], and *ventral posterior nucleus* [**C19**]).

BC20 Superficial dorsal nucleus.
B21 Interventricular foramen (foramen of Monro).
C22 Anterior commissure.
C23 Optic chiasm.
C24 Mamillary body.
C25 Optic tract.
A26 Cut surface of corpus callosum.

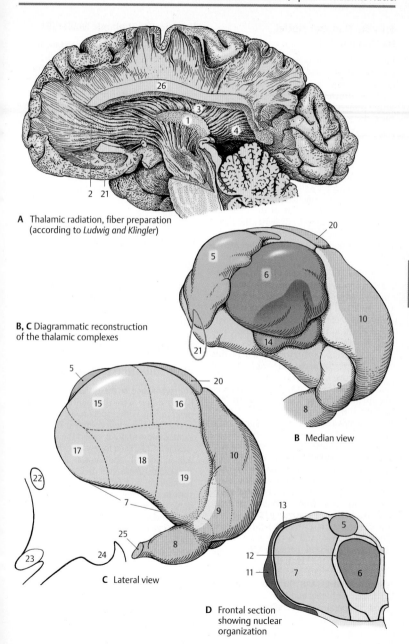

A Thalamic radiation, fiber preparation
(according to *Ludwig and Klingler*)

B, C Diagrammatic reconstruction
of the thalamic complexes

B Median view

C Lateral view

D Frontal section
showing nuclear
organization

Diencephalon

Specific Thalamic Nuclei (continued) (A)

Each nuclear group is connected to a specific region (*field of projection*) in the cerebral cortex; hence the term **specific thalamic nuclei**. Within this system, the nuclei project to their cortical fields and, in turn, the cortical fields project to the respective thalamic nuclei. Thus, there exists a neuronal circuit with a thalamocortical limb and a corticothalamic limb. The neurons of the specific thalamic nuclei transmit impulses to the cerebral cortex and are, in turn, influenced by the respective cortical fields. Hence, the function of a cortical field cannot be examined without the thalamic nucleus belonging to it; likewise, the function of a thalamic nucleus cannot be examined without the cortical field belonging to it.

When neurons of the specific thalamic nuclei become separated from their axon terminals, they respond by retrograde degeneration. Therefore, destruction of circumscribed cortical fields results in neuronal death in the respective thalamic nuclei. The thalamic projection fields of the cerebral cortex can be delineated in this way. The **anterior nuclei** (**A1**) are connected with the cortex of the *cingulate gyrus* (**A2**) and the **medial nuclei** (**A3**) with the cortex of the *frontal lobe* (**A4**). The **lateral nuclei** (**A5**) project to the dorsal and medial cortex of the *parietal lobe* (**A6**), with the lateral dorsal nucleus partly supplying the retrosplenial part of the cingulate gyrus. Among the ventral nuclei, the **ventral anterior nucleus** (**A7**) is connected with the *premotor cortex* (**A8**), the **ventral lateral nucleus** (**A9**) with the *motor precentral area* (**A10**), and the **ventral posterior nucleus** (**A11**) with the *sensory postcentral area* (**A12**). The **pulvinar** (**A13**) projects to the cortical parts of the *parietal and temporal lobes* (**A14**) and to the *cuneus* (**A15**). The **lateral geniculate nucleus** (**A16**) is connected through the visual pathway with the *visual cortex* (striate area) (**A17**), the **medial geniculate nucleus** (**A18**) through the auditory pathway with the *auditory cortex* (transverse temporal gyri, Heschl's convolutions) (**A19**).

Nonspecific Thalamic Nuclei (B)

These nuclei have fiber connections to the *brain stem*, to the *diencephalic nuclei*, and to the *corpus striatum*, but no direct connection to the cerebral cortex has been demonstrated anatomically. Their neurons are not injured by removal of the entire cerebral cortex; they are *cortex-independent*. Two groups of nuclei are distinguished:

- The **median nuclei** (*nuclei of the central thalamic gray matter*) (**B20**), which are small cell clusters located along the wall of the third ventricle
- The **intralaminar nuclei** (**B21**), which are embedded into the internal medullary lamina; the largest of them is the **centromedian nucleus** (**B22**)

Electrical stimulation of these nuclei does not lead to excitation of individual cortical areas but to changes in the electrical activity of the entire cerebral cortex. Hence, they are called **nonspecific nuclei**. The pathways through which cortical activity is influenced are unknown. The ascending pathways of the reticular formation (ascending activating system, p. 146) terminate in the intralaminar nuclei.

Alternative Nuclear Subdivision (C)

The **subdivision of the thalamus according to Hassler** differs from the traditional arrangement mainly with respect to the subdivision of the lateral nuclear complex. The nucleus located most orally is called the *lateropolar nucleus* (**C23**). Then follows the division into dorsal, ventral, and central zones. These three zones are further divided into oral, intermediate, and caudal segments. This results in the following nuclei: dorsally lie the *dorso-oral nucleus* (**C24**), the *dorsointermediate nucleus* (**C25**), and the *dorsocaudal nucleus* (**C26**), and ventrally lie the *ventro-oral nucleus* (**C27**), the *ventrointermediate nucleus* (**C28**), and the *ventrocaudal nucleus* (**C29**).

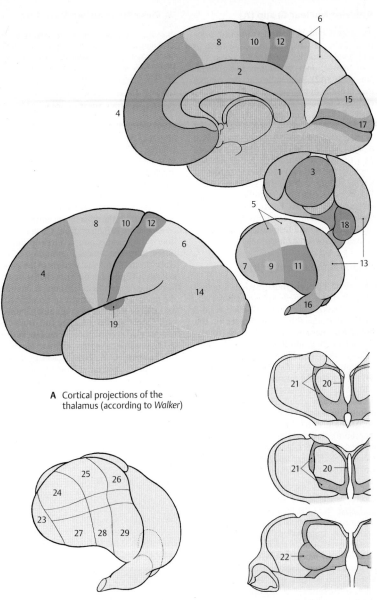

A Cortical projections of the thalamus (according to *Walker*)

C Alternative nuclear subdivision (according to *Hassler*)

B Nonspecific thalamic nuclei

Diencephalon

Anterior Nuclear Group (A)

The anterior nuclear group (anterior complex) (**A1**) consists of a principal nucleus and several smaller nuclei. All these nuclei have bidirectional connections to the **cingulate gyrus** (**A2**), which lies at the medial aspect of the hemisphere directly above the corpus callosum. Predominantly afferent fibers from the mamillary body (**A3**) reach the anterior nucleus as a thick, myelinated bundle, the **mamillothalamic tract** (*Vicq d'Azyr's bundle*) (**A4**). Fibers of the fornix are also thought to terminate in the anterior nucleus. The nucleus is thought to represent a relay station in the limbic system (p. 332). However, its functional significance is not yet precisely known. Electrical stimulation leads to autonomic reactions (changes in blood pressure and respiration rate) due to connections with the hypothalamus.

Medial Nuclear Group (B)

The medial nuclear group (medial complex) (**B5**) consists of a medial magnocellular nucleus, a lateral parvocellular nucleus, and a caudal nucleus. All nuclei project to the **frontal lobe**, namely, to the *premotor cortex*, the *polar cortex*, and the *orbital cortex* (**B6**). Afferent fiber bundles run via the inferior thalamic radiation from the **globus pallidus** (**B7**) and from the *basal nucleus of Meynert* (substantia innominata) (p. 174, A15) to the medial nuclear group. The medial magnocellular nucleus has fiber connections to the hypothalamus (**B8**) (preoptic area and tuber cinereum) and to the amygdaloid body (amygdala). The lateral parvocellular nucleus receives fibers from the adjacent ventral nuclei of the thalamus.

The medial nuclear complex is thought to receive visceral and somatic impulses via pathways from the hypothalamus and ventral nuclei; the impulses are integrated here and then transmitted via the anterior thalamic radiation to the frontal cortex. The *basic affective mood*, which is essentially determined by unconscious stimuli from the visceral and somatic spheres, is thought to enter consciousness in this way.

Clinical Note: It has been observed in patients with severe agitation that incision of the thalamocortical tracts (prefrontal lobotomy) had a calming effect but was also associated with indifference and a regression of personality. The stereotactic destruction of the medial nuclear complex has similar effects.

Centromedian Nucleus (C)

The centromedian nucleus (**C9**) is the *largest of the nonspecific thalamic nuclei* and belongs to the intralaminar nuclei surrounding the medial nuclear complex. It is divided into a ventrocaudal parvocellular part and a dorso-oral magnocellular part. The fibers of the *superior cerebellar peduncle*, which terminate in this nucleus, originate from the *emboliform nucleus of the cerebellum* (**C10**). Apart from these crossed fibers, the nucleus also receives ipsilateral fibers from the *reticular formation* (**C11**). Fibers from the **internal segment of the globus pallidus** (**C12**) branch off from the lenticular fasciculus (Forel's field H2) and radiate into the nucleus. Fibers from the precentral cortex (area 4) are thought to terminate here as well. Efferent fiber bundles run from the magnocellular region to the **caudate nucleus** (**C13**) and from the parvocellular region to the **putamen** (**C14**). These tracts provide a connection between cerebellum and striatum.

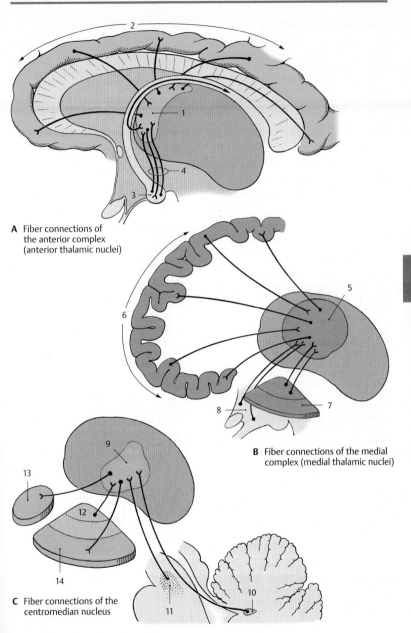

A Fiber connections of the anterior complex (anterior thalamic nuclei)

B Fiber connections of the medial complex (medial thalamic nuclei)

C Fiber connections of the centromedian nucleus

Diencephalon

Lateral Nuclear Group (A)

The lateral nuclei form the dorsal tier of the lateroventral nuclear complex. Neither of the two lateral nuclei, the **lateral dorsal nucleus** (**A1**) and the **lateral posterior nucleus** (**A2**), receives any extrathalamic input; they are connected only with other thalamic nuclei and are therefore viewed as *integration nuclei*. They send their efferent fibers to the parietal lobe (**A3**).

Ventral Nuclear Group (A, B)

Ventral anterior nucleus (VA) (A4). It receives afferent fibers mainly from the internal segment of the globus pallidus (**A5**) (they probably originate in the caudate nucleus [**A6**]) and from the nonspecific thalamic nuclei. Additional fibers are thought to stem from the substantia nigra, the interstitial nucleus of Cajal, and the reticular formation. The nucleus projects to the premotor cortex (**A7**) but depends only partially on the cortex because only half of its neurons die after injury to the cortical region. The ventral anterior nucleus is integrated into the ascending activation system; stimulation within the nuclear zone causes a change in electrical activity of the cortex.

Ventral lateral nucleus (VL) (AB8). The most important afferent system of the nucleus are the fibers of the crossed superior cerebellar peduncle (**A9**). In the anterior segment of the ventral lateral nucleus there terminate fibers from the globus pallidus (thalamic fasciculus) (**A10**). The efferent fibers (**A11**) extend to the cortex of the precentral gyrus (**A12**). The somatotopic organization of the ventral lateral nucleus is obvious in this system (**AB8**); the lateral part of the nucleus is connected to the leg region of the precentral cortex, the adjacent parts to the trunk and arm regions, and the medial part to the head region. Thus, thalamic nucleus and cortical area show a corresponding topical subdivision. Information from the cerebellum (body posture, coordination, muscle tone) reaches the motor cortex via the ventral lateral nucleus; the cerebellum influences voluntary move-

ment in this way. A narrow caudal region of the nucleus is distinguished as *ventral intermediate nucleus* (**B13**). Forel's dorsolateral tegmental fasciculus from the ipsilateral vestibular nuclei terminates here (head and gaze turn to the same side).

Ventral posterior nucleus (VP) (A14). The nucleus is the terminus for the crossed, secondary sensory pathways (**A15**). Originating from the nuclei of the posterior column, the medial lemniscus (p. 140, B) radiates into the lateral segment of the nucleus, the **ventral posterolateral nucleus** (VPL) (**B16**). The fibers of the gracile nucleus lie laterally, those of the cuneate nucleus medially. The resulting somatotopic organization of the nucleus can also be demonstrated electrophysiologically; the lower limb is represented laterally, trunk and upper limb medially. The secondary trigeminal fibers (lemniscus trigeminal) terminate in the medial segment of the nucleus, the **ventral posteromedial nucleus** (VPM) (**B17**). They transmit sensory information from the head and oral cavity, thus completing the homunculus of the contralateral half of the body. The terminus of the secondary gustatory pathway (p. 328) lies most medially. The pathways of the protopathic sensibility (see p. 324), namely, the spinothalamic tract and the trigeminal pain fibers, are thought to terminate bilaterally in the basal areas of the nucleus. The efferent fibers of the nucleus (**A18**) extend to the sensory postcentral area (p. 250), the somatotopic organization of which results from the topical subdivision of the ventral posterior nucleus and its projection to the cerebral cortex.

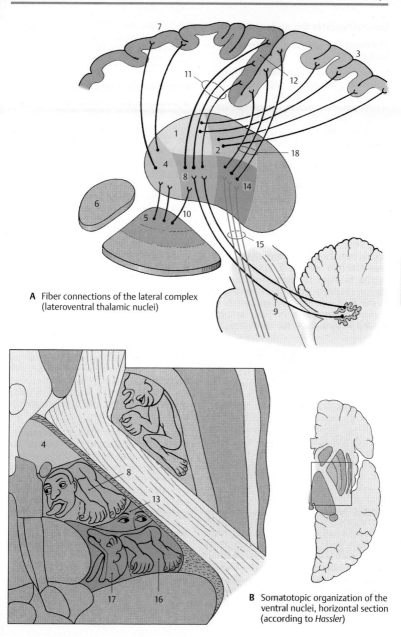

A Fiber connections of the lateral complex (lateroventral thalamic nuclei)

B Somatotopic organization of the ventral nuclei, horizontal section (according to *Hassler*)

Ventral Nuclear Group (continued)

Functional Topography of the Ventral Nuclei (A, B)

Knowledge of the functional organization makes it possible to stop severe pain through stereotactic destruction within the *ventral posterior nucleus* without affecting tactile sensation. By destroying areas within the *ventral lateral nucleus*, motor disturbances (hyperkinesis) can be eliminated without producing concomitant paralysis. The controlled stimulation required for these procedures yields clues regarding the representation of various body regions. As shown in the stimulation diagram (red symbols) (**A**), the representation of the body runs obliquely from dorsolateral (leg region, | ; arm region, –) to mediobasal (head region, (○)). Controlled stimulation of the ventral lateral nucleus (orange symbols) (**B**) results also in involuntary sound production (●) or eruptively uttered sentences (○), thereby revealing which of the two thalami is the dominant one, depending on the dominant hemisphere (p. 262) (in right-handed persons, this is the left thalamus).

Lateral Geniculate Body (C)

This nucleus (**C1**) lies somewhat isolated at the ventrocaudal aspect of the thalamus and is a relatively independent structure. It shows stratification into six cell layers which are separated by the afferent fiber bundles of the **optic tract**. Crossed and uncrossed optic fibers terminate in a regular arrangement (p. 256, A) in each of the two geniculate nuclei. In the left lateral geniculate body, the temporal half of the retina of the left eye and the nasal half of the retina of the right eye are represented; in the right lateral geniculate body, the temporal half of the retina of the right eye and the nasal half of the retina of the left eye are represented (p. 355B). The fibers from the *macula*, which is the region of greatest visual acuity, terminate in a central wedge-shaped area, which extends through all cell layers (p. 190, A9). The neurons of the lateral geniculate nucleus send their axons to the visual cortex,

the **striate area** (**C2**) at the medial hemispheric surface of the occipital lobe (*central optic radiation* or *occipitothalamic radiation*).

Medial Geniculate Body (D)

This nucleus (**D3**) is the *diencephalic relay station of the auditory pathway*. It appears as an externally visible eminence medially to the lateral geniculate body. Its afferent fibers form the **brachium of the inferior colliculus** from the ipsilateral *inferior colliculus* (**D4**). Some fiber bundles of the auditory pathway come from the *nucleus of the trapezoid body* and the ipsilateral cochlear nuclei; however, most fibers originate from the contralateral cochlear nuclei. The efferent fibers of the medial geniculate nucleus extend to the auditory cortex (**D5**), which lies in the **transverse temporal gyri**, or **Heschl's convolutions** (p. 253, C1) of the temporal lobe.

Pulvinar

The pulvinar (p. 179, BC10) occupies the caudal third of the thalamus and is divided into several nuclei. Its functional significance is not understood. Since it does not receive any extrathalamic input, it must be viewed as an *integration nucleus*. Afferent fibers from the lateral geniculate nucleus (collaterals of the optic fibers) and probably also fibers from the medial geniculate nucleus enter the pulvinar.

There are reciprocal fiber connections between the pulvinar and the cortex of the parietal lobe and the dorsal temporal lobe. Hence, the pulvinar is not only integrated into the optic and acoustic systems but is also connected with the cortical areas important for language and symbolic thinking (p. 250).

■ Clinical Note: Injury (or electrical stimulation) of the pulvinar causes speech disorder in humans.

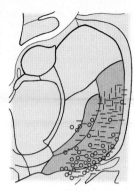

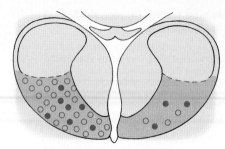

A Motor responses following stimulation of the ventral lateral nucleus (according to *Schaltenbrand, Spuler, Wahren and Rümler*)

B Responses of speech and sound following stimulation of the ventral lateral nucleus (according to *Schaltenbrand, Spuler, Wahren and Rümler*)

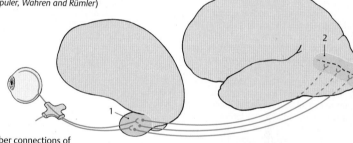

C Fiber connections of the lateral geniculate body

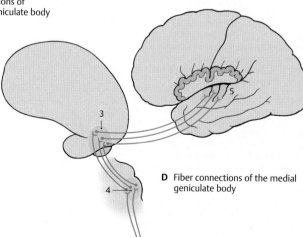

D Fiber connections of the medial geniculate body

Diencephalon

Frontal Section Through the Rostral Thalamus (A – C)

In the myelin staining, the anterior and medial nuclear groups are clearly distinguishable from the lateral nuclear group by their poor and delicate myelination. The dorsally located **anterior nuclear group** (green) (**A – C1**) bulges against the interventricular foramen (foramen of Monro) (**AB2**) and forms the *thalamic eminence*. The **medial nuclear group** (red) is enveloped by the *internal medullary lamina* (**B3**) and the intralaminar nuclei (**C4**) which separate it from the lateral portion. Within the medial nuclear group, a medial magnocellular portion (**AC5**) and a lateral parvocellular portion (**AC6**) surrounding the medial one can be distinguished.

The largest part of the thalamus is formed by the **lateroventral nuclear group** (blue), which surrounds the medial portion like a broad shell. It contains considerably more myelin, and a difference between its dorsal and ventral regions can be recognized in the myelin stained section (**B**). As compared to the dorsal region (lateral dorsal nucleus) (**A – C7**), the ventral nuclear region has more prominent and coarser myelin fibers. Its division into a medial and a lateral segment can be easily recognized in the overview. The section shows the *ventral lateral nucleus*. In its medial segment (**A – C8**) terminate fibers from the midbrain tegmentum. Lateral to it is seen the rostral part of the nucleus (**A – C9**) where the fiber bundles of the superior cerebellar peduncle terminate; its projection to the precentral area (area 4) reveals a somatotopic organization.

The lateral surface of the thalamus is formed by the **reticular nucleus of the thalamus** (**A – C10**). As a narrow layer of cells, this nucleus laterally surrounds the entire thalamus like a shell and extends from the rostral pole, where it is widest, to the pulvinar and to the lateral geniculate nucleus. It is separated from the lateral nuclear complex by a lamella of myelin fibers, the *external medullary lamina* (**B11**). The relationships between cerebral cortex and reticular nucleus vary for different nuclear segments: the frontal cortex is connected with the rostral portion of the nucleus, the temporal cortex with the middle portion, and the occipital cortex with the caudal portion. The functional significance of this nucleus is unknown. Its neurons send many collaterals to the other thalamic nuclei.

Fiber relationships between the thalamic nuclei and certain cortical areas have been established by experimentally destroying the cortical segments or by severing the fibers. The neurons of the respective nuclei undergo retrograde degeneration once their axons have been severed. Neurons of the reticular nucleus, however, are thought to undergo transneuronal degeneration rather than retrograde degeneration, that is, they not degenerate because their axons have been severed but because they have lost afferent fibers terminating on them. This would mean that the cortex projects to the reticular nucleus, while the latter does not project to the cortex.

A – C

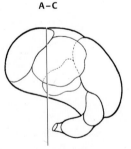

Plane of section

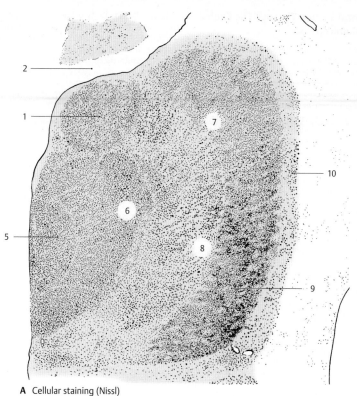

A Cellular staining (Nissl)

A–C Frontal section through the rostral thalamus

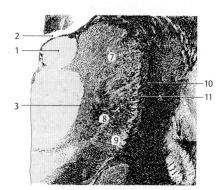

B Fiber staining (myelin)

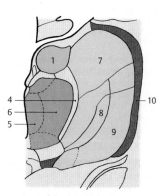

C Nuclear organization (according to *Hassler*)

Diencephalon

Frontal Section Through the Caudal Thalamus (A – C)

At this level, the section again shows the **medial nuclear group** (red) (**A – C1**) and the **lateroventral nuclear group** (blue) (**A – C2**). It includes the most caudal parts of the medial nucleus. Dorsally separated from the *superficial dorsal nucleus* (**A – C3**) by a narrow myelin layer, they are otherwise surrounded by the internal medullary lamina and the intralaminar nuclei. The nonspecific thalamic parts here reach special expansion through the *centromedian nucleus* (**A – C4**).

The most rostral nuclear portions of the **pulvinar** (**A – C5**) lie dorsally between the medial and the lateral nuclear groups. This rostral part of the pulvinar projects to the upper convolutions of the temporal lobe and is thought to receive fibers from the *lateral lemniscus*; it is therefore assumed to be an integration nucleus of the acoustic system.

The *ventral posterior nucleus* (**A – C6**) is seen in the lateroventral area. The *medial lemniscus*, the spinothalamic pathways, and the secondary trigeminal fibers terminate here. The outer portion, which receives the fibers for the limbs and the trunk, is rich in myelinated fibers and has fewer cells than the inner portion, which receives the fibers for the head region. The inner portion is rich in cells and has thinly myelinated fibers. It surrounds the centromedian nucleus ventrally and laterally; it appears as a crescent-shaped figure in the myelin-stained section and, hence, is called the *semilunar nucleus* (**B7**).

The **lateral geniculate body** (**A – C8**) lies slightly apart from the complex of the thalamus at the ventral surface of the diencephalon. It is indented at the base and protrudes laterally (*lateral geniculum*). It is characterized by prominent stratification into six layers of cells and five intercalating layers of fibers. The latter are formed by the fibers of the *optic tract*, which disperse according to a set pattern and terminate at neurons of different cell layers (p.257, A). The upper four of these layers are parvo-

cellular, the lower two layers are magnocellular. In the second, third, and fifth layers terminate the fibers from the retina of the ipsilateral eye (uncrossed optic nerve fibers), while those from the contralateral eye (crossed optic nerve fibers) terminate in the first, fourth, and sixth layers. Fibers from the site of visual acuity, the macula, terminate in the central area (**A9**). When the macula is destroyed, the geniculate cells of this area undergo transneuronal degeneration. The lateral geniculate body is surrounded by a dense capsule of myelinated fibers. These are the dorsally and laterally emerging fibers of the *optic radiation* (*geniculocalcarine tract*) (p.258, C).

Medially to the lateral geniculate body, the section shows the caudal portion of the *medial geniculate body* (**A – C10**). The *reticular nucleus* (**AC11**) forms the lateral capsule. It widens ventrally and also encloses the lateral geniculate body.

A – C

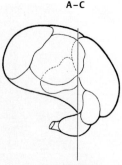

Plane of section

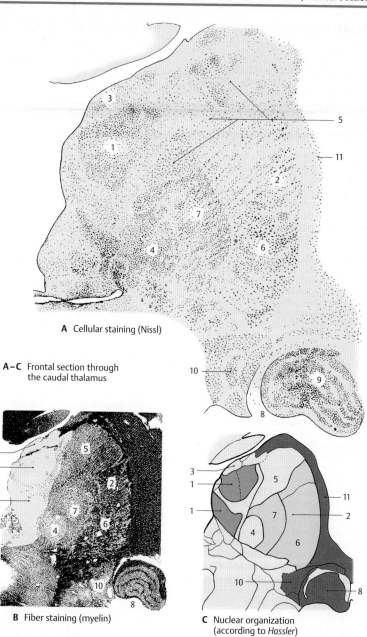

A Cellular staining (Nissl)

A–C Frontal section through
the caudal thalamus

B Fiber staining (myelin)

C Nuclear organization
(according to *Hassler*)

Diencephalon

Subthalamus

Subdivision (A)

The **zona incerta** (**A1**) (p. 174, B25) between *Forel's field H1* (**A2**) and *Forel's field H2* (**A3**) (p. 174, B27, B28) is supposed to be a relay station for descending fibers of the globus pallidus.

The **subthalamic nucleus** (*Luys' body*) (**A4**) (p. 174, B26) between Forel's field H2 and the *internal capsule* (**A5**) has close connections to the pallidum (**A6**), namely, afferent fibers from the outer segment of the globus pallidus and efferent fibers to the inner segment of the globus pallidus. Bidirectional tracts run to the tegmentum and to the contralateral subthalamic nucleus and globus pallidus (supramamillary commissure).

■■ **Clinical Note:** In humans, injury to the subthalamic nucleus leads to hyperkinesis, which may develop into paroxysmal violent flinging movements of the contralateral arm or even the entire body half (*hemiballismus, Luys' body syndrome*). In monkeys, the same symptom complex can be induced by destroying the subthalamic nucleus.

The **globus pallidus** (Pallidium) (**A6**) (p. 174, AB6) is divided by a myelinated fiber lamella into an outer (lateral) and an inner (medial) segment. Both segments are connected by fibers to one another and to the *putamen* (**A7**) and the *caudate nucleus* (**A8**). Bidirectional connections exist to the *subthalamic nucleus* (**A4**), with the *subthalamopallidal fibers* (**A9**) terminating in the inner segment and the *pallidosubthalamic fibers* (**A10**) originating in the outer segment. *Nigropallidal fibers* (**A11**) run from the *substantia nigra* (**A12**) to the inner segment of the pallidum.

The **lenticular fasciculus** (**A13**) emerges at the dorsal margin of the inner segment of the pallidum and forms *Forel's field H2* ventrally to the zona incerta. The **lenticular ansa** (**A14**) (p. 174, A10) emerges from the ventral part of the inner segment and extends in an arch through the inner capsule. Lenticular fasciculus and ansa unite to form the **thalamic fasciculus** (**A15**), which forms *Forel's*

field H1 and radiates into the thalamus (ventral anterior nucleus, ventral lateral nucleus, medial nucleus). Fibers from the inner segment of the pallidum run as pallidotegmental tract (**A16**) into the tegmentum of the midbrain.

■■ **Clinical Note:** Contrary to the old view that *Parkinson's disease* (paralysis agitans) would result from injury to the pallidum, destruction of the pallidum does not lead to motor disorders. Unilateral elimination of the pallidum in patients with Parkinson's disease removes muscle stiffness on the contralateral side and reduces tremor. Bilateral elimination causes mental disorders (postconcussional syndrome: irritability, fatigability, difficulty in concentrating).

Responses to Stimulation of the Subthalamus (B)

Electrical stimulation results in *an increase in muscle tone*, increased *reflex excitability* and promotion of *cortically induced movements*.

Automatic motions can be induced in certain areas. As passing fiber bundles are also affected, the stimulatory effects cannot provide information on the function of individual nuclei. Stimulation at the level of the posterior commissure (fiber region of the interstitial nucleus) results in lowering the head (   ); medial stimulation in the prerubral field results in lifting the head (   ). The field for turning and rolling motions (   ) corresponds to the fiber area of the superior cerebellar peduncle. The field for ipsilateral turning motions (   ) corresponds to the dorsolateral tegmental fasciculus (vestibulothalamic tract). The area for contralateral turning motions (   ) corresponds to the zona incerta.

B17 Fornix.
B18 Mamillothalamic tract.

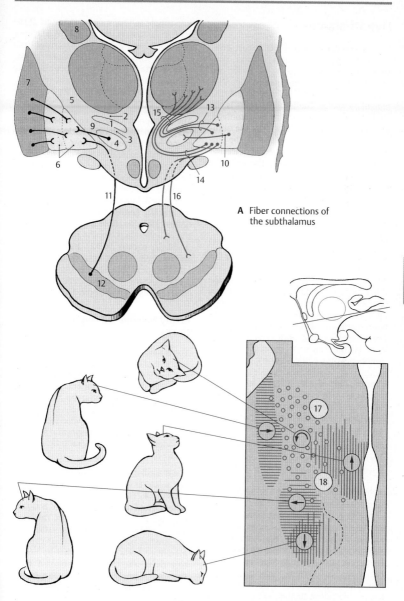

A Fiber connections of the subthalamus

B Motor responses following stimulation of subthalamus and tegmentum; horizontal section through the diencephalon of the cat (according to *Hess*)

Hypothalamus

The hypothalamus forms the lowest layer and the floor of the diencephalon, consisting of the *optic chiasm*, the *tuber cinereum* tapering into the funnel-shaped infundibulum (hypophysial stalk), and the *mamillary bodies*. The hypothalamus is the central region for the control of autonomic functions; it influences not only the autonomic nervous system but also the endocrine system via its connection to the hypophysis, and it coordinates the two. The hypothalamus is divided into two parts, the *poorly myelinated* and the *richly myelinated hypothalamus*.

Poorly Myelinated Hypothalamus (A–C)

This includes the *preoptic region* in front of the optic chiasm, the *tuber cinereum*, the *lateral field* (**A1**) lying dorsolaterally to the tuber cinereum, and the *dorsocaudal area* (**B2**) above the mamillary bodies. The poorly myelinated hypothalamus is the part of the brain that is richest in peptidergic neurons. The different neuropeptides can be demonstrated in diffusely scattered cells (luliberin, cholecystokinin, thyoliberin), in periventricular cell aggregations (somatostatin), and in the nuclear areas (neuropeptides in various compositions). Numerous fiber connections within the hypothalamus and some of the long projection tracts are peptidergic.

The **preoptic region** (**C3**) extends from the *anterior commissure* (**C4**) to the *optic chiasm* (**C5**) as a parvocellular field surrounding the most rostral excavation of the third ventricle, the preoptic recess. The region contains a large number of peptidergic neurons (containing primarily enkephalin). Two prominent magnocellular nuclei lie adjacent to it, namely, the **supraoptic nucleus** (**AC6**) and the **paraventricular nucleus** (**AC7**). The supraoptic nucleus borders on the optic tract (**A8**). Its neurons contain cholecystokinin and dynorphin. The paraventricular nucleus lies close to the wall of the third ventricle, separated from the ependyma

(the simple cell layer lining the walls of the ventricular system, p. 284) only by a layer of glial fibers, and extends as a narrow band obliquely upward into the region of the zona incerta. Its peptidergic neurons have been shown to contain corticoliberin, neurotensin, cholecystokinin, as well as other neuropeptides.

The principal nucleus of the **tuber cinereum** (**A9**) is the **ventromedial nucleus** (**AC10**), a round body occupying most of the tuber cinereum. It contains medium-sized neurons, among them many peptidergic (containing primarily neurotensin), and is surrounded by a delicate fiber capsule formed by the pallidohypothalamic fasciculus. The **dorsomedial nucleus** (**AC11**) is less distinct and contains small neurons. At the base of the infundibulum lies the parvocellular **infundibular nucleus** (**AC12**). Its cells encircle the infundibular recess and reach directly into the ependyma. The neurons of this nucleus contain primarily endorphin and ACTH (adrenocorticotropic hormone).

Richly Myelinated Hypothalamus (B–C)

The **mamillary body** (**BC13**) forms the caudal segment of the hypothalamus. It appears as a round bulge, a myelinated region that is surrounded by a prominent capsule. The latter is formed by afferent and efferent fiber tracts, namely, medially by fibers of the *mamillothalamic fasciculus* (*Vicq d'Azyr's bundle*) and the *mamillotegmental fasciculus* (Gudden's tract), and laterally by fibers of the fornix.

The mamillary body is divided into a medial and a lateral nucleus. The large *medial nucleus* (**B14**) contains small neurons. It is spheroid and makes up the round body that is visible at the base of the brain. The *lateral nucleus* (**B15**) is dorsolaterally attached to the medial nucleus like a small cap. The mamillary body is surrounded by nuclear regions of the poorly myelinated hypothalamus, namely, the *premamillary nucleus* (**C16**) and the *tuberomamillary nucleus* (**B17**).

AC18 Fornix.

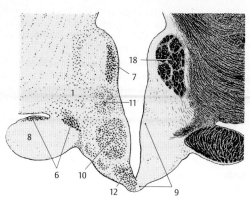

A Frontal section through the tuber cinereum
(poorly myelinated hypothalamus)

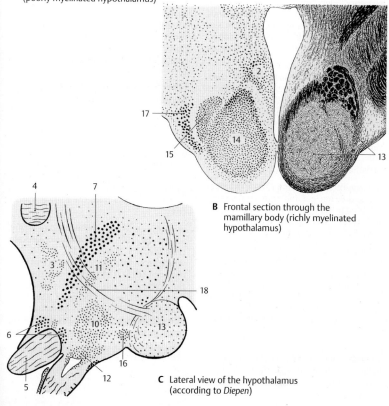

B Frontal section through the
mamillary body (richly myelinated
hypothalamus)

C Lateral view of the hypothalamus
(according to *Diepen*)

Diencephalon

Vascular Supply (A)

The close association of the nervous system with the endocrine system is revealed by the extraordinarily rich vascular supply of individual hypothalamic nuclei. The *supraoptic nucleus* (**A1**) and the *paraventricular nucleus* (**A2**) are about six times more vascularized than the rest of the gray matter. Their neurons have close contact with the capillaries, some of which are even enclosed by the neurons (endocellular capillaries).

Fiber Connections of the Poorly Myelinated Hypothalamus (B)

Numerous connections transmit olfactory, gustatory, viscerosensory, and somatosensory impulses to the hypothalamus. These are usually loose and divergent systems, and almost all of them are bidirectional. The following are the most important ones.

The **medial forebrain bundle** (*medial telencephalic fasciculus*), containing the olfacto-hypothalamotegmental fibers (**B3**), connects almost all hypothalamic nuclei with the olfactory centers and with the reticular formation of the midbrain. The bundle contains a large number of peptidergic fibers (VIP [vasoactive intestinal peptide], enkephalin, somatostatin).

The **terminal stria** (**B4**) runs in an arch around the caudate nucleus and connects the *amygdaloid body* (**B5**) (pp. 226, 228) with the *preoptic region* (**B6**) and the *ventromedial nucleus* (**B7**). It is rich in peptidergic fibers.

The fibers of the **fornix** (**BC8**) (p. 232, B15, C) originate from the pyramidal cells of the hippocampus (**BC9**) and the subiculum (p. 232, A12) and terminate in the mamillary body (**BC10**). A large portion of the fornix fibers are peptidergic. At the level of the anterior commissure, fibers branch off from the fornix to the preoptic region and to the *tuberal nuclei* (**B7**).

The **dorsal longitudinal fasciculus**, or Schütz's bundle (p. 144, B) (**B11**), is the most important component of an extensive periventricular fiber system. At the transition to the midbrain, fibers collect to form a compact bundle that connects the hypothalamus with the nuclei of the brain stem.

Connections with thalamus and pallidum: The hypothalamic nuclei are connected by periventricular fibers with the *medial nuclei of thalamus* (**B12**), the fibers of which project to the frontal cortex. In this way, an indirect connection is established between hypothalamus and frontal cortex. The *pallidohypothalamic fasciculus* runs from the globus pallidus to the tuberal nuclei (ventromedial nucleus).

Commissures: The commissures in the region of the hypothalamus contain almost no fibers from hypothalamic nuclei; fibers from the midbrain and pons cross in the dorsal supraoptic commissure (Ganser's commissure) and in the ventral supraoptic commissure (Gudden's commissure), while fibers from subthalamic nuclei cross in the supramamillary commissure.

Fiber Connections of the Richly Myelinated Hypothalamus (C)

Primarily afferent pathways of the mamillary body (**BC10**) are the **fornix** (**BC8**), the fibers of which terminate primarily in the mamillary body, and the **peduncle of the mamillary body** (**C13**) from the tegmental nuclei of the midbrain, which is thought to contain also gustatory fibers, vestibular fibers, and fibers from the medial lemniscus.

Primarily efferent pathways are represented by the **mamillothalamic fasciculus** (*Vicq d'Azyr's bundle*) (**C14**) (p. 174, B31) which ascends to the *anterior nucleus of the thalamus* (**C15**). The projection of the anterior nucleus to the cingulate gyrus establishes a connection between the hypothalamus and the limbic association cortex. The **mamillotegmental fasciculus** (**C16**) terminates in the tegmental nuclei of the midbrain. All pathways contain a high proportion of peptidergic fibers.

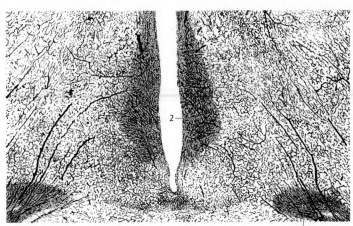

A Vascular supply of the hypothalamus in the rhesus monkey (according to *Engelhardt*)

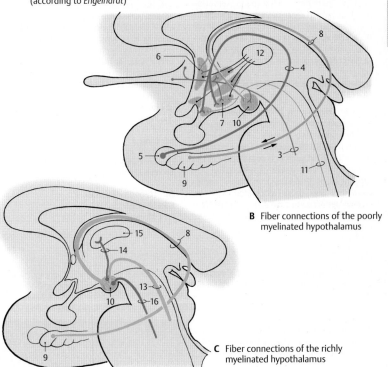

B Fiber connections of the poorly myelinated hypothalamus

C Fiber connections of the richly myelinated hypothalamus

Diencephalon

Functional Topography of the Hypothalamus

The centers of the hypothalamus influence all processes that are important for the *internal environment* of the body; they regulate the performance of organs according to the actual physical load, that is, they control the balance of temperature, water, and electrolytes; the activity of the heart; circulation and respiration; metabolism; and the rhythm of sleeping and waking.

Vital functions (such as food intake, gastrointestinal activity and defecation, fluid intake and urination) are controlled from here, and so are the processes essential for the preservation of the species (procreation and sexuality). These vital activities are triggered by physical needs that are perceived as hunger, thirst, or sexual drive. Instinctive activities serving the preservation of the organism are usually accompanied by a strong affective component, such as desire or aversion, joy, anxiety, or anger. Hypothalamic excitation plays an important role in the creation of these emotions.

Dynamogenic and Trophotropic Zones (A–C)

Electric stimulation of the hypothalamus evokes autonomic reactions that can be divided into two groups (**A**): those associated with **regeneration** and metabolic processes (blue symbols: contraction of the pupils ; slowing down of respiration ; drop in blood pressure ; urination ; defecation), and those associated with **increased performance** in response to the environment (red symbols: dilation of the pupils ; accelerated respiration ; increase in blood pressure). Stimulation responses from specific regions have established a dorsocaudal and lateral region for the *dynamogenic mechanisms*, the **dynamogenic zone** (**B1**), and a ventro-oral region for the *mechanisms promoting regeneration*, the **trophotropic zone** (**B2**). The two regions correspond to the subdivision of the peripheral autonomic nervous system into a sympathetic component (dynamogenic) and a parasympathetic component (trophotropic) (p. 292). In humans, controlled stimulation of the caudal hypothalamus (**C**) yielded similar results, namely, dilation of pupils (**C3**), increase in blood pressure (**C4**), and accelerated respiration (**C5**).

Stimulation and Lesion Experiments (D–F)

Experiments of these types reveal the significance of defined regions for the regulation of certain processes; destruction of the tuber cinereum in the region of the infundibular nucleus (**D6**) in juvenile animals leads to *atrophy of the gonads*. On the other hand, lesions between optic chiasm and tuber cinereum in infantile rats result in *premature sexual development*. Estrous cycle and sexual behavior are affected by lesions to the hypothalamus. Lesions to the caudal hypothalamus between the tuber cinereum and the premamillary nucleus (**E7**) lead to *adipsia* (cessation of spontaneous drinking). More dorsal lesions result in *aphagia* (refusal to eat), while dorsal stimulation (**F8**) results in *hyperphagia* (compulsive eating). In the anterior hypothalamus at the level of the optic chiasm lies a region responsible for the *control of body temperature*. Stimulation in the vicinity of the fornix triggers fits of rage and aggressive behavior (*perifornical zone of rage*).

Clinical Note: In humans, pathological processes in the hypothalamus cause similar changes (*premature sexual development, bulimia, anorexia nervosa*).

A9 Hypophysial stalk.
B10 Hypophysis.
B11 Mamillary body.

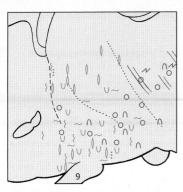

A Trophotropic and dynamogenic zones in the diencephalon of the cat, sagittal section, stimulation pattern (according to *Hess*)

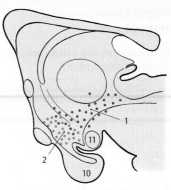

B Trophotropic and dynamogenic zones in the cat, sagittal section

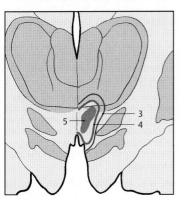

C Responses following stimulation in the caudal hypothalamus in humans, frontal section (according to *Sano, Yoshioka, Ogishiwa, et al.*)

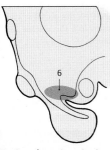
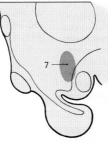
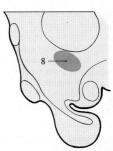

D Gonadotropic region in cats and rabbits (according to *Spatz, Bustamante, and Weisschedel*)

E Adipsia, experimental lesion in the rat (according to *Stevenson*)

F Hyperphagia, stimulation experiment in the cat (according to *Brügge*)

Hypothalamus and Hypophysis

Development and Subdivision of the Hypophysis (A, B)

The **hypophysis**, or *pituitary gland* (see vol. 2), consists of two parts, namely, the **adenohypophysis** (*anterior lobe*) (**AB1**), which develops from an evagination of the roof of the primitive foregut (Rathke's pouch), and the **neurohypophysis** (*posterior lobe)* (**AB2**), which represents an evagination of the floor of the diencephalon. The adenohypophysis is an endocrine gland, while the neurohypophysis is a part of the brain composed of nerve fibers, a capillary bed, and a unique type of glia, the pituicytes. The two parts of the hypophysis adjoin each other, bringing the nervous system into close association with the endocrine system.

Infundibulum (A – D)

The tuber cinereum tapers at its base to form the **infundibulum** (*hypophysial stalk*) (**A3**) and the **infundibular recess** (**A4**). The funnel-shaped descent of the infundibulum, where the contact between nervous and endocrine systems is established, is also called the **median eminence of the tuber**. A thin tissue layer of the adenohypophysis reaches to the tuber cinereum and covers the anterior side of the infundibulum (**infundibular part of the adenohypophysis**) (**A5**); several islets of tissue cover also its back. Thus, we can distinguish a proximal part of the hypophysis bordering on its tuber cinereum (infundibulum and infundibular part of the adenohypophysis) and a distal part lying in the sella turcica (adenohypophysis with the **intermediate part** [**A6**, **hypophysial cavity** [**A7**], and neurohypophysis). The *proximal contact surface* is of special significance for the interconnection of the nervous and endocrine systems. The outer *glial fiber layer* (**A8**), which seals off the rest of the brain surface, is absent here, and *portal capillaries* (**C**) enter from the adenohypophysis into the infundibulum. These are vascular loops with an afferent and an efferent limb, which take a convoluted and sometimes spiral course (**D**).

Blood Vessels of the Hypophysis (C – E)

A prominent vascular supply ensures the coupling of the nervous and endocrine parts of the hypophysis. The afferent vessels, the **superior hypophysial artery** (**E9**) and the **inferior hypophysial artery** (**E10**), branch off from the internal carotid artery. The two superior hypophysial arteries form an arterial ring around the proximal part of the infundibulum, from where small arteries extend through the adenohypophysial cover into the infundibulum and disperse into **portal capillaries** (**E11**). The recurrent limbs of the latter collect in the **portal veins** (**E12**), which transport the blood to the capillary bed of the adenohypophysis. The **trabecular arteries** (**E13**) extend to the adenohypophysis, ascend caudally, and supply the distal segment of the infundibulum. The blood then runs from the capillary bed of the adenohypophysis into the veins.

The two inferior hypophysial arteries supply the neurohypophysis; with several branches in the area of the intermediate part, they also form special vessels (**E14**); from here, the blood also runs via short portal vessels into the capillary bed of the adenohypophysis.

Thus, the adenohypophysis receives no direct arterial supply. The latter goes into the infundibulum and into the neurohypophysis, from where the blood flows into the adenohypophysis via the portal vessels and, only then, drains into the venous side (**E15**) of the circulation.

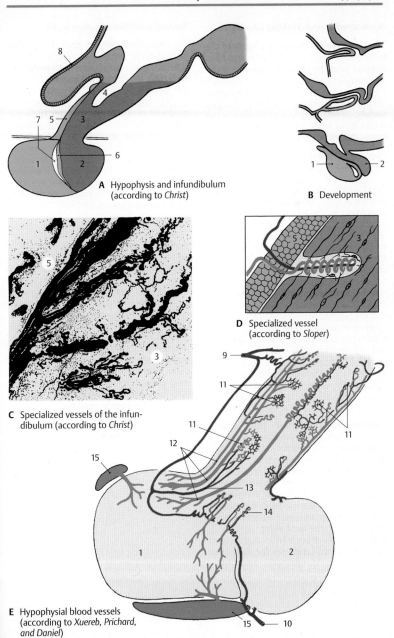

A Hypophysis and infundibulum (according to *Christ*)

B Development

C Specialized vessels of the infundibulum (according to *Christ*)

D Specialized vessel (according to *Sloper*)

E Hypophysial blood vessels (according to *Xuereb, Prichard, and Daniel*)

Neuroendocrine System (A – D)

The hypophysis is under the control of the hyopothalamic centers. Unmyelinated fiber bundles from the hypothalamus run in the hypophysial stalk and in the hypophysial posterior lobe. Severing the hypophysial stalk leads to retrograde cellular changes, namely, to extensive loss of neurons in the nuclei of the tuber cinereum when transection is carried out at a high level.

Neurons in the hypothalamus produce substances that migrate within the axons into the hypophysis and enter there into the bloodstream. This endocrine function of neurons is called **neurosecretion**. The substances are produced in the perikarya and appear there as small secretory droplets (**B1**). The cells, which are genuine neurons with dendrites and axons, represent a transitional stage between neurons and secretory cells. Both these cell types are of ectodermal origin and are closely related with respect to physiology and metabolism. Both of them produce a specific substance that they secrete in response to a nervous or humoral stimulus: the neurons (**A2**) release transmitter substances (neurotransmitters) and the secretory cells (**A3**) release their secretion. Transitional forms between the two cell types are the neurosecretory cells (**A4**) and the endocrine cells **A5**), both of which release their secretion into the bloodstream.

Hypothalamohypophysial fiber tracts.

Corresponding to the structure of the hypophysis with an anterior lobe (adenohypophysis) and a posterior lobe (neurohypophysis), there are two different fiber systems extending from the hypothalamus to the hypophysis, namely, the *tuberoinfundibular tract* and the *hypothalamohypophysial tract*. In both, the coupling of the neural system with the endocrine system is achieved by the sequential arrangement of nerve fibers and capillaries (*neurovascular chain*).

Tuberoinfundibular System (C, D)

The **tuberoinfundibular tract** (**D**) consists of thin nerve fibers that originate in the tuberal nuclei, namely, the **ventromedial nucleus** (**D6**), the **dorsomedial nucleus** (**D7**), and the **infundibular nucleus** (**D8**), and extend into the hypophysial stalk. The substances produced in the perikarya enter from the axon terminals into the portal capillaries (**D9**) and pass through the portal veins (**D10**) into the capillary bed of the adenohypophysis. They are stimulating substances, *releasing factors*, which cause the release of glandotropic hormones (messengers that affect other endocrine glands) by the adenohypophysis (see vol. 2).

The production of specific releasing factors cannot be attributed to individual hypothalamic nuclei. The regions from which an increased secretion can be induced by electrical stimulation (**C**) do not correspond to the tuberal nuclei. Stimulation of the preoptic region (**C11**) results in an increased secretion of *luteotropic hormone*. Stimulation caudally to the optic chiasm (**C12**) leads to the release of *thyrotropic hormone*, and stimulation of the ventral hypothalamus (tuber cinereum to mamillary recess) (**C13**) leads to the release of *gonadotropic hormone*. In addition to releasing factors, inhibiting factors, inhibiting factors have been identified that block the release of hormones in the adenohypophysis.

CD14 Optic chiasm.
CD15 Mamillary body.

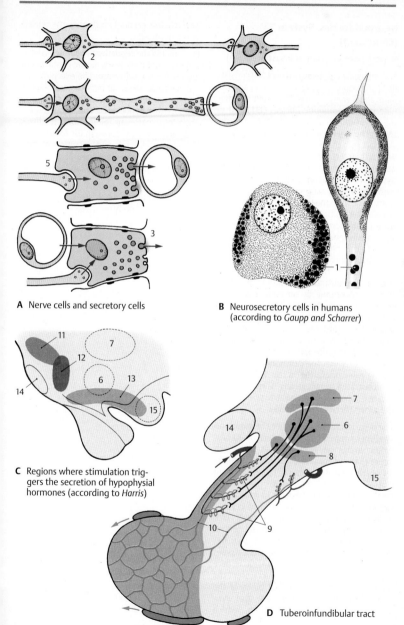

A Nerve cells and secretory cells

B Neurosecretory cells in humans
(according to *Gaupp and Scharrer*)

C Regions where stimulation triggers the secretion of hypophysial
hormones (according to *Harris*)

D Tuberoinfundibular tract

Neuroendocrine System (continued)

Hypothalamohypophysial System (A–D)

The **hypothalamohypophysial tract** (**D**) consists of the *supraopticohypophysical tract* and the *paraventriculohypophysial tract* which originate in the *supraoptic nucleus* (**D1**) and in the *paraventricular nucleus* (**D2**), respectively. The fibers run through the hypophysial stalk into the hypophysial posterior lobe where they terminate at the capillaries. The hormones produced by the neurons of both hypothalamic nuclei migrate along this pathway to the axon terminals and enter from here into the bloodstream. *Electrical stimulation* of the supraoptic nucleus (**C3**) leads to an increased secretion of *vasopressin* (antidiuretic hormone), while stimulation of the paraventricular nucleus (**C4**) leads to an increased secretion of *oxytocin*. In this system, the neurons do not release stimulating substances that affect the secretion of a hormone by an endocrine gland (such as the glandotropic hormones or releasing factors of the tuberoinfundibular system), but they themselves produce hormones that have a direct effect on the target organs (effector hormones). The carrier substances to which the hormones are bound during their migration in the axons can be demonstrated histologically. These Gomoripositive substances often cause swellings of the axons (**Herring bodies**) (**B5**).

The neurosecretory substances in axons and swellings appear in the electron-microscopic image as granules that are much larger than synaptic vesicles. At the capillaries of the neurohypophysis, the axons form club-shaped endings (**AD6**) containing small, clear synaptic vesicles in addition to the large granules. At the sites of contact with axon terminals, the capillary walls lack the glial covering layer that, in the central nervous system, forms the boundary between ectodermal and mesodermal tissues and envelops all vessels (p. 44). It is here that the neurosecretory product enters the bloodstream. At the terminal bulbs of the neurosecretory cells, there are also synapses (**A7**) of unknown origin, which nevertheless certainly influence the release of the hormones.

Presumably the **regulation of neurosecretion** is achieved not only via synaptic contacts but also via the bloodstream. The exceptionally rich vascularization of hypothalamic nuclei and the existence of endocellular capillaries support this hypothesis. This arrangement provides a pathway for humoral feedback and forms a regulatory circuit for controlling the production and secretion of hormones, consisting of a neural limb (supraopticohypophysial tract) and a humoral limb (circulation).

CD8 Optic chiasm.
CD9 Mamillary body.

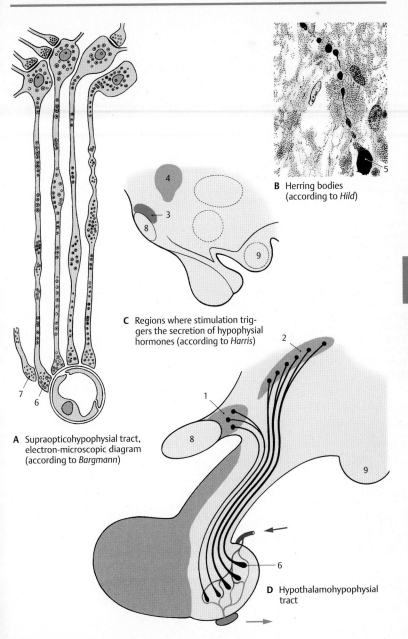

B Herring bodies
(according to *Hild*)

C Regions where stimulation triggers the secretion of hypophysial hormones (according to *Harris*)

A Supraopticohypophysial tract,
electron-microscopic diagram
(according to *Bargmann*)

D Hypothalamohypophysial
tract

Diencephalon

Telencephalon

Overview

Subdivision of the Hemisphere (A, B)

The embryonic hemispheric vesicle (**A**) clearly shows the subdivision of the telencephalon into four parts, some of which develop early (phylogenetically old portions), while others develop late (phylogenetically new portions). The four parts are the *paleopallium*, the *striatum*, the *neopallium*, and the *archipallium*.

The hemispheric wall is called the **pallium**, or **brain mantle**, because it covers the diencephalon and brain stem and envelops them like a mantle.

The **paleopallium** (blue) (**AB1**) is the oldest portion of the hemisphere. It forms the floor of the hemisphere and corresponds, with the *olfactory bulb* (**A2**) and adjacent **paleocortex** (p. 224 ff), to the olfactory brain, or *rhinencephalon*, in the narrower sense. The **neostriatum** (deep yellow) (**AB3**) (p. 236) develops above the paleopallium; it, too, is part of the hemispheric wall, although it does not appear on the outer aspect of the hemisphere.

The largest area is made up by the **neopallium** (light yellow). Its outer aspect, the **neocortex** (p. 240 ff) (**AB4**), develops very late and encircles ventrally a transitional area to the paleocortex that lies over the striatum; this is the **insula** (p. 238) (**B14**).

The medial hemispheric wall is formed by the **archipallium** (red) (**AB5**), an old portion of the brain; its cortical band, the **archicortex** (p. 230 ff), later curls up to form the hippocampus (*Ammon's horn*).

The relationships in the mature brain are determined by the massive expansion of the neocortex, which pushes the paleocortex and the transitional cortex of the insula into the deeper parts of the brain. The archicortex becomes displaced caudally and appears on the surface of the corpus callosum only as a thin layer (**B5, F10**).

Rotation of the Hemisphere (C–F)

The hemispheric vesicle does not expand evenly in all directions during its development but widens primarily in caudal and basal directions. The temporal lobe is formed in this way, and it finally turns rostrally in a circular movement (**C**); to a lesser degree, such a rotation can also be observed with the frontal lobe. The axis around which the hemispheric vesicle rotates is the insular region; like the *putamen* (**E6**) lying beneath it, the insula does not participate in the movement. Other structures of the hemisphere, however, follow the rotation and end up having an arched shape in the mature brain. The *lateral ventricle* (**D7**) forms such an arch with its anterior and inferior horns. The lateral portion of the striatum, the *caudate nucleus* (**E8**), participates in the rotation as well and follows precisely the arched shape of the lateral ventricle. The main part of the archipallium, the *hippocampus* (**F9**), moves from its original dorsal position in basal direction and comes to lie in the temporal lobe. The remnants of the archipallium on the dorsal aspect of the corpus callosum, the *indusium griseum* (**F10**), and the *fornix* (**F11**) reflect the arched expansion of the archipallium. The *corpus callosum* (**F12**) also expands in caudal direction but follows the rotation only partially as it develops only late toward the end of this process.

D13 Third ventricle.

A, B Subdivision of the hemispheres

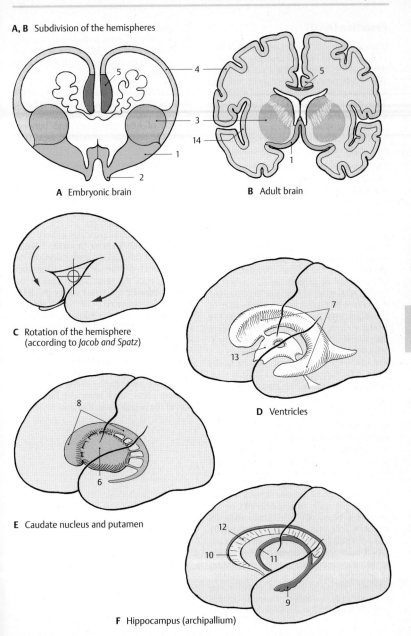

A Embryonic brain

B Adult brain

C Rotation of the hemisphere
(according to *Jacob and Spatz*)

D Ventricles

E Caudate nucleus and putamen

F Hippocampus (archipallium)

Telencephalon

Evolution (A – D)

During primate evolution, the telencephalon has undergone changes similar to those taking place during human embryonic development; it developed late and then overgrew the other parts of the brain. Thus, the cerebellum (**A1**) is still completely exposed in the brain of primitive mammals (hedgehog), while it becomes more and more covered by the hemispheres of the telencephalon during primate evolution.

The **paleopallium** (rhinencephalon) (blue) (**A – C2**) with *olfactory bulb* (**A – C3**) and *piriform lobe* (**A – C4**) forms the largest part of the hemisphere in the primitive mammalian brain (**A**), and the **archipallium** (red) (**A – D5**) still has its original dorsal position above the diencephalon. These two old components of the hemisphere then become overgrown by the **neopallium** (yellow) (**A – D6**) during the course of evolution. The paleopallium of prosimians (**C**) is still of considerable size. In humans (**D**), however, it becomes displaced deep into the base of the brain and no longer appears in the lateral view of the brain. The archipallium (*hippocampus*), which lies above the diencephalon in the hedgehog (**A5**), appears as a part of the temporal lobe at the base of the brain in humans (**D5**). Only a narrow remnant remains above the corpus callosum (indusium griseum).

The positional changes largely correspond to the rotation of the hemisphere during embryonic development; they also lead to the formation of the temporal lobe (**B – D7**). While still absent from the brain of the hedgehog (**A**), the temporal lobe is already recognized as a ventrally directed projection in the brain of the tree shrew (Tupaia), the most primitive of primates (**B**). In the prosimian brain (**C**), a caudally directed temporal lobe has developed that finally turns rostrally in the human brain (**D**). In addition, sulci and gyri develop in the region of the neopallium. Whereas the neopallium of primitive mammals is smooth (*lissencephalic brains*), a relief of convolutions develops only in higher mammals (*gyrencephalic brains*). The development of sulci and gyri considerably enlarges the surface of the cerebral cortex. In humans, only one-third of the cortical surface lies at the surface of the hemispheres, two-thirds lie deep in the sulci.

Two types of cortical areas can be distinguished on the neocortex: the primary **areas of origin** (light red) and **termination areas** (green) of long pathways, and between them the secondary **association areas** (yellow).

The area of origin of motor pathways, the *motor cortex* (**A – D8**), constitutes the entire frontal lobe in the hedgehog. An association area (**B – D9**) appears for the first time in primitive primates (Tupaia) and achieves extraordinary expansion in the human brain. The termination area of sensory pathways, the *sensory cortex* (**A – D10**), borders caudally on the motor cortex. Owing to the enlargement of the adjacent association area, most of the termination area of the visual pathway, the *visual cortex* (**A – D11**), becomes displaced to the medial hemispheric surface in humans. The termination area of the acoustic pathway, the *auditory cortex* (**CD12**), becomes displaced deep into the lateral sulcus (fissure of Sylvius) by the expansion of the temporal association areas. Thus, the association areas expand much more during evolution than the primary areas; they represent the largest part of the neocortex in humans.

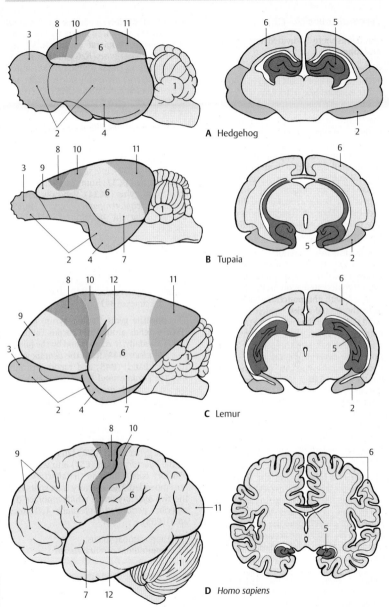

A Hedgehog

B Tupaia

C Lemur

D *Homo sapiens*

A–D Evolution of the telencephalon (modified after *Edinger, Elliot Smith, and Le Gros-Clark*)

Telencephalon

Cerebral Lobes (A – C)

The hemisphere is divided into four **cerebral lobes**:

- The **frontal lobe** (red) (p. 246)
- The **parietal lobe** (light blue) (p. 250)
- The **temporal lobe** (dark blue) (p. 252)
- The **occipital lobe** (purple) (p. 254)

The hemispheric surface consists of *grooves*, or **sulci**, and *convolutions*, or *gyri*. We distinguish *primary*, *secondary*, and *tertiary sulci*. The primary sulci appear first and are equally well developed in all human brains (central sulcus, calcarine sulcus). The secondary sulci are variable. The tertiary sulci appear last, being irregular and different in each brain. Thus, each brain has its own surface relief as an expression of individuality, like the features of the face.

The **frontal lobe** extends from the *frontal pole* (**AC1**) to the **central sulcus** (**AB2**), which together with the *precentral sulcus* (**A3**) defines the **precentral gyrus** (**A4**). The latter is grouped with the **postcentral gyrus** (**A5**) to form the **central region**, which spreads beyond the *edge of the hemisphere* (**AB6**) to the *paracentral gyrus* (**B7**). Furthermore, the frontal lobe exhibits three major convolutions: the *superior frontal gyrus* (**A8**), *middle frontal gyrus* (**A9**), and the *inferior frontal gyrus* (**A10**); they are separated by the *superior frontal sulcus* (**A11**) and the *inferior frontal sulcus* (**A12**). Three parts are distinguished at the inferior frontal gyrus that define the lateral sulcus (sulcus of Sylvius) (**AC13**): the *opercular part* (**A14**), the *triangular part* (**A15**), and the *orbital part* (**A16**).

The **parietal lobe** adjoins the frontal lobe with the postcentral gyrus (**A5**) which is defined caudally by the *postcentral sulcus* (**A17**). This is followed by the *superior parietal lobule* (**A18**) and the *inferior parietal lobule* (**A19**), which are separated by the *intraparietal sulcus* (**A20**). The end of the lateral sulcus is surrounded by the *supramarginal gyrus* (**A21**); the **angular gyrus** (**A22**) lies ventrally to it. The medial surface of the parietal lobe is formed by the *precuneus* (**B23**).

The **temporal lobe** includes the *temporal pole* (**AC24**) and three major convolutions: the *superior temporal gyrus* (**A25**), the *middle temporal gyrus* (**A26**), and the *inferior temporal gyrus* (**AC27**), which are separated by the *superior temporal sulcus* (**A28**) and the *inferior temporal sulcus* (**A29**). The *transverse temporal gyri* (Heschl 's convolutions) of the dorsal aspect of the temporal lobe lie in the depth of the lateral sulcus (p. 252, C). On the medial surface is the *parahippocampal gyrus* (**BC30**) which merges rostrally into the *uncus* (**BC31**) and caudally into the *lingual gyrus* (**BC32**). It is separated by the *collateral sulcus* (**BC33**) from the *middle occipitotemporal gyrus* (**BC34**). Ventrally lies the *lateral occipitotemporal gyrus* (**BC35**), delimited by the *occipitotemporal sulcus* (**BC36**).

The **occipital lobe** includes the *occipital pole* (**A – C37**) and is crossed by the *transverse occipital sulcus* (**A38**) and the deep **calcarine sulcus** (**B39**). Together with the *parieto-occipital sulcus* (**B40**), the latter defines the *cuneus* (**B41**).

The **cingulate gyrus** (limbic gyrus) (green) (**B42**) extends around the corpus callosum (**B43**). Caudally, it is separated by the *hippocampal sulcus* (**B44**) from the *dentate gyrus* (dentate band) (**B45**) and tapers rostrally into the *paraterminal gyrus* (**B46**) and into the *subcallosal area* (parolfactory area) (**B47**). *Isthmus of cingulate gyrus* (**B48**).

Base of the brain. The basal aspect of the frontal lobe is covered by the *orbital gyri* (**C49**). Along the edge of the hemisphere runs the *gyrus rectus* (**C50**), laterally defined by the *olfactory sulcus* (**C51**) into which the *olfactory bulb* (**C52**) and the *olfactory tract* are embedded. The olfactory tract splits into the two *olfactory striae* which embrace the *anterior perforated substance* (olfactory area) (**C53**).

C54 Hippocampal sulcus.
C55 Longitudinal cerebral fissure.

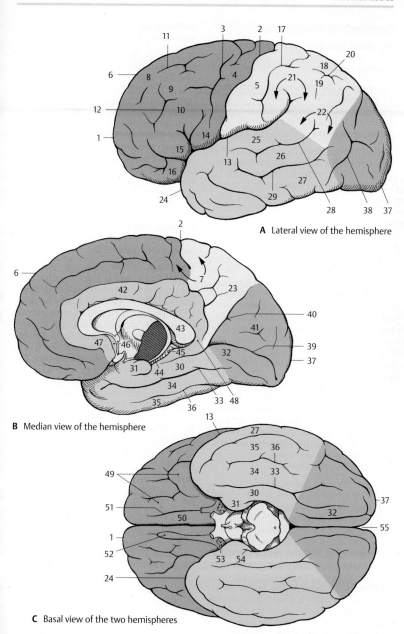

A Lateral view of the hemisphere

B Median view of the hemisphere

C Basal view of the two hemispheres

Sections Through the Telencephalon

Frontal Sections

The posterior cut surface is shown for each brain section.

Section at the Level of the Exit of the Olfactory Tract (A)

The cut surface shows the two hemispheres separated by the *cerebral longitudinal fissure* (**AB1**); the gray matter (cortex and nuclei) is easily distinguished from the white matter (myelinated fiber masses). The *corpus callosum* (**AB2**) connects the two hemispheres. The section shows the *cingulate gyrus* (**AB3**) above the corpus callosum.

The lateral field of the section shows the deep *lateral sulcus* (**AB4**). Dorsally to it lies the frontal lobe with the *superior frontal gyrus* (**AB5**), the *middle frontal gyrus* (**AB6**), and the *inferior frontal gyrus* (**AB7**). They are separated by the *superior frontal sulcus* (**AB8**) and the *inferior frontal sulcus* (**AB9**). Ventrally to the lateral sulcus lies the temporal lobe with the *superior temporal gyrus* (**AB10**), the *middle temporal gyrus* (**AB11**), and the *inferior temporal gyrus* (**AB12**). The temporal gyri are separated by the *superior temporal sulcus* (**AB13**) and *inferior temporal sulcus* (**AB14**). The lateral sulcus expands deep into the *lateral fossa* (*fossa of Sylvius*) (**AB15**), on the inner surface of which is the insula. The insular cortex extends basally almost to the exit of the *olfactory tract* (**A16**). It represents a transitional area between paleocortex and neocortex.

In the depth of the hemisphere lies the *neostriatum* which is divided by the *internal capsule* (**AB17**) into the *caudate nucleus* (**AB18**) and the *putamen* (**AB19**). The section shows the *anterior horn* (**AB20**) of the lateral ventricle. The lateral wall of the ventricle is formed by the caudate nucleus, while its medial wall is formed by the *septum pellucidum* (**AB21**) containing the *cavity of the septum pellucidum* (**AB22**). At the lateral aspect of the putamen lies a narrow, cup-shaped layer of gray matter, the *claustrum* (**AB23**). It is separated from the putamen by the *external capsule* (**AB24**) and from the insular cortex by the *extreme capsule* (**AB25**).

Section at the Level of the Anterior Commissure (B)

At this level, the section shows the central regions of the frontal lobe and the temporal lobe. The lateral fossa is closed, and the insula is covered by the frontal operculum (**AB26**) and the temporal operculum (**AB27**). The ventral regions of both hemispheres are connected by the *anterior commissure* (**B28**) where fibers of the paleocortex and the temporal neocortex cross. Above the commissure appears the globus pallidus (**B29**) (part of the diencephalon), and close to the midline lies the *septum pellucidum* (**AB21**), or more specifically, its wide ventral segment containing the septal nuclei (also known as *peduncle of the septum pellucidum*). The mediobasal aspect of the hemisphere is covered by the *paleocortex*, the olfactory cortex (**B30**).

Claustrum. In the past, the claustrum (**AB23**) was either grouped together with the striatum to form the so-called *basal ganglia* or was assigned to the insular cortex as an additional cortical layer. Developmental studies and comparative anatomical investigations, however, suggest that it consists of cell clusters of the *paleocortex* which have become displaced during development. The claustrum merges with its wide base into paleocortical regions (namely, the prepiriform cortex and the lateral nucleus of the amygdaloid body). Unmyelinated fibers from the cortices of parietal, temporal, and occipital lobes are thought to terminate in the claustrum in a topical arrangement. The function of the claustrum is largely unknown.

B31 Optic chiasm.

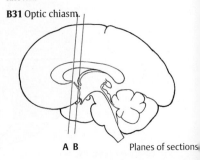

A B   Planes of sections

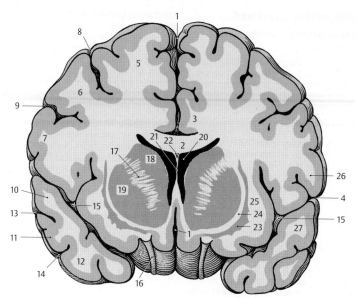

A Frontal section at the exit of the olfactory tract

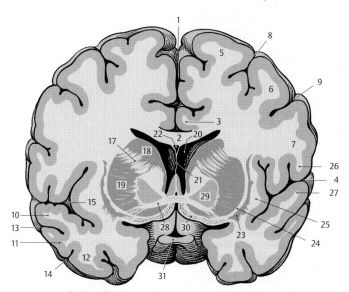

B Frontal section at the level of the anterior commissure

Telencephalon

Frontal Sections (continued)

Section at the Level of the Amygdaloid Body (A)

At this level the **central sulcus** (**AB1**), which runs obliquely from dorsocaudal to ventrorostal, has been cut in the more rostral part; the *frontal lobe*, which is dorsal to it, therefore occupies a far larger part of the section than the parietal lobe, which is ventral to it. The convolution above the central sulcus is the **precentral gyrus** (**AB2**); the convolution below it is the **postcentral gyrus** (**AB3**). Deep in the temporal lobe appears the **amygdaloid body** (amygdala) (**A4**). It reaches the surface at the medial aspect of the temporal lobe and might therefore be regarded partly as cortex, partly as nucleus, or rather as a transition between the two structures. Since not only the surrounding periamygdalar cortex but also its corticomedial half belong to the primary olfactory centers, the amygdaloid body can be assigned to the paleocortex, despite its nuclear features. The *claustrum* (**AB5**) ends above this region with a wide base.

Between the hemispheres lies the diencephalon with *thalamus* (**AB6**), *globus pallidus* (**AB7**), and *hypothalamus* (**A8**). Laterally to the diencephalic nuclei border the *neostriatum* with *putamen* (**AB9**) and *caudate nucleus* (**AB10**). Below the *corpus callosum* (**AB11**) lies a strong fiber bundle, the *fornix* (**AB12**). Also seen are the *longitudinal cerebral fissure* (**AB13**), the *lateral cerebral sulcus* (**AB14**), the *lateral fossa* (**AB15**), the *optic tract* (**A16**), and the *infundibulum* (**A17**).

Section at the Level of the Hippocampus (B)

Once the more caudally cut sections no longer show the amygdaloid body, the **hippocampus** (**B18**) appears in the medial area of the temporal lobe. This most important portion of the archicortex is a cortical formation that has curled up and projects against the inferior horn of the lateral ventricle (**B23**). The section also shows the caudal part of the *lateral fossa* (**B15**). The

inner surface of the temporal operculum exhibits prominent convolutions; these are the obliquely cut **transverse temporal gyri** (**B19**), or Heschl's convolutions, representing the auditory cortex. In the ventral region of the diencephalon lie the *subthalamic body* (**B20**), the *mamillary body* (**B21**), and the *substantia nigra* (**B22**), which is a part of the midbrain.

Basal Gaglia. The gray nuclear complexes deep in the hemisphere are collectively known as basal ganglia. Some authors use the term only for the striatum and the pallidum, while others include the amygdaloid body and the claustrum, some even the thalamus. As this term is vague and ill-defined, it is not used in the present description. Earlier anatomists viewed the pallidum and the putamen as parts of the *lentiform nucleus* (a concept still surviving as lenticular ansa and lenticular fasciculus), a term that is no longer used.

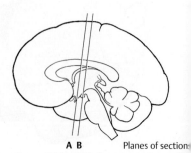

A B   Planes of section

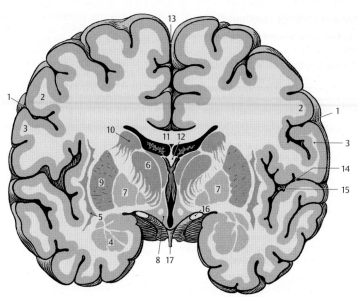

A Frontal section at the level of the amygdaloid body

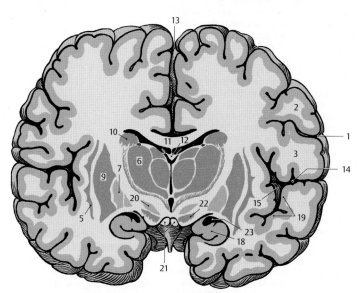

B Frontal section at the level of the hippocampus

Frontal Sections (continued)

Section at the Level of Midbrain and Pons (A)

The caudal portion of the *lateral fossa* (**A1**) is open to the lateral aspect of the hemisphere. Dorsally to the *lateral sulcus* (**A2**) lies the parietal lobe, ventrally the temporal lobe. The dorsal convolutions of the latter, which lie deep in the lateral sulcus and represent the *transverse temporal gyri* (**A3**) (p. 252, C1), are cut obliquely. At the bottom of the lateral fossa lies the insular cortex, which rests on the caudal extensions of *claustrum* (**A4**) and *putamen* (**A5**). The *caudate nucleus* (**A6**) appears at the lateral wall of the lateral ventricle (**A7**). At the medial aspect of the temporal lobe, concealed by the *parahippocampal gyrus* (**A8**), the cortex curls up to form the hippocampus (*Ammon's horn*) (**A9**). Corpus callosum (**A10**) and fornix (**A11**) are seen above the choroid plexus.

The field between the hemispheres represents the transition between diencephalon and midbrain. The section shows the caudal nuclear regions of the thalamus (**A12**). Separated from the main complex lies the *lateral geniculate body* (**A13**), and medially to the ventricular wall lies the *habenular nucleus* (**A14**). The plane of section has been oriented according to Forel's axis (p. 4, B), thus showing telencephalon and diencephalon in frontal section, while the structures of midbrain and pons (Meynert's axis; p. 4, B) have been cut obliquely. Ventral to the aqueduct (**A15**) lies the *decussation of the superior cerebellar peduncle* (**A16**). A narrow strip of dark cells, the *substantia nigra* (**A17**), extends ventrally on both sides. The *cerebral peduncles* (**A18**) are seen laterally to it; the course of their fiber masses can be traced from the internal capsule to the pons (**A19**).

Section at the Level of the Splenium of the Corpus Callosum (B)

In this section, the dorsal part of the hemisphere belongs to the parietal lobe and the ventral part to the temporal lobe; at this plane of section, the latter is merging into the occipital lobe. The boundary between parietal lobe and temporal lobe lies in the region of the *angulate gyrus* (**B20**). The lateral sulcus and the lateral fossa are no longer present in the section. The cut surface of the corpus callosum is particularly wide at the level of the *splenium* (**B21**) (p. 220, A6; p. 260, E14). Dorsally and ventrally to it lies the *cingulate gyrus* (**B22**), which encircles the splenium in an arch. The *parahippocampal gyrus* (**B23**) adjoins ventrally. Neither the hippocampus nor the calcarine sulcus are present in the section; hence, the section lies behind the hippocampus but in front of the calcarine sulcus. The two lateral ventricles are remarkably wide, each representing the most anterior part of the posterior horn at the transition into inferior horn and central part (see p. 281, BC7 – 9).

The lower aspects of the hemispheres border on the cerebellum. The medulla oblongata appears in the middle, the oblique section shows the fourth ventricle (**B24**), the *olives* (**B25**), and the *pyramids* (**B26**).

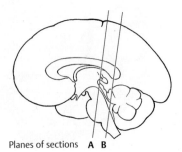

Planes of sections   **A B**

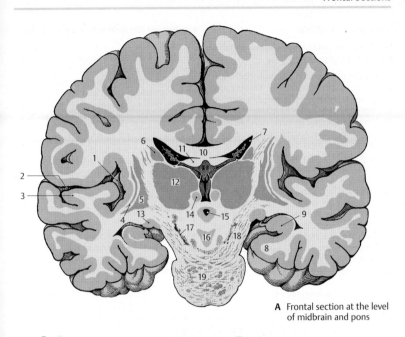

A Frontal section at the level of midbrain and pons

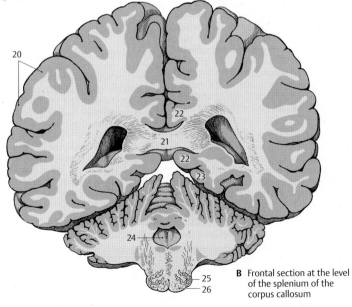

B Frontal section at the level of the splenium of the corpus callosum

Horizontal Sections

Superior Aspect of Corpus Callosum and Lateral Ventricles (A)

The horizontal section through the brain has been cut above the corpus callosum, and the superior aspect of the corpus callosum and the lateral ventricles have been exposed by removal of deeper portions of white matter. The section shows the frontal lobes (**A1**) at the top, the temporal lobes (**A2**) on both sides, and the occipital lobes (**A3**) at the bottom. The superior surface of the corpus callosum (**A4**) belongs to the free brain surface lined by the pia mater and arachnoidea. Lying deep in the brain, it is covered by the convolutions of the medial walls of the hemispheres. Rostrally, the superior surface of the corpus callosum turns in ventral direction and forms the *genu of the corpus callosum* (**A5**) (p. 260, E11); caudally, it forms the *splenium of the corpus callosum* (**A6**) (p. 260, E14). On the superior aspect of the corpus callosum extend four myelinated fiber ridges: one *lateral longitudinal stria* (**A7**) and one *medial longitudinal stria of Lancisi* (**A8**) run along each half of the corpus callosum (see p. 230). Their fiber tracts extend from the hippocampus to the subcallosal area. Between the two longitudinal striae lies a thin layer of gray matter consisting of a narrow layer of neurons, the *indusium griseum*. This is a cortical portion of the archicortex that regressed as a result of the extensive development of the corpus callosum (p. 7, E) and subsequent displacement of the archicortex into the inferior horn of the lateral ventricle (see p. 209, F).

The anterior horns (**A9**) of the lateral ventricles (p. 280, A1) are opened in the area of the frontal lobes, and the posterior horns (**A10**) in the area of the occipital lobes. The protruding *hippocampus* (**A11**) forms the floor of the inferior horn. The central part and the inferior horn of the lateral ventricle contain the *choroid plexus* (**A12**) (p. 282).

Exposure of the Roof of the Diencephalon (B)

This is an oblique horizontal section below the corpus callosum, which has been completely removed. Upon opening the two lateral ventricles, the dorsal aspect of the *caudate nucleus* (**B13**) and, bordering medially, the dorsal aspect of the *thalamus* (**B14**) become visible. Parts of the diencephalon become exposed as well, namely, the *pineal gland* (**B15**) and both *habenulae* (**B16**) which are connected to it. The two *fornices* (**B17**) between the heads of the two caudate nuclei have been cut in their rostral part (columns of fornix). The *septum pellucidum* (**B18**) extends from there to the corpus callosum.

The lateral wall of the hemisphere contains a particularly wide medullary layer between the cortex and the ventricle, the *semioval center* (**B19**). The *central sulcus* (**B20**) cuts into it and separates the frontal lobe (at the top of the figure) from the parietal lobe (bottom). Starting from the central sulcus, the *precentral gyrus* (**B21**) and the *postcentral gyrus* (**B22**) can be located.

Caudally in the *longitudinal cerebral fissure* (**AB23**), the *cerebellum* (**B24**) is visible. The caudal portion of the hemisphere is formed by the occipital lobe. The **striate area** (**B25**), the visual cortex, lies in this region and occupies primarily the **calcarine sulcus** (**B26**) at the medial aspect of the occipital lobe, while extending only a short distance onto the occipital pole. It can be distinguished even by the naked eye from the rest of the cortex through a white streak, the **line of Gennari** (**B27**), which divides the cortex into two gray bands. Gennari's line is a wide band of myelinated nerve fibers corresponding to the slightly narrower external band of Baillarger in the other areas of the neocortex (see p. 240, A16; p. 254).

B28 Mesencephalic tectum.

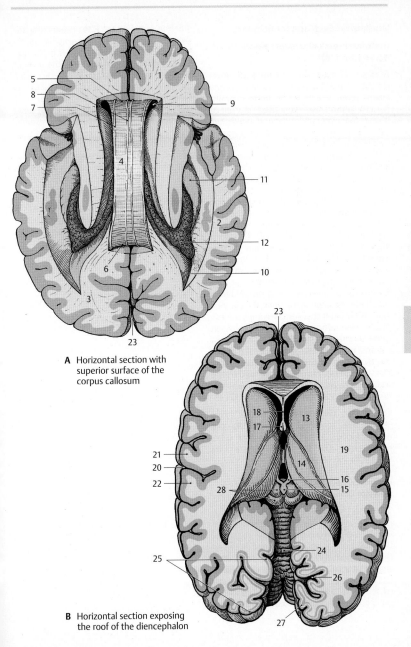

A Horizontal section with superior surface of the corpus callosum

B Horizontal section exposing the roof of the diencephalon

Horizontal Sections (continued)

Horizontal Section through the Neostriatum (A)

At this level, the *lateral cerebral fossa* (**AB1**) is exposed in its longitudinal expansion. The *lateral sulcus* (**A2**) is found more rostrally, with the *frontal operculum* (**AB3**) in front of it and the elongated *temporal operculum* (**AB4**) caudally to it. The longitudinal expansion is also apparent in the deep structures of the telencephalon, the *claustrum* (**AB5**) and the *putamen* (**AB6**). The arched structures have been cut twice; the *corpus callosum* (**A7**) appears rostrally with its anterior part, the *genu of the corpus callosum*, and caudally with its end, the *splenium*. The *caudate nucleus* has been cut twice as well; the *head of the caudate nucleus* (**AB8**) is seen rostrally and the *tail of the caudate nucleus* (**AB9**) caudolaterally to the *thalamus* (**AB10**). The thalamus is separated from the *globus pallidus* (**AB11**) by the *internal capsule* which, in horizontal sections, exhibits the shape of a hook made up of the *anterior limb* (**AB12**) and the *posterior limb* (**AB13**). Also the lateral ventricle has been exposed twice. Its *anterior horn* (**A14**) has been cut in the area of the frontal lobe and, caudally, in the transition to the *posterior horn* (**A15**). The two anterior horns are separated by the septum pellucidum (**A16**), which spans between corpus callosum and fornix (**A17**).

The section also shows the frontal lobes (**AB18**), the temporal lobes (**AB19**), the occipital lobes (**A20**), the longitudinal cerebral fissure (**AB21**), and the striate area (visual cortex) (**A22**).

Horizontal Section at the Level of the Anterior Commissure (B)

While the section still shows the entire frontal lobe and temporal lobe, the occipital lobe has only been cut in its anterior part at the transition to the temporal lobe. Between the two hemispheres appears the cone-shaped dorsal aspect of the cerebellum (**B23**). The anterior horn of the lateral ventricle and the corpus callosum are no longer seen in this section. Instead there is the

anterior commissure (**B24**) connecting the two hemispheres. The two *columns of the fornix* (**B25**), lying close together in the previous section, are separated at the level of the anterior commissure. While the *posterior limb of the internal capsule* (**AB13**) retains its usual width, the *anterior limb* (**AB12**) is only indicated by some fiber bundles. As a result, the *head of the caudate nucleus* (**AB8**) is no longer separated from the putamen (**AB6**), and the striatum is seen as uniform nuclear complex. In the area of the temporal lobe, the curled-up cortical band of the hippocampus (Ammon's horn) (**B26**) is almost covered by the *parahippocampal gyrus* (**B27**).

B28 Mesencephalic tectum.

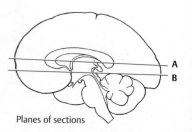

Planes of sections

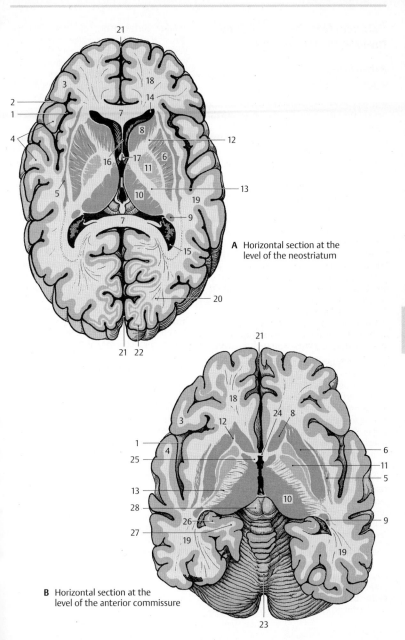

A Horizontal section at the level of the neostriatum

B Horizontal section at the level of the anterior commissure

Telencephalon

Paleocortex and Amygdaloid Body

Paleocortex

Subdivision (A, B)

The paleocortex (blue) is the oldest cortical area of the telencephalon. Together with the olfactory bulb and the olfactory tract it forms the *olfactory brain*, or **rhinencephalon**. In primitive mammals (hedgehog) (**A**), this is the largest part of the telencephalon. The large, compact *olfactory bulb* (**A1**) lies rostrally and, adjacent to it, the *olfactory tubercle* (**A2**), or olfactory cortex. The rest of the base of the brain is occupied by the *piriform lobe* (**A3**) with the *uncus* (**A4**). The piriform lobe contains various cortical areas, namely, laterally the *prepiriform area* (**A5**), medially the *diagonal band of Broca* (bandeletta diagonalis) (**A6**), and caudally the *periamygdalar area* (**A7**). The caudal part of the piriform lobe is occupied by the *entorhinal area* (**A8**), a transitional area (orange) between archicortex (red) and neocortex. Medially appears a portion of the hippocampal formation, the uncus with the superficial *dentate gyrus* (dentate band) (**A9**).

The enormous expansion of the neocortex in humans (**B**) has displaced the paleocortex into the depth where it represents only a small part of the base of the brain. The slender *olfactory bulb* (**B10**) is connected by the *olfactory tract* (**B11**) with the olfactory cortex. The fibers of the tract divide at the *olfactory trigonum* (**B12**) into two (but often into three or more) bundles: the *medial olfactory stria* (**B13**) and the *lateral olfactory stria* (**B14**). They enclose the *olfactory tubercle* which, in humans, has sunk into the depth as *anterior perforated substance* (**B15**). It is delimited caudally by the *diagonal band of Broca* (**B16**) which contains afferent fibers for the olfactory bulb.

The rotation of the hemisphere in humans has displaced the other parts of the *piriform lobe* mainly to the medial aspect of the temporal lobe, where they form the *ambient gyrus* (**B17**) and the *semilunar gyrus* (**B18**). The ambient gyrus is occupied by the *pre-piriform cortex* (**B19**), and the semilunar gyrus by the *periamygdalar cortex* (**B20**). Ventrocaudally to it the *uncus* (**B21**) bulges with the superficial end of the *dentate gyrus* known as *Giacomini's band*. It merges into the *parahippocampal gyrus* (**B22**) which is covered by the *entorhinal cortex* (**B23**).

Olfactory Bulb (C)

The olfactory bulb has regressed in humans, who belong to the *microsmatic mammals*. Mammals with a highly developed sense of smell (*macrosmatic mammals*) possess a large olfactory bulb of complex structure (p. 211, AB3). In the human olfactory bulb we distinguish a *glomerular layer* (**C24**), a *mitral layer* (**C25**), and a *granular layer* (**C26**). The mitral cells of the glomerular layer form synaptic contacts with the terminals of the olfactory nerves (p. 228, A). The axons of the mitral cells run through the olfactory tract to the primary olfactory centers. The olfactory tract contains a discontinuous aggregation of medium-sized neurons along its entire length, the *anterior olfactory nucleus*. Their axons join the fibers of the olfactory tract and partly cross to the contralateral olfactory bulb.

Anterior Perforated Substance (D)

The anterior perforated substance, which is characterized by numerous vascular perforations (**D27**), is covered externally by an irregular layer of small pyramidal cells, the *pyramidal layer* (**D28**), and internally by the loose *multiform layer* (**D29**) with individual clusters of dark cells, the *islands of Calleja* (**D30**). Olfactory bulb, olfactory tract, and anterior perforated substance contain large numbers of peptidergic neurons (corticoliberin, enkephalin, and other peptides).

D31 Nucleus of the diagonal band.
D32 Longitudinal cerebral fissure.
D33 Lateral ventricle.
D34 Paraterminal gyrus.

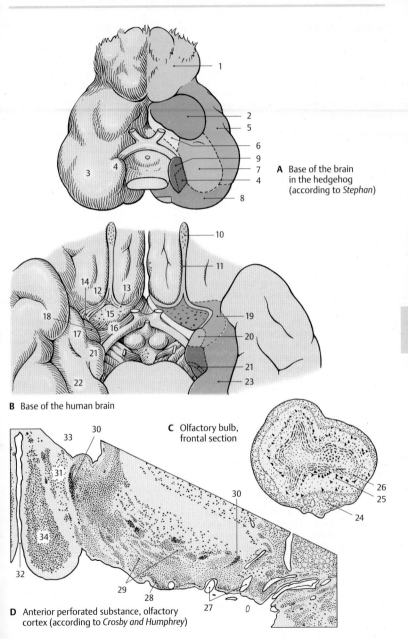

A Base of the brain
in the hedgehog
(according to *Stephan*)

B Base of the human brain

C Olfactory bulb,
frontal section

D Anterior perforated substance, olfactory
cortex (according to *Crosby* and *Humphrey*)

Amygdaloid Body

The **amygdaloid body** (amygdala) lies at the medial aspect of the temporal lobe (**B**). It consists of a cortical part, the *cortical nucleus*, and a nuclear part lying in the depth; hence, it must be viewed as a transition between cortex and nucleus. The nuclear complex is covered by the *periamygdalar cortex* (**A1**).

Subnuclei (A – D)

The complex is divided into several subnuclei, namely, the superficial *cortical nucleus* (**ACD2**), the *central nucleus* (**ACD3**), the *basal nucleus* (**CD4**) consisting of a parvocellular *medial part* (**A5**) and a magnocellular *lateral part* (**A6**), and the *lateral nucleus* (**ACD7**). The assignment of the medial nucleus (**A8**) to the amygdala complex is questionable. The amygdaloid body is rich in peptidergic neurons. Primarily enkephalin and corticoliberin can be demonstrated in the central nucleus and VIP in the lateral nucleus.

The subnuclei form two groups: the phylogenetically old **corticomedial group** (*cortical nucleus, central nucleus*) and the phylogenetically younger **basolateral group** (*basal nucleus, lateral nucleus*). The corticomedial group receives fibers of the olfactory bulb and is the area of origin of the stria terminalis. The basolateral group has fiber connections with the prepiriform area and the entorhinal area. Electrophysiological recordings have demonstrated that only the corticomedial group receives olfactory impulses, while the basolateral group receives optic and acoustic impulses.

Functional Organization (C – E)

Electrical stimulation of the amygdala and its surroundings induces *autonomic and emotional responses*. Anger (■) or flight reaction (□) with the corresponding autonomic phenomena (dilatation of pupils, rise in blood pressure, increase in cardiac and respiratory rates) can be triggered by stimulation of the collecting area of the stria terminalis fibers (**C**). Other sites produce *reactions of alertness* associated with turning the head. Stimulation may induce *chewing* (○), *licking* (●), or *salivation* (▲) (**D**). It may also result in food uptake, secretion of gastric juice, and increased intestinal motility or bulimia. Hypersexuality may occur as a result of stimulation but may also be produced by lesions to the basolateral group of nuclei. *Urination* (△) or defecation may be induced as well.

The stimulation responses are difficult to arrange topically; many fibers run through the nuclear complex, and the stimulation responses may originate not only from the site of stimulation but also from affected fiber bundles of other nuclei. The medial part of the basal nucleus has been assigned to the corticomedial group of nuclei, and an attempt has been made to correlate the two nuclear groups with the different responses; the corticomedial group (**E9**) is thought to promote *aggressive behavior, sexual drive, and appetite*, while the lateral group (**E10**) has an inhibitory effect.

▬ **Clinical Note:** Stimulation of the amygdaloid body in humans (a diagnostic measure in the treatment of severe epilepsy) may trigger anger or anxiety, but also a feeling of tranquillity and relaxation. The patients may feel "transformed" or "in a different world". The response will essentially be influenced by the emotional state at the onset of the stimulation.

A–E11 Optic tract.
A12 Hypothalamus.
A13 Claustrum.

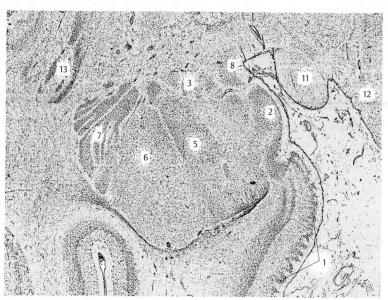

A Subdivision of the amygdaloid body, frontal section, semi-diagram

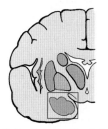

B Location of section in **A**

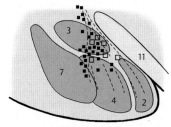

C Fight-or-flight reaction, stimulation experiment
in the cat (according to *de Molina and Hunsperger*)

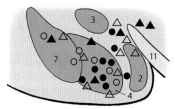

D Autonomic reactions, stimulation experiments
in the cat (according to *Ursin and Kaada*)

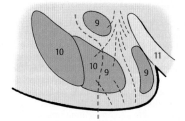

E Functional organization
(according to *Koikegami*)

Fiber Connections

Olfactory Bulb (A)

The bundled axons of the olfactory cells (**A1**) (p. 331, C) pass as *olfactory nerves* (1st neuron) through the openings of the cribriform lamina (**A2**) into the olfactory bulb (**A3**). Here they terminate on the dendrites of the **mitral cells** (**A4**) with which they form *glomeruli* (**A5**). In this glomerular system, one mitral cell is in contact with numerous sensory cells. Other cell types, such as *granule cells*, *periglomerular cells*, and *tufted cells*, belong to the integration center of the olfactory bulb. The axons of the mitral cells (2nd neuron) pass through the **olfactory tract** (**A6**) to the primary olfactory centers. Medium-sized neurons are scattered along the olfactory tract; they constitute the **anterior olfactory nucleus** (**AC7**). The axons, or their collaterals, of the mitral cells terminate here. The neuronal processes partly cross through the anterior commissure to the contralateral olfactory bulb, where they form the *medial olfactory stria* (**B8**).

Lateral Olfactory Stria (B)

All fibers of the mitral cells extend in the lateral olfactory stria to the primary olfactory centers, namely, the *anterior perforated substance* (olfactory area) (**BC9**), the *prepiriform area* (**B10**), and the *periamygdalar area* (**B11**) including the cortical nucleus of the amygdaloid body. The prepiriform area and the periamygdalar area are thought to be the olfactory cortex proper for the conscious perception of olfactory stimuli. The medial olfactory stria is thought to receive exclusively fibers running from the olfactory cortex to the olfactory bulb.

Fiber systems extend from the olfactory cortex (olfactory impulses for the search for food, food uptake, and sexual behavior) to the entorhinal area (**B12**), to the basolateral nuclear group of the amygdaloid body (**BC13**), to the anterior and lateral portions of the hypothalamus (**B14**), and to the magnocellular nucleus of the medial thalamic nuclei (**B15**). A connection to the centers of

the brain stem is established through fibers running to the habenular nuclei (**B16**) (p. 176, A). These association pathways do not directly belong to the olfactory system.

Amygdaloid Body (B)

The basolateral nuclear group receives fibers from the premotor, prefrontal, and temporal cortices; from the magnocellular nucleus of the medial thalamic nuclei; and from nonspecific thalamic nuclei. The most important efferent fiber system of the *amygdaloid body* is the **stria terminalis** (**BC17**). It arches in the sulcus between caudate nucleus and thalamus and runs below the thalamostriate vein (p. 171, C14; p. 175, AB2) as far as the anterior commissure. Its fibers terminate in the *septal nuclei* (**B18**), in the *preoptic area* (**B19**), and in the nuclei of the *hypothalamus*. Fiber bundles cross from the stria terminalis into the medullary stria (**B20**) and extend to the habenular nuclei. Other efferent bundles from the basolateral portion of the amygdaloid body extend as ventral **amygdalofugal fibers** (**B21**) to the *entorhinal area*, to the *hypothalamus*, and to the *medial thalamic nuclei*, from where additional connections lead to the frontal lobe. The stria terminalis is rich in peptidergic fibers.

Anterior Commissure (C)

In the **anterior part** of the anterior commissure, fibers of the olfactory tract (anterior olfactory nucleus) (**AC7**) and fibers of the olfactory cortex (**BC9**) cross to the contralateral side. The anterior part is poorly developed in humans. The main part is formed by the **posterior part**, where fibers of the temporal cortex (**C22**) cross; they are primarily from the cortex of the medial temporal gyrus. Furthermore, the posterior part contains crossing fibers from the amygdaloid bodies (**BC13**) and the striae (terminales) (**BC17**).

B23 Optic chiasm.

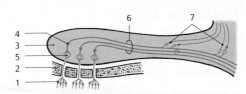

A Olfactory bulb

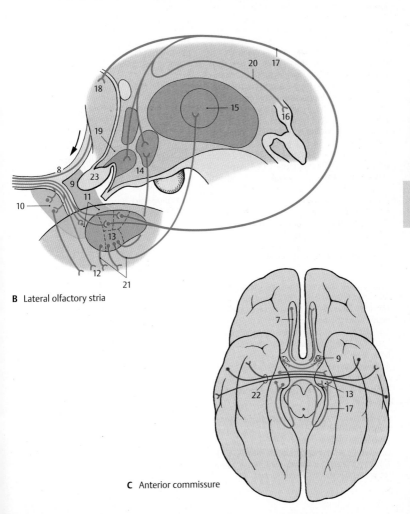

B Lateral olfactory stria

C Anterior commissure

Telencephalon

Archicortex

Subdivision and Functional Significance (A – D)

The **hippocampus** (**A – D1**) is the main part of the archicortex. It lies at the medial aspect of the temporal lobe in the depth and is largely covered by the parahippocampal gyrus. The left hemisphere has been removed in the preparation, showing the cut surface of the corpus callosum (**A2**) with only the left hippocampus being left intact. The latter looks like a paw with claws, the *digitations*. The temporal lobe of the right hemisphere in the background illustrates the position of the hippocampus in the temporal lobe. The hippocampus extends to the caudal end of the corpus callosum. Here, it becomes reduced to a thin layer of gray matter, the **indusium griseum** (**A3**), which extends along the superior surface of the corpus callosum to its rostral end in the region of the *anterior commissure* (**A4**). Two narrow fiber bundles, the **lateral** and the **medial longitudinal striae** of Lancisi (p. 220, A7 and A8) also run here bilaterally. On the dorsal surface of the hippocampus lies a thick fiber band, the **fimbria of hippocampus** (**A – D5**), which separates from the hippocampus beneath the corpus callosum and continues as **fornix** (**A6**), arching down to to *mamillary body* (**A7**).

In a horizontal section through the temporal lobe, the *inferior horn* (**BC8**) and the *posterior horn* (**B9**) of the lateral ventricle are exposed, and the protrusion of the hippocampus into the ventricle is visible. Medially, already at the outer aspect of the temporal lobe, lies the *fimbria* and, beneath it, the **dentate gyrus** (fascia dentata) (**B – D10**), separated from the *parahippocampal gyrus* (entorhinal area) (**B – D11**) by the *hippocampal sulcus* (**BC12**).

In a frontal section the hippocampal cortex forms a curled band, **Ammon's horn**, which protrudes against the ventricle and is covered by a layer of fibers, the **alveus** (**C13**). Ammon's horn shows considerable variations at different planes of section (**D**).

In the past, the hippocampus has been assigned to the rhinencephalon, but it has no direct relationship with the olfactory sense. In reptiles, which do not have a neocortex, the telencephalon is the highest integration organ. Electrical recordings from the hippocampus of mammals show that it receives optic, acoustic, tactile, and visceral input, but only a few olfactory impulses. It is an *integration organ* influencing *endocrine, visceral, and emotional processes* via its connections to the hypothalamus, septal nuclei, and cingulate gyrus. Furthermore, the hippocampus plays a major role in processes of learning and memory.

▬ **Clinical Note:** Bilateral removal of the hippocampus in humans (treatment of severe epileptic seizures) leads to a loss in memory. While old memories are retained, new information can be remembered only for a few seconds. Such a short-term memory may persist for years. The hippocampal neurons possess a very low absolute threshold for convulsive discharges. Thus, the hippocampus is of special importance for the origin of epileptic seizures and memory deficits.

C14 Optic tract.
C15 Choroid plexus.

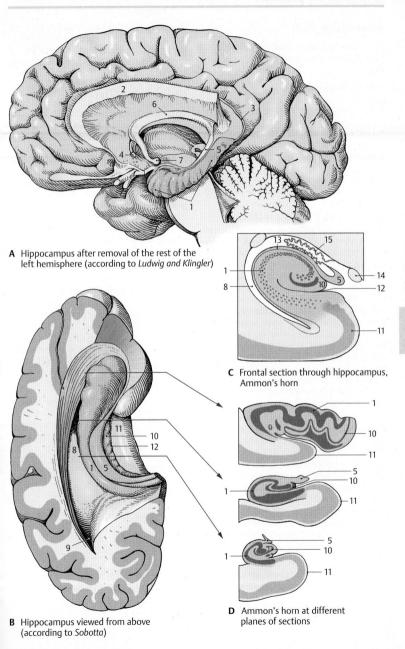

A Hippocampus after removal of the rest of the left hemisphere (according to *Ludwig and Klingler*)

C Frontal section through hippocampus, Ammon's horn

B Hippocampus viewed from above (according to *Sobotta*)

D Ammon's horn at different planes of sections

Ammon's Horn (A)

The hippocampus is subdivided into four parts according to width, cell size, and cell density:

- *Field CA1* (**A1**) contains small pyramidal cells.
- *Field CA2* (**A2**) is characterized by a narrow, dense band of large pyramidal cells.
- *Field CA3* (**A3**) is characterized by a wide loose band of large pyramidal cells.
- *Field CA4* (**A4**) forms the loosely structured inner zone. Recently, it has been called into question whether a separate CA4 region can be delimited from the CA3 region.

The narrow band of densely packed granule cells of the **dentate gyrus (fascia dentata)** (**A5**) surrounds the ending band of pyramidal cells. The dentate gyrus is fused with the surface of the curled-up Ammon's horn and appears only partially at the surface of the brain. It is separated by the *hippocampal sulcus* (**A6**) from the *parahippocampal gyrus* (**A7**) and by the *fimbriodentate sulcus* (**A8**) from the *fimbria of the hippocampus* (**A9**). The inner layer bordering on the ventricle is the *alveus of the hippocampus* (**A10**), in which the efferent fibers collect before leaving the hippocampus via the fimbria. The transitional area between Ammon's horn and the bordering entorhinal cortex (**A11**) is called the **subiculum** (**A12**).

Fiber Connections (B, C)

Afferent Pathways (B)

The fiber bundles from the **entorhinal area** (**B13**) are thought to be the most important afferent system, where the pathways from the primary olfactory centers (prepiriform area), from the amygdaloid body, and from various regions of the neocortex terminate. Direct connections between olfactory bulb and hippocampus have not been demonstrated.

The fibers from the cingulate gyrus collect in the **cingulum** (**B14**) and extend primarily to the subiculum.

The **fornix** (**B15**) contains bundles from the *septal nuclei* (**B16**) but above all fibers from the hippocampus and the entorhinal area of the contralateral hemisphere (via the *commissure of the fornix*).

Efferent Pathways (B)

Apart from a few fibers leaving the hippocampus via the *longitudinal stria* (**B17**), the fornix contains all other efferent pathways. It is divided into a precommissural part and a postcommissural part. The fibers of the **precommissural fornix** (**B18**) terminate in the septum, in the preoptic area (**B19**), and in the hypothalamus (**B20**). The fibers of the **postcommissural fornix** (**B21**) terminate in the *mamillary body* (**B22**) (predominantly in the *medial nucleus of the mamillary body*), in the *anterior thalamic nucleus* (**B23**), and in the *hypothalamus*. Some fibers of the fornix extend to the central gray matter of the midbrain.

A large neuronal circuit can be recognized in this system of pathways. Hippocampal impulses are conducted via the fornix to the anterior thalamic nucleus. The latter is connected with the cingulate gyrus, from where there is feedback via the cingulum to the hippocampus (Papez circuit) (p. 332, C).

Fornix (C)

At the inferior surface of the corpus callosum, the two *limbs of the fornix* (**C24**) unite to form the *commissure of the fornix* (*psalterium*) (**C25**) and the *body of the fornix* (**C26**), which then divides again into the two *columns of the fornix* (**C27**) above the foramen of Monro.

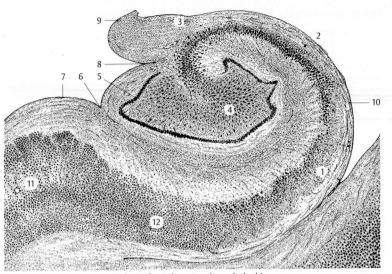

A Ammon's horn, frontal section through the hippocampus

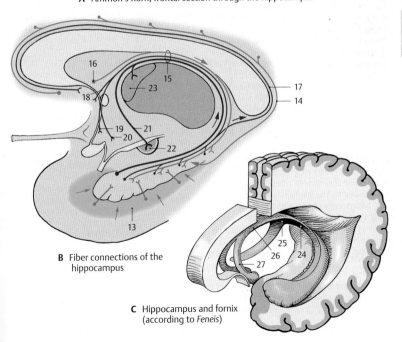

B Fiber connections of the hippocampus

C Hippocampus and fornix (according to *Feneis*)

Hippocampal Cortex (A, B)

The structure of the archicortex is simpler than that of the neocortex, and its neuronal circuits are therefore easier to elucidate. The hippocampal cortex belongs to those brain regions where inhibitory and excitatory neurons have been identified both histologically and electrophysiologically.

Fields CA1 (**A1**), *CA2* (**A2**), and *CA3* (**A3**) show differences with respect to organization and fiber connections. The majority of afferent fibers enter Ammon's horn via the **perforant path** (**A4**), and only a few do so via the alveus of hippocampus. They terminate on the dendrites of the **pyramidal cells** (**AB5**). Many of the fibers (**AB6**) extend to the **granule cells** (**AB22**) of the dentate gyrus (fascia dentata); their axons, *mossy fibers* (**AB7**), too, have synaptic contacts with the dendrites of pyramidal cells. However, mossy fibers run only to *field CA3*; they are absent from *fields CA1* and *CA2*.

The pyramidal cells are the efferent elements; their axons collect in the **alveus** (**AB8**) and leave the cortex through the **fimbria** (**A9**). The axons of the CA3 pyramidal cells give off recurrent collaterals (*Schaffer collaterals*) (**AB10**) that form synapses with dendrites of the CA1 pyramidal cells. The efferent fibers running to the septum originate in *CA3*, the fibers for the mamillary body and the anterior thalamic nucleus originate in *CA1*. Many of the efferent fibers of the hippocampus, however, run to the subiculum.

Organization of layers. Ammon's horn consists of the following layers: the *alveus* (**AB8**) with the efferent fibers lies inside and is followed by the *stratum oriens* (**B11**) with the **basket cells** (**B12**), the axons of which split up and fill the pyramidal layer with a dense fiber network (**B13**). The fibers envelope the pyramidal cell bodies and form synaptic contacts (axosomatic synapses) with them. Basket cells are inhibitory neurons that are excited by the axon collaterals of the pyramidal cells and cause pyramidal cell inhibition following pyramidal cell discharge. The pyramidal cells form the

stratum pyramidale (**B14**). Their apices are oriented toward the subsequent *stratum radiatum* (**B15**), their bases toward the *stratum oriens*. They send dense dendritic trees in both directions. The long apical dendrite reaches with its branches into the *stratum lacunosum-moleculare* (**B16**). In the CA3 region, one can also distinguish a stratum lucidum (**B20**) where the mossy fibers run.

The **afferent fibers** originating from different regions run in different layers. Many of the commissural fibers from the contralateral hippocampus extend into the *stratum oriens* (**B11**) and the stratum radiatum (**B15**). The fibers of the entorhinal area (**B5**) extend into the *stratum lacunosum-moleculare* (**B16**) and form contacts with the outermost branches of the apical dendrites (**B17**). Schaffer collaterals (**B10**) have contact with distal segments of the apical dendrites of the CA1 pyramidal cells, while the mossy fibers (**B7**) have contact with proximal segments of the CA3 pyramidal cells. The dendrites of granule cells of the dentate gyrus are contacted in a similar way; entorhinal fibers terminate on distal dendritic segments, while commissural fibers terminate on proximal segments of the dendrites. In addition to the principal cells—pyramidal cells and granule cells—the afferent fibers of the hippocampus also form synaptic contacts with inhibitory GABAergic interneurons (feed-forward inhibition of principal neurons, p. 35, C). Apart from the basket cells mentioned above (**B12**), which form axosomatic synapses, GABAergic cells have been found in recent years that form synaptic contacts on the initial segment of the axon (axo-axonal cells or chandelier cells) (**B18**) or on the dendrites (**B19**) of the principal cells. From the course of the fibers and from electrophysiological studies, the following impulse flow emerges in the hippocampus: glutamatergic, entorhinal afferent fibers activate granule cells which, in turn, activate CA3 pyramidal cells via mossy fibers. These then activate CA1 pyramidal cells via Schaffer collaterals (trisynaptic excitatory pathway of the hippocampus).

B21 Hilus of dentate gyrus.

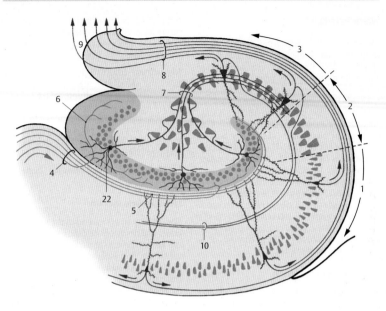

A Organization of the hippocampus (according to *Cajal*)

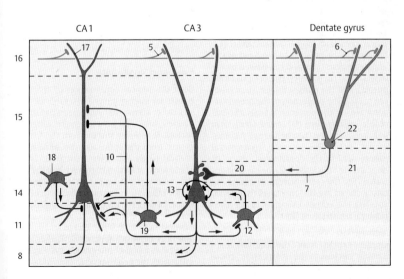

B Neuronal circuits in the hippocampus

Neostriatum

The neostriatum (or *striatum*) is the *highest integration site of the extrapyramidal motor system* (p. 310). It is a large, gray complex in the depth of the cerebral hemisphere and is divided into two parts by the internal capsule (**ABD1**), namely, the **caudate nucleus** (**ABD2**) and the **putamen** (**ABD3**) (p. 214, AB18 and AB19; p. 216, AB9 and AB10). The caudate nucleus consists of the large **head of the caudate nucleus** (**A4**), the **body of the caudate nucleus** (**A5**), and the **tail of the caudate nucleus** (**A6**). Immunohistochemical assays for neurotransmitter substances yield a spotty, mosaic-like structure created by the terminals of various fiber tracts. The spots form a system of interconnected fields (*striosomes*) that stand out from the rest of the tissue because of their content of a specific neurotransmitter.

Afferent Pathways (B–D)

Corticostriate fibers (**B8**). Fibers extend from all areas of the neocortex to the neostriatum. They are the axons of medium-sized and small pyramidal cells of the fifth layer (see p. 240). However, there are no fiber connections extending from the striatum to the cortex. The corticostriate projection reveals a topical organization (**C**): the frontal lobe projects to the head of the caudate nucleus (red) and is followed by the parietal lobe (light blue), the occipital lobe (purple), and the temporal lobe (dark blue) (see p. 213). The projection of the precentral motor area in the putamen reveals a somatotopic organization (**D**): head (red), arm (light red), and leg (hatched area). A somatotopic projection of the postcentral sensory area to the dorsolateral region of the caudate nucleus has been demonstrated. The fibers from areas adjoining the central sulcus are the only ones that partly cross via the corpus callosum to the contralateral neostriatum (**B9**).

Centrostriate fibers (**B10**). These fiber bundles extend from the centromedian thalamic nucleus to the neostriatum; those for the caudate nucleus originate in the dorsal part, those for the putamen in the ventral part of the nucleus. Impulses from the cerebellum and from the reticular formation of the midbrain reach the neostriatum via these fibers.

Nigrostriate fibers (**B11**). Fibers extending from the substantia nigra to the neostriatum can be traced by fluorescence microscopy. They are the axons of dopaminergic neurons, and they cross the inner capsule in groups. They run without interruption through the globus pallidus to the neostriatum (p. 136, B16).

Serotoninergic fiber bundles from the raphe nuclei.

Efferent Pathways (B)

The efferent fibers extend to the globus pallidus. The fibers of the caudate nucleus terminate in the dorsal parts of the two segments of the pallidum (**B12**), while the fibers of the putamen terminate in the ventral parts (**B13**). Here, they synapse with the pallidofugal system, namely, with the pallidosubthalamic fibers, the lenticular ansa, the lenticular fasciculus, and the pallidotegmental fibers (p. 192, A16).

Strionigral fibers (**B14**). Fibers of the caudate nucleus terminate in the rostral part and fibers of the putamen in the caudal part of the substantia nigra (p. 136, B12, B14).

Functional Significance

Both the topical organization of the corticostriate fiber systems and its mosaic-like structure show that the neostriatum is divided into many functionally different sectors. It receives stimuli from the frontal cortex, from the optic, acoustic, and tactile cortical fields and their association areas. These areas are thought to have an effect on the motor system via the stratum (sensory motor integration, cognitive function of the neostriatum). The neostriatum has no direct control over elementary motor processes (its destruction does not lead to an appreciable loss of motor functions). Rather, it is viewed as a higher integration system that influences the behavior of an individual.

A7 Amygdaloid body.

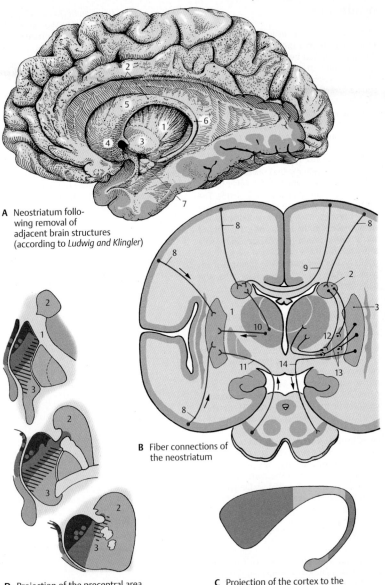

A Neostriatum following removal of adjacent brain structures (according to *Ludwig and Klingler*)

B Fiber connections of the neostriatum

D Projection of the precentral area onto the putamen in the monkey (according *Künzle*)

C Projection of the cortex to the caudate nucleus in the monkey (according to *Kemp and Powell*)

Insula

The **insula** is the region at the lateral aspect of the hemisphere that lags behind during development and becomes covered by the more rapidly growing adjacent regions of the hemisphere. The parts of the hemisphere overlapping the insula are called **opercula**. They are named according to the cerebral lobe they belong to: the **frontal operculum** (**A1**), the **parietal operculum** (**A2**), and the **temporal operculum** (**A3**). In diagram **A**, the opercula have been moved apart to expose the insula. They normally leave only a cleft, the **lateral cerebral sulcus** (*fissure of Sylvius*, p. 10, A4), which widens over the insula into the **lateral fossa** (p. 216, AB15). The insula has roughly the shape of a triangle and is bordered at its three sides by the **circular sulcus of the insula** (**A4**). The *central sulcus of the insula* (**A5**) divides the insula into a rostral and a caudal part. At its lower pole, the **limen of insula** (**A6**), the insular region merges into the olfactory area, the paleocortex.

The insular cortex represents a *transitional region between paleocortex and neocortex.* The lower pole of the insula is occupied by the **prepiriform area** (**B7**) (blue) which belongs to the paleocortex. The upper part of the insula is covered by the **isocortex** (neocortex; see p. 244) (**B8**) (yellow) with the familiar six layers (p. 240). Between both parts lies a transitional region, the **mesocortex** (proisocortex, see p. 244) (**B9**) (hatched area). Unlike the paleocortex, it has six layers; however, these are only poorly developed as compared to the neocortex. The fifth layer (**C10**) is characteristic for the mesocortex by standing out as a distinct narrow, dark stripe in the cortical band. It contains small pyramidal cells that are densely packed like palisades, a feature otherwise found only in the cortex of the cingulate gyrus.

Stimulation responses (D). Stimulation of the insular cortex is difficult because of the hidden position of the region; it has been carried out in humans during surgical treatment of some specific forms of epilepsy. It caused an increase (+) or decrease (−) in the peristaltic movement of the stomach. Nausea and vomiting (●) were induced at some stimulation sites, while sensations in the upper abdomen or stomach region (×) or in the lower abdomen (○) were produced at other sites. At several stimulation sites, taste sensations were induced (▲). Although the stimulation chart does not show a topical organization of these effects, the results do indicate *viscerosensory and visceromotor functions* of the insular cortex. Experiments with monkeys yielded not only salivation but also motor responses in the muscles of the face and the limbs. In humans, surgical removal of the insular region does not lead to any functional losses.

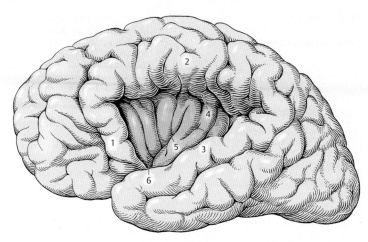

A Insula with the opercula moved
apart (according to *Retzius*)

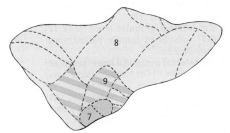

B Cortical areas of the insula
(according to *Brockhaus*)

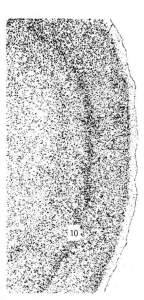

C Mesocortex

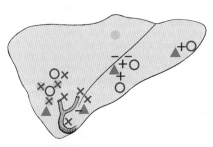

D Stimulation map of the human insular
cortex (according to *Penfield and Faulk*)

Neocortex

Cortical Layers (A – C)

The neocortex (isocortex) exhibits a *stratification into six layers* running parallel to the surface of the hemisphere. The stratification can be demonstrated by silver impregnation (**A1**), cellular staining according to Nissl (**A2**), myelin staining (**A3**), and pigment staining (**B**). The layers are distinguished according to the different shapes, sizes and numbers of their neurons and by the different densities of myelinated nerve fibers.

Cellular staining (**A2**) reveals the following features:

- The outermost layer, the **molecular layer** (layer I) (**A4**), contains few cells.
- The **external granular layer** (layer II) (**A5**) is densely packed with small granule cells.
- The **external pyramidal layer** (layer III) (**A6**) contains predominantly medium-sized pyramidal cells.
- The **internal granular layer** (layer IV) (**A7**) consists of densely packed small granule cells.
- The **internal pyramidal layer** (*ganglionic layer*) (layer V) (**A8**) contains large pyramidal cells.
- The **multiform layer** (layer VI) (**A9**) completes the stratification with a loose mixture of different cell types.

Silver impregnation (**A1**), which shows the neuron with all its processes (p. 18), makes it possible to identify the granule cells of layer II as small pyramidal cells and stellate cells, and the granule cells of layer IV predominantly as stellate cells. The pyramidal cell (**C**) is the typical neuron of the neocortex. Its axon (**C10**) takes off from the base of the cell, where the basal dendrites (**C11**) branch off at the margins. One long, thick dendrite, the apical dendrite (**C12**), ascends to the surface of the cortex. The dendrites have thousands of spines at which other neurons synapse.

Myelin staining (**A3**) of the nerve fibers reveals the following layers based on the different densities of tangential fibers:

- The *tangential layer* (**A13**).
- The *dysfibrous layer* (**A14**).
- The *suprastriate layer* (**A15**).
- The *external* (**A16**) and *internal* (**A17**) *Baillarger's bands* of high fiber density, the external band being created by branches of afferent fibers, the internal band by axon collaterals of pyramidal cells.
- The *substriate layer* (**A18**) completes the stratification.
- In addition, there are the vertical bundles of *radial fibers* (**A19**).

Pigment staining (**B**). The various neurons differ in their degree of pigmentation. The different pigment contents cause the characteristic stratification of the cortex, usually with two unpigmented bands corresponding to the two Baillarger's bands.

Vertical Columns (D)

The basic functional units of the neocortex are vertical cell columns that reach through all layers and have a diameter of 200 – 300 μm. Electrophysiological studies have shown that, in the cortical projection areas, each cell column is connected to a defined peripheral group of sensory cells. Stimulation of the peripheral field always yields a response from the entire column.

Fiber tracts connect the cortical columns with each other (**D**): the fibers of a column (**D20**) run either to columns of the ipsilateral hemisphere (association fibers, see p. 260) or via the corpus callosum to mostly symmetrically localized columns of the contralateral hemisphere (commissural fibers, see p. 260). Branches of individual fibers terminate in different columns (**D21**). It is estimated that the neocortex is made up of 4 million columns.

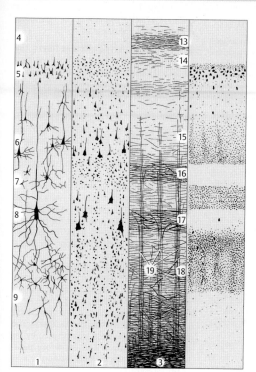

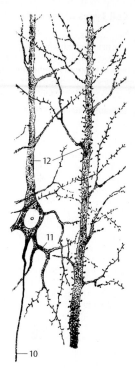

A Layers of the neocortex: 1, silver impregnation; 2, cellular staining; 3, myelin staining (according to *Brodmann*)

B Pigment staining

C Pyramidal cell and apical dendrite (according to *Cajal*)

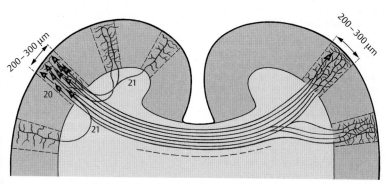

D Connection of vertical columns in the neocortex (*Szentágothai* according to *Goldman and Nauta*)

Cell Types of the Neocortex (A)

In principle, we distinguish between **projection neurons with long axons** (excitatory glutamatergic *pyramidal cells*) and **interneurons with short axons** (inhibitory GABAergic *interneurons*).

The *pyramidal cell* (**A1**) is characterized by one apical dendrite (**A2**), which ascends to the molecular layer and branches there, and numerous basal dendrites (**A3**). Its descending axon gives off numerous recurrent collaterals (**A4**). The cell-deficient molecular layer (layer I) contains *Cajal – Retzius cells* (**A5**) with tangentially running axons. The different types of granule cells or stellate cells are predominantly interneurons and are found in all layers at various densities. They include *Martinotti's cells* (**A6**), the vertically ascending axons of which ramify in various cortical layers and reach as far as the molecular layer. The *cellules à double bouquet dendritique* of Cajal, cells with two vertically oriented dendritic trees (**A7**) (primarily in layers II, III, and IV), possess long ascending or descending axons. The axon of some stellate cell types arborizes after a short course (**A8**), or it bifurcates and terminates with basketlike networks (*basket cells*) (**A9**) on adjacent pyramidal cells. Axon bifurcations may run horizontally and terminate on distant pyramidal cells (**A10**). Their inhibitory function has been confirmed by detection of GABA in the synapses of basket cells.

The Module Concept (B)

The results of histological and electrophysiological studies have made it possible to design models in which the described cell types are organized in a functional group. The vertical column is conceived as a module, that is, as a group of elements forming a functional unit.

The **efferent elements** of the column are the *pyramidal cells* (**B11**). Their axons either run to other cortical columns, where their terminal ramifications end at the spines of other pyramidal cells, or they run to subcortical groups of neurons. The numerous axon collaterals (**A4**) terminate at the pyramidal cells of nearby columns.

There are two kinds of **afferent fibers**: the association fibers from other columns (p. 240, D) and the specific sensory fibers from peripheral sensory areas. In every layer the **association fibers** (**B12**) give off branches that terminate at the spines of pyramidal cells. They ascend to the molecular layer, where they branch into horizontally running fibers. The latter have synaptic contacts with apical dendrites within a radius of 3 mm. The excitation transmitted by them reaches far beyond the column; however, it remains weak because the number of synaptic contacts is limited. The **specific fibers** (**B13**) terminate in layer IV on interneurons (**B14**), primarily on the cells with two dendritic trees (**B15**). The axons of the latter ascend vertically along the apical dendrites of pyramidal cells and form synapses with their spines (**B16**). These series of synapses result in powerful transmission. The *basket cells* (**B17**), which are inhibitory interneurons, send their axons to the pyramidal cells of adjacent columns and inhibit them, thereby restricting the excitation. The basket cells themselves are activated by recurrent collaterals of the excitatory pyramidal cells. The axons of Martinotti's cells (**B18**) ascend to the molecular layer where they form branches.

The number of neurons per column is estimated to be 2500, approximately 100 of which are pyramidal cells. It should be considered, however, that a vertical column is not a clearly defined histological entity. Possibly, it does not represent a permanent morphological unit but rather a functional unit, which forms and disintegrates according to the level of excitation.

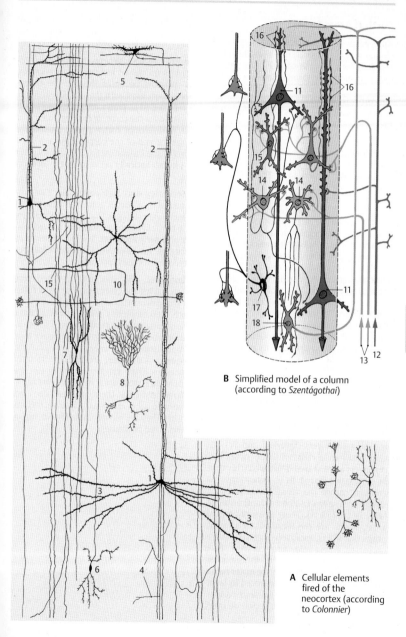

B Simplified model of a column
(according to *Szentágothai*)

A Cellular elements
fired of the
neocortex (according
to *Colonnier*)

Cortical Areas (A, B)

All regions of the neocortex develop in a similar way. First, a wide cell layer, the *cortical plate*, forms on the surface of the hemisphere which then divides into six layers. Because of this similar development, the neocortex is also known as *isogenetic cortex*, or briefly, **isocortex**, or as *homogenetic cortex*.

Nevertheless, the adult neocortex exhibits considerable regional variations, and we distinguish several regions of different structure, the **cortical areas**. The individual layers may vary considerably in these areas: wide or narrow, with densely or loosely packed cells. The cells may vary in size, or a specific cell type may predominate. The definition of individual areas according to such criteria is called **architectonics**. Depending on the staining method employed (p. 240A), the following terms are used: *cytoarchitectonics*, *myeloarchitectonics*, or *pigment architectonics*. A map of cortical areas can be reconstructed on the surface of the hemisphere, similar to a geographic map. The cytoarchtectonic map of cortical areas established by Korbinian Brodmann (**A**, **B**) has been confirmed repeatedly and is generally accepted.

Types of cortices. A special feature of **projection areas** (terminals of ascending sensory pathways) is the prominent development of their granular layers. In the sensory cortex (area 3) as well as in the auditory cortex (areas 41 and 42, temporal transverse gyri), the granular layers (layers II and IV) are wide and rich in cells, while the pyramidal cell layers are less well developed. This type of cortex is referred to as *koniocortex*, or *granular cortex*. The visual cortex (area 17, striate area) even exhibits a duplication of layer IV (p. 255, A). These sensory cortical areas, which are the terminals of afferent projection fibers, are involved in associative processes, where relay neurons with short axons play an important role. On the other hand, the granular layers in the **motor cortex** (areas 4 and 6) are largely reduced in favor of the pyramidal layers (*agranular cortex*). The pyramidal cells are projection neurons

with long axons; for example, in the motor cortex, their axons form the corticospinal tract (pyramidal pathway).

Border zones. Wherever the isocortex borders on the archicortex or paleocortex, its structure becomes simpler. The transitional formation with the simpler structure is referred to as **proisocortex**. The proisocortex includes the *cortex of the cingulate gyrus*, the *retrosplenial cortex* (lying around the posterior end of the corpus callosum), and *parts of the insular cortex* is phylogenetically older than the neocortex.

Paleocortex and archicortex are also surrounded by a border zone, the structure of which approximates that of the neocortex. The border zones are known as **periarchicortex** and **peripaleocortex**. The periarchicortex includes, for example, the *entorhinal area* bordering on the hippocampus.

Allocortex. The allocortex is often contrasted with the isocortex. The term refers to the paleocortex and the archicortex. As both are completely different parts of the telencephalon, genetically as well as structurally and functionally, this collective term is not justified. The only thing they have in common is that they are different from the isocortex. Their simpler structure (for example, no stratification into six layers as it is characteristic for the neocortex) does not justify to label them as primitive regions. Rather, they are phylogenetically old, highly specialized structures.

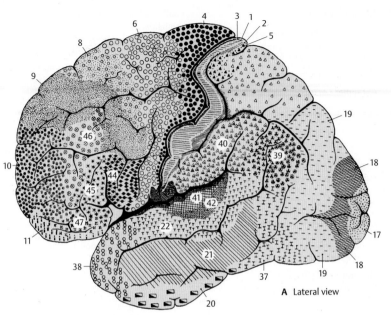

A, B Cortical areas of the hemisphere
(according to *Brodmann*)

A Lateral view

B Median view

Telencephalon

Frontal Lobe (A – C)

We distinguish the precentral area (the motor cortex proper) and the premotor, the prefrontal, and the orbitofrontal cortical areas.

Agranular Cortex

The cortex of the **precentral area** (red), consisting of *primary motor cortex* (area 4) and *premotor cortex* (area 6) (**C**), is characterized by the reduction or *loss of granular layers* and a general *increase in pyramidal cells*. Also typical are the exceptional thickness of the cortex and its gradual transition into the white matter. These features are especially prominent in the cortex of *area 4* (**A**), where certain regions of layer V contain **giant pyramidal cells** (**Betz's cells**) (**A1**). The latter possess the thickest and longest axons in the nervous system, reaching as far as the sacral spinal cord.

The prefrontal cortex (area 9, light red) is shown for comparison (**B**). It is not only narrower and better delimited against the white matter by a distinct layer VI, but also possesses well-developed granular layers (II and IV).

The agranular cortex (areas 4 and 6) is the *principal site of origin of the pyramidal tract* and is regarded as prototype of the **motor cortex**. Nevertheless, it also receives afferent fibers: following the stimulation of skin at the extensor and flexor sides of the limbs, electrical potentials can be recorded at the precentral area. They are probably afferent systems for controlling and fine-tuning the motor system. On the other hand, motor responses can be induced by enhanced stimulation at some points of the postcentral area of the parietal lobe (somatosensory area, blue) (**C**) and the premotor area of the frontal lobe. Accordingly, pyramidal cells with long axons are found in layer V of these areas (△). Physiologists therefore speak of a *motosensory area*, Ms I (predominantly motor) and of a *sensorimotor area*, Sm I (predominantly sensory). However, these findings do not affect the basic fact that the precentral area represents the motor cortex, while the postcentral area represents the somatosensory (tactosensory) cortex.

Granular Cortex

The **prefrontal** and **orbitofrontal cortices** exhibit well-developed granular layers (**B**).

▨ **Clinical Note:** Injury to the granular frontal cortex results in severe changes in personality. This affects the formal intellectual capacity less than it does initiative, ambition, concentration, and judgment. The patients show a silly self-satisfied euphoria, are only interested in everyday trivia, and are unable to plan ahead.

Similar changes have been observed in patients with *prefrontal lobotomy*. This is a complete surgical severing of the frontal fiber connections, which was performed as a form of treatment of raving, mentally deranged patients and in patients with the most severe conditions of pain (this treatment has now become obsolete thanks to psychopharmaceuticals). The operation achieved permanent calming and indifference of the patients. A characteristic change was observed in the emotional sphere: the patents still felt the pain, but they did not perceive it as troublesome; previously intolerable pain had become immaterial. Radical changes in character also occur following *injury of the orbitofrontal area* (*basal neocortex*). Previously cultured and educated people exhibit a disintegration of decency, tact, and sense of shame, which may lead to serious social offenses.

Telencephalon

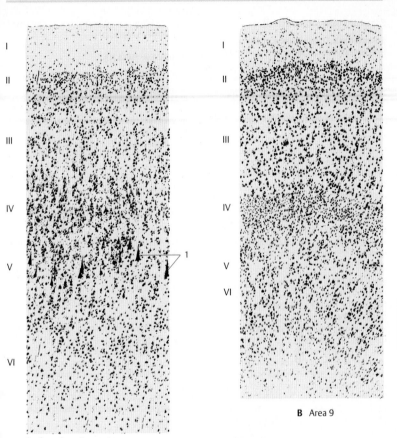

I

II

III

IV

V

VI

1

A Area 4, motor cortex

I

II

III

IV

V

VI

B Area 9

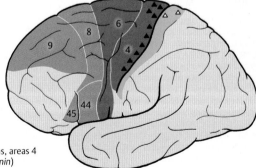

C Precentral cortical areas, areas 4 and 6 (according to *Bonin*)

Frontal Lobe (continued)

Somatotopic Organization of the Precentral Area (A, B)

Electrical stimulation of individual cortical regions of the precentral area (**A, B1**) causes muscle contraction in specific body regions. The area exhibits a somatotopic organization. The head region lies above the lateral sulcus, with the lowest parts representing the throat (**A2**), the tongue (**A3**), and the lips. The areas for hand, arm, trunk, and leg follow dorsally, with the leg area reaching beyond the edge of the pallium to the median surface of the hemisphere. This creates an inverted *motor homunculus*. The areas for individual body parts differ in size; those parts of the body where muscles have to perform differentiated movements are represented by particularly large areas. The fingers and the hand occupy the largest areas, and the trunk the smallest.

Each body half is represented on the contralateral hemisphere, that is, the left body half on the right hemisphere and the right body half on the left hemisphere. Unilateral stimulation yields *bilateral* responses of masticatory, laryngeal, and palatal muscles, and partly also of the trunk muscles. The muscles of face and limbs yield *strictly contralateral* responses. Representation of the limbs is organized in such a way that areas for distal parts of the limbs lie deep in the central sulcus, while those for the proximal parts lie more rostrally on the precentral gyrus (**B1**).

Supplementary Motor Areas (B)

Apart from the precentral area (Ms I), there are two additional motor areas. Their somatotopic organization has been demonstrated in monkeys but not in humans. The *second motosensory area*, Ms II (**B4**), lies on the medial surface of the hemisphere above the cingulate gyrus adjacent to areas 4 and 6. The *second sensorimotor area*, Sm II (**B5**), which is predominantly a tactosensory area rather than a motor area, lies above the lateral sulcus and roughly corresponds to area 40. The functional significance of these areas within the entire motor system has not yet been elucidated.

Frontal Eye Field (C)

Conjugated eye movements can be induced by electrical stimulation of the precentral area and, above all, of area 8. This is the frontal eye field for voluntary eye movements. In general, stimulation causes the gaze to turn to the opposite side, sometimes with simultaneous turning of the head. The fibers of area 8 do not terminate directly at the nuclei of the eye-muscle nerves. Their impulses are probably relayed in the interstitial nucleus of Cajal.

Broca's Motor Speech Area (D)

Injury to the region in the lower frontal gyrus (areas 44 and 45) of the dominant hemisphere (p. 262) causes **motor aphasia**. The patients are no longer able to form and articulate words, even though the speech muscles (lips, tongue, larynx) are not paralyzed. Speech comprehension is not affected.

It is certainly impossible to localize the faculty of speech to a defined cortical area ("speech center"). Speech is one of the highest cerebral functions and involves large regions of the cortex. However, Broca's motor speech area is without doubt a crucial relay station in the complex neuronal foundation of speech (*sensory aphasia*, p. 262).

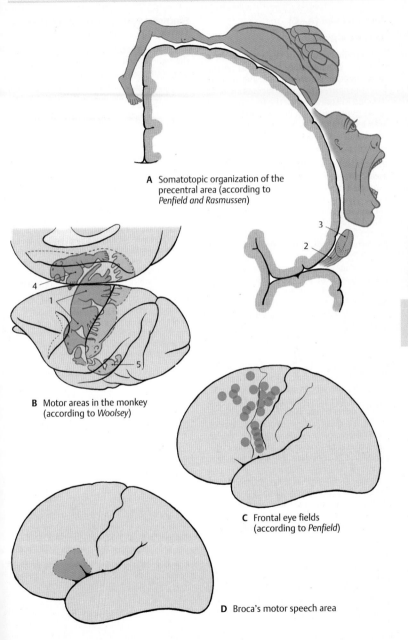

A Somatotopic organization of the precentral area (according to *Penfield and Rasmussen*)

B Motor areas in the monkey (according to *Woolsey*)

C Frontal eye fields (according to *Penfield*)

D Broca's motor speech area

Telencephalon

Parietal Lobe (A – C)

Postcentral Area

The terminal station of the sensory pathways, the **somatosensory cortex**, lies on the most anterior convolution of the parietal lobe, the *postcentral gyrus*. It contains areas 3, 1, and 2. Area 3 lies on the frontal surface of the gyrus and deep in the central sulcus; area 1 covers the top of the gyrus as a narrow band; and area 2 covers the posterior surface (p. 245, A).

In contrast to the motor cortex, the cortex of area 3 (**A**) is extremely narrow and clearly separated from the white matter. The pyramidal layers (III and V) are narrower and contain few cells, while the granular layers (II and IV) are much wider. Hence, the cortex of area 3 is part of the *koniocortex*, or *granular cortex* (p. 244). The cortex of area 40 is shown for comparison (**B**); it covers the *supramarginal gyrus* and may be regarded as the prototype of the parietal cortex. Both granular and pyramidal layers are well developed, and the radial striation running through all layers is clearly visible.

The somatosensory cortex receives its afferent fibers from the ventral posterior thalamic nucleus in a somatotopic organization that results in representation of the contralateral body parts in specific cortical areas. The areas for throat and oral cavity lie above the lateral sulcus (**C1**), and the areas for face, arm, trunk, and leg are superior to it. The leg area reaches beyond the edge of the pallium to the median surface, where the representations of bladder, colon, and genitals (**C2**) complete the sequence. This creates a *sensory homunculus*. The cutaneous regions of highly differentiated sensibilities, such as hand and face, are represented by particularly large cortical areas. The areas for distal parts of the limbs are usually larger than those for proximal parts.

According to clinical and electrophysiological studies, the *superficial skin sensibility* is represented by area 3, and the *deep sensibility* by area 2 (predominantly impulses from joint receptors). In area 2, position and movement of the limbs are constantly registered.

Functional Significance of the Parietal Cortex

The function of this area has become known through mental deficiencies following injury of the parietal lobe.

Clinical Note: Various types of *agnosia* may occur. Although sensory impressions are perceived, the importance and characteristics of objects are not recognized. Such disturbances may also affect tactile, optic, or acoustic sensations. There may be disturbances in symbolic thinking when the parietal lobe (*angular gyrus*) of the dominant hemisphere (p. 262, A) is affected: the loss of understanding of letters or numbers makes reading and writing, counting and calculating impossible.

Furthermore, disturbances of the *body scheme* may be observed. They may involve the inability to distinguish between left and right. Also, one's own paralyzed or nonparalyzed limbs may be sensed as foreign limbs, for example, one's own arm may be perceived as a heavy iron bar lying on the chest. The disturbance may affect an entire body half, which is then perceived as a different person, "my brother" (*hemidepersonalization*).

The parietal cortex, which lies between the tactile and the optic cortices and is closely linked to both by fiber connections, is thought to have a special importance for bringing about three-dimensional *space perception*. Injury to the parietal lobe may result in the destruction of this sense.

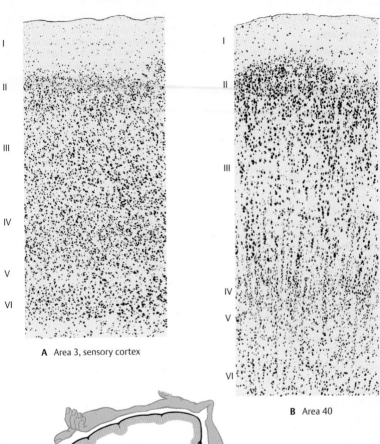

I

II

III

IV

V

VI

A Area 3, sensory cortex

I

II

III

IV

V

VI

B Area 40

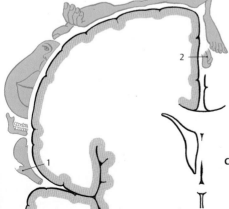

2

1

C Somatotopic organization of the postcentral area (according to *Penfield and Rasmussen*)

Telencephalon

Temporal Lobe (A – C)

Auditory Cortex

The main convolutions on the lateral surface of the temporal lobe run mostly longitudinally. However, two convolutions on the superior surface, the **transverse temporal gyri**, or *Heschl's convolutions* (**C1**), run transversely. They lie deep in the lateral sulcus and become visible upon removal of the overlying parietal operculum. The cortex of the anterior transverse convolution is the terminal station of the *acoustic radiation* (p. 380) originating from the medial geniculate body. The cortical areas of the two transverse convolutions correspond to area 41 (**A**) and area 42; they represent the **auditory cortex**. Like all receptive cortical areas, they are part of the *koniocortex*, or *granular cortex* (p. 244). The external granular layer (II) and, even more so, the internal granular layer (IV) are rich in cells and very wide. The pyramidal layers (III and V), on the other hand, are narrow and contain only small pyramidal cells. The cortex of area 21 (**B**) is shown for comparison; it covers the *medial temporal gyrus* and represents a typical temporal cortex with prominent granular layers, wide pyramidal layers, and distinct radial striation.

Electrical stimulation of area 22 close to the transverse convolutions induces acoustic sensations, such as humming, buzzing, and ringing. The auditory cortex is organized according to tone frequencies (*tonotopic organization*, p. 381C). In the human auditory cortex, the highest frequencies are assumed to be registered medially and the lowest frequencies laterally.

Functional Significance of the Temporal Cortex

Electrical stimulation of the remaining parts of the temporal lobe (performed during surgical treatment of temporal lobe epilepsy) induces *hallucinations* involving fragments of past experiences. The patients hear the voices of people familiar to them in their youth. They relive momentary episodes of their own past. These are mainly acoustic hallucinations and less often visual ones.

During stimulation of the temporal lobe, however, *misinterpretation* of the current situation may occur as well. A new situation may thus appear as an old experience (*déjà vu*). Surrounding objects may move away or come closer. The entire surroundings may take on an uncanny or threatening character.

Such phenomena occur only upon stimulation of the temporal lobe and cannot be elicited from any other part of the cortex. It is therefore assumed that the temporal cortex plays a special role in the conscious and unconscious *availability of one 's own past* and of things experienced in the past. Only if we are continuously aware of past experiences are we able to *judge and interpret new situations correctly*. Without this ability, we would not find our way around in our surroundings. Therefore, the temporal cortex has also been called the *interpretative cortex*.

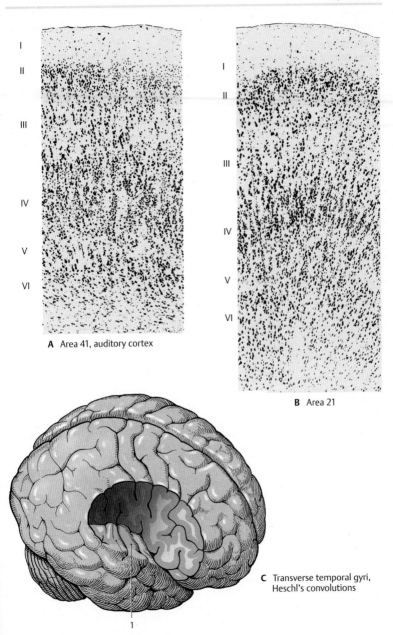

A Area 41, auditory cortex

B Area 21

C Transverse temporal gyri,
Heschl's convolutions

1

Occipital Lobe (A – C)

The medial surface of the occipital lobe is crossed by the horizontally running **calcarine sulcus** (**BC1**); its deepest point corresponds to an eminence on the ventricular wall, the **calcar avis** (**B2**). Frontal sections through the occipital lobe clearly show a fiber plate in the white matter, the **tapetum** (**B3**). It contains commissural fibers of the corpus callosum that run through the splenium and radiate in an arch into the occipital lobe (*forceps major*, see p. 260, F16).

Visual Cortex (Striate Area)

Area 17 (**AC4**) is the terminal station of the *optic radiation* and represents the **visual cortex**. It lies on the medial surface of the occipital lobe and spreads slightly to the convex part at the pole. The cortex lines the calcarine sulcus and extends further to its dorsal and ventral lips. Area 17 is surrounded by area 18 (**A5**) (left side of figure, the arrow shows the boundary between areas 17 and 18) and area 19, which represent the *optic integration areas*.

Like all receptive cortical areas, the cortex of area 17 is characterized by reduced pyramidal layers and well-developed granular layers. The cortex is very thin and set apart from the white matter by a cell-rich layer VI. The internal granule layer (IV) is divided by a zone deficient in cells. In myelin-stained sections, this zone corresponds to *Gennari's line* (**B6**). Because of the striate appearance, the visual cortex is therefore also called *striate cortex*. In the cell-deficient zone (IVb) are remarkably large cells, the giant stellate cells or *Meynert's cells*. The two cellular layers of the internal granular layer (IVa and IVc) contain very small granule cells. These are the layers with the highest cell density within the entire cerebral cortex. Area 18 exhibits a uniform granular layer made up of large granule cells. Area 19 forms a transition to the parietal and temporal cortices.

Functional Organization of the Visual Cortex

Electrophysiological studies on the visual cortex of experimental animals have shown that there are two main types of neurons in the striate area: *simple cells* and *complex cells*. A **simple cell** receives impulses from a cell group of the retina. Its response is strongest to narrow lines of light, to dark lines against a light background, or to straight lines of light/dark boundaries. Orientation of the lines is crucial: some cells respond only to horizontal lines of light, others only to vertical or oblique lines.

Complex cells also respond to lines of light of a specific orientation. However, whereas a simple cell becomes excited only by its receptive field, a complex cell responds to mobile lines of light that move over the retina: each complex cell is stimulated by a large number of simple cells. It is assumed that the axons of numerous simple cells terminate on one complex cell. The internal granular layers consist almost entirely of simple cells, while the complex cells aggregate in the external granular layer. More than one half of all neurons in areas 18 and 19 are complex or **hypercomplex cells**. It is assumed that they play a special role in *shape recognition*.

Electric stimulation of the visual cortex (area 17) causes the sensation of light sparks or flashes. Stimulation of the parastriate and peristriate areas (areas 18 and 19) is thought to produce figures and shapes; it also causes the gaze to turn (**occipital eye field**). Eye movements induced by the occipital lobe are purely reflex, in contrast to the voluntary movements directed by the *frontal eye field* (p. 248, C).

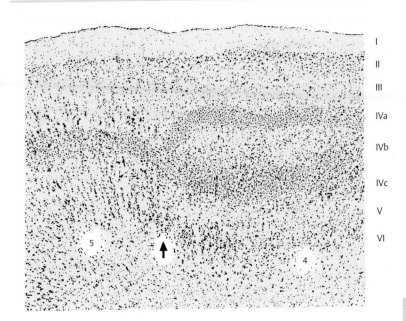

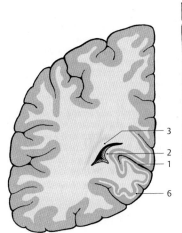

B Occipital lobe, frontal section

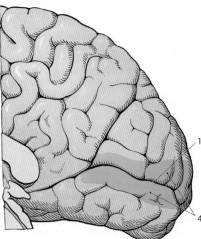

C Median view of hemisphere, showing area 17

Telencephalon *(side margin)*

Occipital Lobe (continued)

Functional Organization of the Visual Cortex (continued) (A, B)

Subdivision of the visual cortex into columns. In addition to the structural subdivision into cell layers, the visual cortex shows a functional subdivision into columns; these run vertically to the cell layers and through the entire width of the cortex; they are 0.3 – 0.5 mm in diameter. Each column is connected to a defined field of the retina. When the sensory cells of such a field are stimulated, all neurons of the respective column respond. Each column is connected to a peripheral field of only one of the two retinas. In the visual cortex (**A1**), columns for the right retina alternate with columns for the left retina (*ocular dominance columns*). Hence, the impulses from each retina are segregated along the entire visual pathway.

The nerve fibers from the two corresponding halves of the retina terminate in the **lateral geniculate body**: the fibers from the left retinal halves of both eyes (right halves of the visual fields) terminate in the left geniculate body (**A2**), and the fibers from the right retinal halves (left halves of the visual fields) terminate in the right geniculate body. The fibers from corresponding fields of both retinal halves terminate in different cell layers of the geniculate body: the uncrossed fibers (**A3**) from the ipsilateral retina extend to the second, third, and fifth layers, while the crossed fibers (**A4**) from the contralateral retina run to the first, fourth, and sixth layers. The neurons receiving the optic fibers from corresponding points of the two retinas lie along a line that runs through all cell layers (*projection column*). Their axons project via the optic radiation (**A5**) to the visual cortex. Each geniculate fiber is thought to branch extensively and to terminate on several thousands of stellate cells of cortical layer IV. Fibers conducting the excitation of the ipsilateral retina extend to other columns than the fibers for excitation of the contralateral retina.

The **organization of the visual cortex into vertical columns** has been visualized by administering ^{14}C-deoxyglucose to experimental animals and determining the varying distribution of the substance by autoradiography. Excited neurons possess an increased metabolism and rapidly take up ^{14}C-deoxyglucose, whereas resting cells do not.

The visual cortex of an experimental animal with both eyes open (rhesus monkey) showed a bandlike distribution of the radioactive label, corresponding to the familiar stratification of cells (**B6**). The distribution of the label indicated that layers I, II, III, and V have a low glucose content, while layer VI has a higher content and layer IV the highest. When both eyes of the test animal were closed, there were no significant differences between individual layers; instead, low-grade labeling of relatively uniform distribution throughout the cortex (**B7**) was found. If one eye was open while the other one was closed, this method yielded a series of columns running perpendicularly to the cell layers and showing intensely dark columns alternating with pale ones (**B8**). The pale columns, where the neurons did not take up the label, represented the retina of the closed eye. The dark columns contained newly incorporated ^{14}C-deoxyglucose because they received input from the retina of the open eye. Again, layer IV was most intensely labeled. The columns were absent in a small area (**B9**) representing the monocular zones of the retina, namely, the outermost margin of the retina and the blind spot.

In addition to the ocular dominance columns, periodically arranged cell populations have been demonstrated that respond in a characteristic fashion to the orientation of lines in the visual field (*orientation columns*). By demonstrating the enzyme cytochromoxidase in sections cut parallel to the cerebral surface, periodically arranged spots have been found that represent neurons responding to color stimuli mediated by one eye (*color columns*).

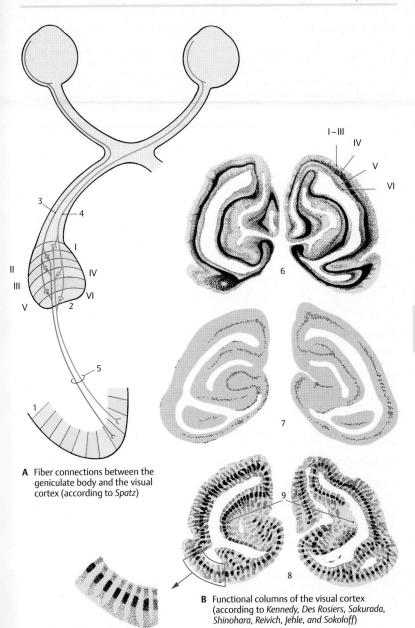

A Fiber connections between the geniculate body and the visual cortex (according to *Spatz*)

B Functional columns of the visual cortex (according to *Kennedy, Des Rosiers, Sakurada, Shinohara, Reivich, Jehle,* and *Sokoloff*)

Telencephalon

Fiber Tracts (A – C)

A broad layer of **white matter** lies between the cerebral cortex and the gray nuclei in the depth. It consists of fiber bundles originating from the neurons of the cortex or those extending to the cortex and terminating on the cortical neurons. There are three different types of fiber systems:

• Projection fibers
• Association fibers
• Commissural fibers

Projection fibers provide connections between the cerebral cortex and the subcortical centers, either as ascending systems terminating in the cortex or as descending systems extending from the cortex to the deeper-lying centers. *Association fibers* provide connections between different cortical areas of the same hemisphere. *Commissural fibers* connect the cortices of both hemispheres; they are really nothing else but interhemispheric association fibers. However, commissural fibers often connect identical regions of the two hemispheres.

Projection Fibers

The pathways descending from different cortical areas merge and form a fanlike structure known as the internal capsule. The ascending fibers pass through the internal capsule and then radiate outward like a fan. In this way, ascending and descending fibers form a radiating crown of fibers beneath the cortex, the **corona radiata** (**A1**).

The **internal capsule** (**A2**, **B**) appears in horizontal sections as an angle consisting of an **anterior limb** (**B3**), bordered by the head of the caudate nucleus (**B4**), globus pallidus (**B5**) and putamen (**B6**), and a **posterior limb** (**B7**), bordered by the thalamus (**B8**), globus pallidus and putamen (see p. 222, A). Between both limbs lies the **genu of the internal capsule** (**B9**). The different fiber tracts pass through specific parts of the internal capsule. The *frontopontine tract* (**B10**) (red lines) and the *anterior thalamic radiation* (**B11**) (blue lines) pass through the anterior limb. The *corticonuclear fibers* supplying the

cranial nerve nuclei run through the genu of the internal capsule. The fibers of the *corticospinal tract* (red dots) pass through the posterior limb in a somatotopic organization: upper extremity, trunk, and lower extremity. The thalamocortical fibers leading to area 4 and the corticorubral and corticotegmental fibers coming from area 6 pass through the same region. The caudal part of the posterior limb is occupied by the fibers of the *central thalamic radiation* (blue dots) (**B12**) which extend to the postcentral area. The fibers of the *posterior thalamic radiation* (**B13**) (light blue dots) and of the *temporopontine tract* (light red dots) (**B14**) pass obliquely through the caudal part.

The most important projection pathways include the acoustic radiation and the optic radiation. The fibers of the **acoustic radiation** originate in the medial geniculate body, extend across the lateral geniculate body, and cross the internal capsule at the inferior margin of the putamen. In the white matter of the temporal lobe, they ascend almost vertically to the anterior transverse gyrus (Heschl's convolution) (pp. 252, 380). The **optic radiation** originates in the lateral geniculate body. The fibers fan out into a wide medullary lamina (**A15**) and run to the temporal lobe, where they form the temporal genu (**C16**) of the visual pathway. They then pass in an arch around the inferior horn of the lateral ventricle and through the white matter of the occipital lobe to the calcarine sulcus (**C17**).

A18 Corpus callosum.
A19 Cerebral peduncle.

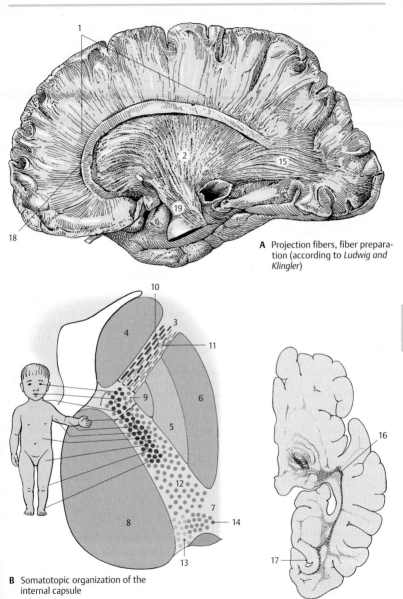

A Projection fibers, fiber preparation (according to *Ludwig and Klingler*)

Telencephalon

B Somatotopic organization of the internal capsule

C Optic radiation during myelin maturation (according to *Flechsig*)

Fiber Tracts (continued)

Association Fibers (A–D)

The connections between different cortical areas are very different in length. For the purpose of simplification, we distinguish short and long association fibers.

The **short association fibers**, or *arcuate fibers* (**B**), provide connections within a cerebral lobe (**B1**) or from one convolution to the next (**B2**). The shortest fibers connect adjacent cortical parts. They reenter the cortex after running just a short distance through the white matter. The layer of arcuate fibers lies immediately beneath the cortex.

The **long association fibers** connect different cerebral lobes and form compact bundles that can be recognized with the naked eye. The cingulum (**D3**) is a strong system of shorter and longer fibers lying underneath the cingulate gyrus; it follows the entire course of the cingulate gyrus. The long fibers extend from the parolfactory area and the rostrum of the corpus callosum to the entorhinal area. The *subcallosal fasciculus* (*superior occipitofrontal fasciculus*) (**CD4**) lies dorsolaterally to the caudate nucleus below the radiation of the corpus callosum. Its fibers connect the frontal lobe with the temporal lobe and the occipital lobe. Some of the fibers run to the insula, others connect the frontal lobe with the caudate nucleus. The *superior longitudinal fasciculus* (**ACD5**) lying dorsolaterally to the putamen is a strong association bundle between frontal lobe and occipital lobe with fibers branching to the parietal lobe and temporal lobe. The *inferior occipitofrontal fasciculus* (**ACD6**) passes through the ventral part of the extreme capsule from the frontal lobe to the occipital lobe. The *inferior longitudinal fasciculus* (**C7**) extends between occipital lobe and temporal lobe. The *uncinate fasciculus* (**AC8**) connects the temporal cortex with the frontal cortex. Its ventral part provides a connection between the entorhinal area and the orbital area of the frontal lobe. Other fiber bundles are the *vertical occipital fasciculus* (**AC9**) and the *orbitofrontal fasciculus* (**C10**).

Commissural Fibers (E, F)

The *interhemispheric association fibers* pass through the corpus callosum, the anterior commissure (p. 228, C), and the commissure of the fornix (p. 232, C25) to the contralateral hemisphere. The most important commissure of the neocortex is the **corpus callosum** (**E**). Its curved rostral part is the *genu of the corpus callosum* (**E11**) with the pointed *rostrum* (**E12**). It is followed by the middle part, the *trunk of the corpus callosum* (**E13**), and the thickened end, the *splenium of the corpus callosum* (**E14**). The fibers of the corpus callosum spread through the white matter of both hemispheres and form the radiation of the corpus callosum. The arched fibers passing through the genu of the corpus callosum and connecting both frontal lobes are called *minor forceps* (**F15**), those passing through the splenium and connecting both occipital lobes are called *major forceps* (**F16**).

We distinguish between homotopic and heterotopic interhemispheric fibers. **Homotopic fibers** connect the same cortical areas in both hemispheres, while **heterotopic fibers** connect different areas. The vast majority of fibers of the corpus callosum is homotopic. Not all areas are connected to the same extent with their counterpart in the other hemisphere. The hand and foot parts of both somatosensory areas, for example, possess no interhemispheric fiber connections; nor are the two visual cortices connected to each other. However, very strong fiber connections exist between both areas 18, which are regarded as optic integration areas.

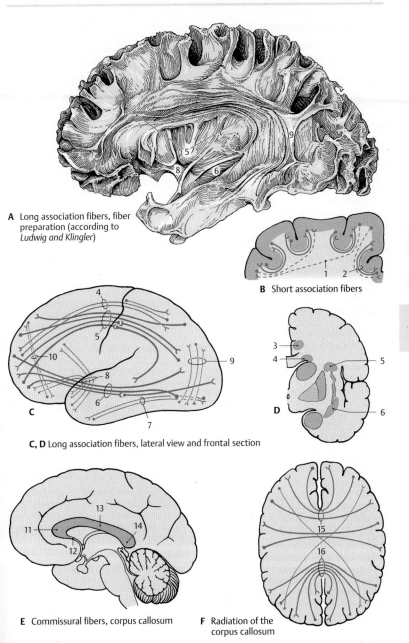

A Long association fibers, fiber preparation (according to *Ludwig and Klingler*)

B Short association fibers

C, D Long association fibers, lateral view and frontal section

E Commissural fibers, corpus callosum

F Radiation of the corpus callosum

Telencephalon

Hemispheric Asymmetry (A, B)

Consciousness is dependent on the cerebral cortex. Only those sensory stimuli reach consciousness that are transmitted to the cerebral cortex.

The faculty of speech is unique to humans. Internal speech is the prerequisite for thought, just as the spoken word is the basis for communication and writing is the information transmitted over thousands of years. In the individual person, speech depends on the integrity of specific cortical areas that usually lie only in one hemisphere. This hemisphere is called the **dominant hemisphere** and is normally the left one in right-handed persons. In left-handed persons, it may be the right or the left hemisphere, or the faculty may be represented in both hemispheres. Thus, *handedness* is not a reliable indication for dominance of the contralateral hemisphere.

In the posterior region of the superior temporal gyrus of the dominant hemisphere lies **Wernicke's speech center** (**A1**). Injury to this area results in disturbed word comprehension (*receptive aphasia*, or *sensory aphasia*). It is an *integration area* that is indispensable for the continuous availability of learned word patterns and for the interpretation of heard or spoken words. Patients with sensory aphasia utter a senseless word salad (*schizophasia*), and the speech of other persons sounds to them like an incomprehensible foreign language. Injury to the *angular gyrus* (**A2**), which borders on the supramarginal gyrus (**A3**) (see p. 212, A21), results in the loss of the abilities to write (*agraphia*) and to read (*alexia*). Stimulation of adjacent areas (**A5**), especially of the middle temporal gyrus, causes disturbance in spontaneous speech or writing. Broca 's area for motor speech coordination (**A4**) lies in the inferior frontal gyrus (p. 248, D).

Injury to the right (nondominant) hemisphere may cause disturbance of the visual and spatial orientation or the appreciation of music (*amusia*). Although speech is preserved, the melody of the language and the emotional timbre of speech are affected.

Different ways of thinking have often been assigned to one of the two hemispheres: the left, dominant hemisphere is thought to work in a logical, rational, and analytical way, while the right hemisphere is supposed to be integrative, synthetic, and intuitive. These generalizations are largely speculative.

Transection of the corpus callosum (split brain)**.** There are no changes in personality or intelligence following the transection of the corpus callosum. The patients are completely normal in everyday life. Only special tests of the tactile and visual systems will reveal any shortcomings (**B**).

Touch sensation of the left hand is registered in the right hemisphere, that of the right hand in the left hemisphere (in right-handed persons, the dominant hemisphere controls the ability to speak). Visual stimuli affecting the left halves of each retina are transmitted to the left hemisphere, while the stimuli for the right halves are transmitted to the right hemisphere (p. 256). An examination shows that right-handed persons with a severed corpus callosum can only read with the left halves of the retinas. They cannot name objects perceived with the right halves of the retina. However, they can illustrate the use of these objects through movements with their hands. The same phenomenon occurs when such persons have their eyes covered and receive an object in their left hand: they are unable to describe it verbally but can indicate its use through gestures. Objects perceived with the right hand, or with the left halves of the retina which are connected to the "speaking" hemisphere, are immediately named.

Movements performed with one limb cannot be repeated with the contralateral limb, because the one hemisphere has not been informed about the impulses been sent out by the other hemisphere.

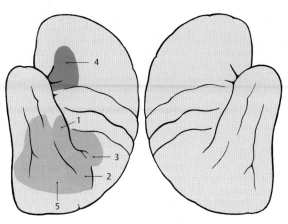

A Areas for speech and writing in a right-handed person

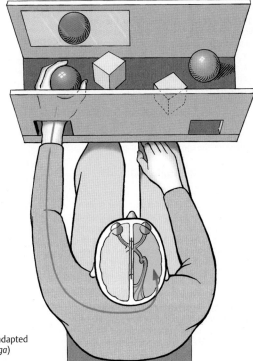

B Split brain experiment (adapted from *Sperry and Gazzaniga*)

Imaging Procedures

The development of imaging methods for clinical diagnostics has been extremely rapid during the last 20 years. The currently most important methods for visualizing the central nervous system are:

- *Contrast medium-assisted projection radiography* for the visualization of blood vessels and the ventricular system
- *Computed tomography* (CT)
- *Magnetic resonance imaging* (MRI) (p. 266)

Conventional radiography is mainly used for the visualization of bony structures. Ultrasound is not suitable for the central nervous system because sonic waves do not penetrate the cranial bones in adults. However, examination by ultrasound does play a major role in diagnostics during early childhood. In addition to purely anatomical imaging methods, there is also the possibility of visualizing functional parameters following the injection of a suitable radioactive nuclide. The methods of nuclear medicine—*single-photon emission computed tomography* (SPECT) and *positron emission tomography* (PET) (p. 266)—can be used for studying cerebral blood flow and metabolic activity as well as for mapping specific receptor systems. In the following section, some areas of application are briefly described for the most important procedures.

Contrast Radiography (A)

Radiation emitted by an X-ray source penetrates the body. Absorbing structures (mainly the bones) appear as silhouettes on the recording screen. The screen consists either of a light-sensitive film or, to an increasing extent, of digital receiver arrangements. For the *visualization of blood vessels*, a contrast medium is injected into the patient's vessels between two recordings. By subtracting the images obtained prior to and after the injection of contrast medium, the (unchanged) bony structures are eliminated. The resulting subtraction image selectively shows the blood vessels filled with contrast medium. The contrast medium is usually administered to a selected vessel by introducing a catheter into the vascular system of interest. The result is a selective high-resolution image of the corresponding vascular segments (**A**). However, *intra-arterial* **digital subtraction angiography** (DSA) subjects the patient to considerable stress due to the introduction of the arterial catheter. *Intravenous DSA*, where contrast medium is injected into a brachial vein, produces images of lower contrast because the contrast medium is diluted by the blood volume; furthermore, it results in nonspecific coloration of the vessels. Hence, it has not yet gained acceptance in the practice.

The same recording principle is used for **myelography**, in which contrast medium is administered directly into the *cerebrospinal fluid* following puncture of the lumbar spine.

Computed Tomography (B)

Computed tomography (CT), or *computed axial tomography* (CAT), also produces images by means of X-rays. However, the imaging principle is not based on the creation of silhouettes as in the projection method. Rather, individual transmissions through a thin layer of tissue ("slice") defined by the geometry of the X-ray are recorded. After obtaining the absorption profiles at different projection angles (in Figure **B** shown as red, green, and blue areas), a two-dimensional image of the slice penetrated by the X-rays is computed by means of the filtered **back projection**. Only digital photocells of very high sensitivity are used as detectors. These allow the detection of even soft tissue structures with great sensitivity and precision, in addition to bony structures.

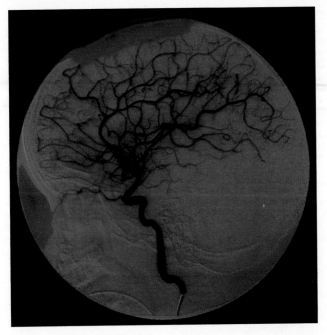

A Digital subtraction angiography (DSA): Selective visualization of the internal carotid artery, lateral view (Courtesy of *M. Orszagh*, Department of Radiodiagnostics, University of Freiburg, Germany)

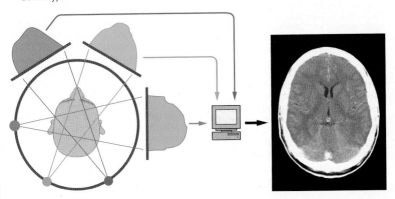

B Principle of computed tomography (CT): After recording the absorption profiles (red, green, and blue areas) of fan-shaped X-rays, the image is reconstructed by the computer and displayed on the monitor (Courtesy of *J. Lautenberger*, Department of Radiodiagnostics, University of Freiburg, Germany)

Computed Tomography (continued)

The main area of CT application for the CNS is above all the examination of patients with *head trauma*: apart from high-precision imaging of bony displacements, CT is also very sensitive for the visualization of hemorrhages. Furthermore, all types of space-occupying processes can be examined by means of CT because of the highly detailed precision of the anatomical imaging. However, these areas of application are now shifting increasingly to magnetic resonance imaging owing to its higher sensitivity.

Magnetic Resonance Imaging (A, B)

Magnetic resonance imaging (MRI) is based on the magnetic properties of the atomic nuclei in the body (mainly hydrogen nuclei) and uses radiofrequency resonance technique to measure the magnetic moment induced by an external magnetic field. Variations imposed on the magnetic field allow spatial visualization of the T1 and T2 relaxation times, which are nuclear magnetic properties governed by the chemical form of the hydrogen atoms containing the nuclei in the tissues. These relaxation times, which mainly reflect the ratio of free to bound water molecules, produce a contrast mechanism that delineates not only different tissues but also pathological tissue changes. The images in **A** and **B** display the same anatomic section and illustrate the exquisite contrast achievable by MRI, which makes it an excellent method for the *identification of pathological processes*. The static magnetic and electromagnetic fields used with MRI have not been found to be hazardous to patients. Furthermore, MRI has the advantage over CT that the images can be reconstructed in any plane. The spatial resolution of MRI is in the range of 0.7 – 1.0 mm for a section thickness of 5 mm. It takes several minutes to acquire and process the data for the axial, coronal, and sagittal sections constituting a standard cranial MRI.

PET and SPECT (C, D)

Both of these imaging methods of nuclear medicine are based on detecting the radiation of radioactive nuclides administered prior to the investigation. **Single-photon emission computed tomography** (SPECT) uses gamma-emitters. Detection of the generated radiation is carried out by means of photocells in a similar way as in CT. The geometry of the detected radiation, however, is not predetermined by the arrangement of transmitter and receiver; instead, both intensity and direction of the received gamma rays must be determined. This yields a spatial resolution of about 1 cm, clearly less favorable than when using computed tomography. Hence, high-resolution anatomical images cannot be obtained; nevertheless, SPECT is suitable for the *determination of cerebral blood flow* by means of time-resolved kinetics of the detected signals.

Positron emission tomography (PET) uses emitters of beta particles (positrons). The positrons almost instantly combine with electrons in the immediate surrounding, with both particles undergoing mutual annihilation and giving rise to two gamma rays of 511 keV energy emitted in opposite directions (**C**). The geometry of these gamma rays can be much better defined than in the case of SPECT, namely, by recording the coincidence of the two gamma photons received in the detector ring. Therefore, the spatial resolution of PET lies at 2 – 4 mm, though this is clearly still not as good as in the case of CT and MRI. Through incorporation of suitable radioisotopes into biologically active molecules, PET also allows us to carry out *metabolic studies* in addition to *measurements of blood flow* (**D**), or to *selectively visualize receptor systems* (for example, dopamine receptors).

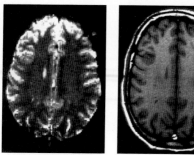

A Spin–lattice relaxation, T1 **B** Spin–spin relaxation, T2

A, B Magnetic resonance imaging (MRI): Two different scans of the head at the level above the ventricles

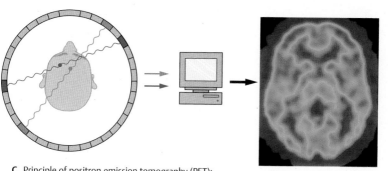

C Principle of positron emission tomography (PET): Detection of gamma rays emitted simultaneously in opposite directions during the decay of a radioisotope (Courtesy of *F. Jüngling*, Department of Radiodiagnostics, University of Freiburg, Germany)

Telencephalon

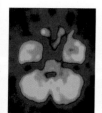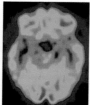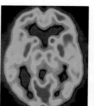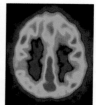

D PET scans of four parallel slices (intensity of glucose utilization in the brain)

Cerebrovascular and Ventricular Systems

Cerebrovascular System

Arteries

The brain is supplied by four large arteries: the two *internal carotid arteries* and the two *vertebral arteries* (see vol. 2).

The **internal carotid artery** (**A1**) (p. 272) passes through the dura mater medially to the anterior clinoid process of the sphenoid bone. Between the subarachnoid and the pia mater, it gives off the *superior hypophysial artery* (p. 200, E9), the *ophthalmic artery*, the *posterior communicating artery* (**A16**), and the *anterior choroidal artery* (**A2**). It then divides into two large terminal branches, the *anterior cerebral artery* (**A4**) and the *middle cerebral artery* (**A7**).

The **anterior choroidal artery** (**A2**) runs along the optic tract to the choroid plexus (**A3**) in the inferior horn of the lateral ventricle. It gives off fine branches that supply the optic tract, the temporal genu of the optic radiation, the hippocampus, the tail of the caudate nucleus, and the amygdaloid body.

The **anterior cerebral artery** (**A4**) runs on the medial surface of the hemisphere across the corpus callosum. The two anterior cerebral arteries are interconnected by the *anterior communicating artery* (**A5**). The long central artery (recurrent artery of Heubner) (**A6**) branches off shortly after the communicating artery. It passes through the anterior perforated substance into the brain and supplies the anterior limb of the internal capsule, the adjacent region of the head of the caudate nucleus, and the putamen.

The **middle cerebral artery** (**A7**) runs laterally toward the lateral sulcus; above the perforated substance it gives off 8 – 10 *striate branches* that enter the brain. At the entrance of the lateral fossa, it divides into several large branches that spread over the lateral surface of the hemisphere.

The two **vertebral arteries** (**A8**) arise from the two subclavian arteries and enter the cranial cavity through the foramen magnum; they unite at the upper margin of the medulla oblongata to form the unpaired

basilar artery (**A9**). The latter ascends along the ventral surface of the pons and bifurcates at the upper margin of the pons into the two **posterior cerebral arteries** (**A10**). The vertebral artery gives off the *posterior inferior cerebellar artery* (**A11**), which supplies the lower surface of the cerebellum and the choroid plexus of the fourth ventricle. The basilar artery gives off the *anterior inferior cerebellar artery* (**A12**), which also supplies the lower surface of the cerebellum as well as the lateral parts of medulla and pons. The *labyrinthine artery* (**A13**) runs as a fine branch together with the facial nerve and the vestibulocochlear nerve through the internal acoustic meatus into the inner ear. It may arise from the basilar artery or from the anterior inferior cerebellar artery. Numerous small branches reach as *pontine arteries* (**A14**) directly into the pons. The *superior cerebellar artery* (**A15**) runs along the upper margin of the pons and extends deep into the cisterna ambiens around the cerebral peduncles to the dorsal surface of the cerebellum.

The circle of Willis. The *posterior communicating arteries* (**A16**) connect on both sides the posterior cerebral arteries with the internal carotid arteries so that the blood flow of the vertebral arteries can communicate with the carotid circulation. The anterior cerebral arteries are interconnected through the anterior communicating artery. This way, a closed *cerebral arterial circle* is created at the base of the brain. However, the anastomoses are often so thin that they do not allow for a significant exchange of blood. Under conditions of normal intracranial pressure, each hemisphere is fed by the ipsilateral internal carotid artery and the ipsilateral posterior cerebral artery.

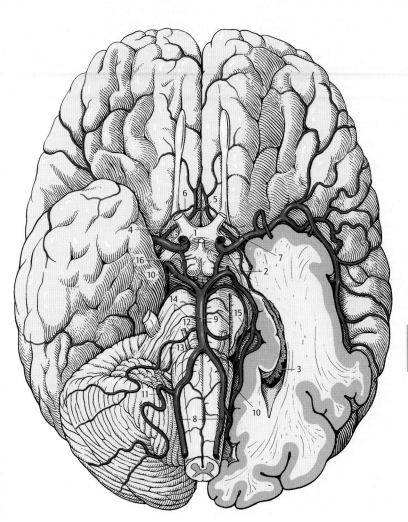

A Arteries of the base of the brain

Internal Carotid Artery (A – C)

The **internal carotid artery** (**C1**) can be subdivided into a *cervical part* (between the carotid bifurcation and the base of the skull); a *petrosal part* (in the carotid canal of the petrous bone); a *cavernous part* (within the cavernous sinus, p. 104, A7); and a *cerebral part*. The cavernous and cerebral segments of the artery form an S-shaped curve (*carotid siphon*) (**C2**). The *inferior hypophysial artery* (p. 200, E10) comes off in the cavernous part, followed by small branches to the dura and to cranial nerves IV and V. After giving off the *superior hypophysial artery* (p. 200, E9), the *ophthalmic artery*, and the *anterior choroidal artery* in the cerebral part, the internal carotid artery divides into two large terminal branches, the *anterior cerebral artery* and the *middle cerebral artery*.

The **anterior cerebral artery** (**BC3**) turns toward the longitudinal cerebral fissure after giving off the anterior communicating artery. The *postcommunical part* of the artery (*pericallosal artery*) (**BC4**) runs on the medial hemispheric wall around the rostrum and the genu of the corpus callosum (**B5**) and along the superior surface of the corpus callosum toward the parieto-occipital sulcus. It gives off branches to the basal surface of the frontal lobe (*medial frontobasal artery*) (**B6**). The other branches spread over the medial surface of the hemisphere; they are the *frontal branches* (**BC7**), the *callosomarginal artery* (**BC8**), and the *paracentral artery* (**B9**) which supplies the motor area for the leg.

The **middle cerebral artery** (**AC10**) extends laterally to the bottom of the lateral fossa, where it divides into several groups of branches. The artery is divided into three segments. The *sphenoidal part* gives off the *central arteries* (fine branches for the striatum, thalamus, and internal capsule), and the *insular part* gives rise to the short *insular arteries* (**C11**) for the insular cortex, the *lateral frontobasal artery* (**A12**), and the *temporal arteries* (**A13**) for the cortex of the temporal lobe. The last segment, the *terminal part*, is formed by the long branches for

the cortex of the central region and the parietal lobe (**AC14**). There are considerable variations regarding the bifurcation and course of individual arteries.

The **posterior cerebral artery** (**BC15**) develops as a branch of the internal carotid artery. In adults, however, it lies relatively far caudally. It is only connected with the internal carotid artery through the thin posterior communicating artery. Because most of its blood comes from the vertebral arteries, it is regarded as part of their supply area; the latter consists of the subtentorial parts of the brain (brain stem and cerebellum) and the supratentorially located occipital lobe, the basal part of the temporal lobe, and the caudal parts of the striatum and thalamus (tentorium of the cerebellum, p. 288, B5). All parts of the frontal brain lying in front of it receive their blood from the internal carotid artery.

The posterior cerebral artery ramifies on the medial surface of the occipital lobe and on the basal surface of the temporal lobe. It gives off the posterior choroidal artery for the choroid plexus of the third ventricle as well as fine branches for the striatum and thalamus. The ophthalmic artery (**C16**) runs from the internal carotid artery to the orbit.

Carotid angiogram (arterial phase). A diagram of a carotid angiogram is shown in C. For diagnostic purposes, a contrast medium is injected into the internal carotid artery. Within a few seconds, the contrast medium spreads through the supply area of the artery. A radiograph *taken immediately* illustrates the arterial vascular tree. When viewing arteriographic images, it must be taken into consideration that all blood vessels are seen in a single plane (see contrast radiography, p. 264, A).

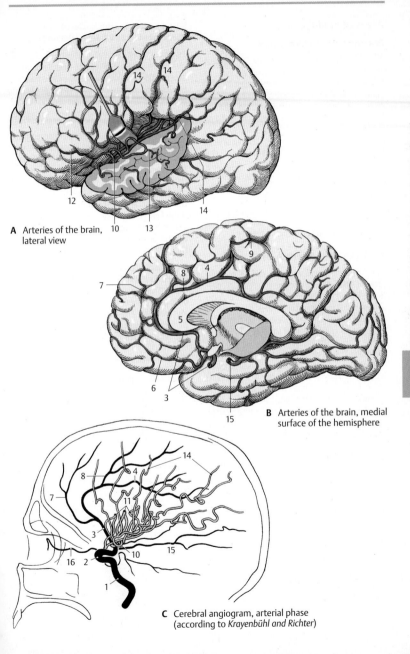

A Arteries of the brain, lateral view

B Arteries of the brain, medial surface of the hemisphere

C Cerebral angiogram, arterial phase (according to *Krayenbühl and Richter*)

Areas of Blood Supply

Anterior Cerebral Artery (A, B)

The short branches of the **anterior cerebral artery** (**AB1**) extend to the optic chiasm, to the septum pellucidum, to the rostrum and genu of the corpus callosum, while the long *recurrent artery of Heubner* (long central artery) runs to the medial part of the head of the caudate nucleus and to the anterior limb of the internal capsule. The cortical branches supply the medial parts of the base of the frontal brain and the olfactory lobe, and also the frontal and parietal cortex on the medial surface of the hemisphere and the corpus callosum as far as the splenium. The supply area of the artery extends beyond the edge of the pallium to the dorsal convolutions of the convexity.

Middle Cerebral Artery (A, B)

The *striate branches* of the **middle cerebral artery** (**AB2**) terminate in the globus pallidus, in parts of the thalamus, in the genu and parts of the anterior limb of the internal capsule. The branches given off by the insular arteries ramify in the insular cortex and in the claustrum and reach into the external capsule. The area supplied by the cortical branches includes the lateral surface of frontal, parietal, and temporal lobes and also a large part of the central region and the temporal pole. The branches not only supply the cortex but also the white matter as far as the lateral ventricle, including the central part of the optic radiation.

Posterior Cerebral Artery (A, B)

The **posterior cerebral artery** (**AB3**) gives off fine short branches that supply the cerebral peduncles, pulvinar, geniculate bodies, quadrigeminal plate, and splenium of the corpus callosum. The cortical supply area occupies the basal part of the temporal lobe and the occipital lobe with the visual cortex (striate area); however, the latter is also reached in the region of the occipital pole by the most inferior branches of the middle cerebral artery.

Blood Supply for the Nuclei of Diencephalon and Telencephalon (C, D)

The recurrent artery of Heubner and the *striate branches* (**D4**) of the middle cerebral artery (**D5**) supply the head of the caudate nucleus, putamen, and internal capsule. The role of the *anterior choroidal artery* (**C6**) in the supply of deep structures varies; its branches extend not only to the hippocampus and amygdaloid body but also to parts of the pallidum and thalamus. The rostral part of the thalamus receives a branch from the posterior communicating artery (**C7**), the thalamic branch (**C8**). The middle and caudal parts of the thalamus are supplied by the basilar artery (**C9**), from where direct branches (**C10**) may run to the thalamus. Other fine thalamic branches are given off by the posterior choroidal artery (**C11**) and the posterior cerebral artery (**C12**).

Vascularization. The large cerebral vessels lie without exception on the surface of the brain. They give off small arteries and arterioles that penetrate vertically into the brain and ramify. The capillary network is very dense in the gray matter but far less so in the white matter.

▄▄▄ **Clinical Note:** Sudden obstruction of an artery by a thrombus, air bubble, or fat droplet in the bloodstream (*embolism*) causes the death of brain tissue in the supply area of the affected artery. The anastomoses between the vascular areas are not adequate to supply the affected region through nearby areas in case of a sudden obstruction. Especially affected are the *middle cerebral artery* and its branches.

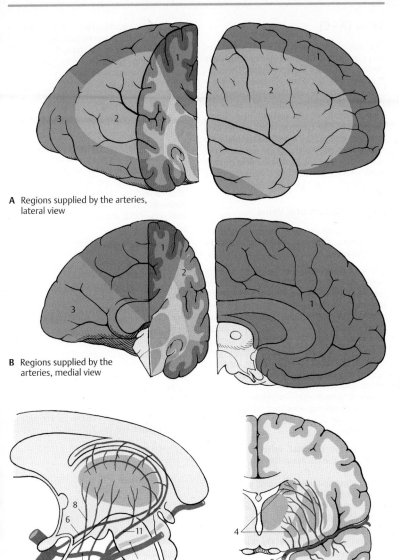

A Regions supplied by the arteries, lateral view

B Regions supplied by the arteries, medial view

C Arterial supply of the thalamus (according to *Van den Bergh and Van der Eeken*)

D Arterial supply to the neostratum

Veins (A – C)

The major veins lie on the surface of the brain in the subarachnoid space; some deep veins run beneath the ependyma. The cerebral veins do not possess valves. They exhibit considerable variations with respect to course and drainage. Quite often there are several small vessels instead of a single major vein. The cerebral veins are divided into two groups: the **superficial cerebral veins** which drain the blood into the sinuses of the dura mater (see vol. 2), and the **deep cerebral veins** which drain the blood into the *great cerebral vein* (great vein of Galen).

Superficial Cerebral Veins

We distinguish between the group of *superior cerebral veins* and the group of *inferior cerebral veins*.

The **superior cerebral veins** (**AC1**), totaling about 10 – 15 veins, collect the blood from the frontal and parietal lobes and carry it into the *superior sagittal sinus* (**BC2**). They run within the subarachnoid space and empty into the *lateral lacunae* (**BC3**), pouchlike cavities of the superior sagittal sinus. For a short distance, they pass through the subdural space. Here, the thin-walled veins can easily rupture during head injury and bleed into the subdural space (*subdural hematoma*). Strangely, the veins empty into the superior sagittal sinus at an oblique angle against the dominating blood flow in the sinus.

The **inferior cerebral veins** receive the blood from the temporal lobe and from the basal regions of the occipital lobe; they empty into the *transverse sinus* and the *superior petrosal sinus*. The largest and most consistent of these veins is the *superficial middle cerebral vein* (**AC4**) located in the lateral sulcus; it often consists of several venous trunks. It drains the blood from most of the lateral aspect of the hemisphere into the cavernous sinus (p. 104, A7).

The superior and inferior cerebral veins are interconnected only by a few anastomoses. The most important one is the *superior anastomotic vein* (Trolard's vein) (**AC5**); it empties into the superior sagittal sinus and is connected with the superficial middle cerebral vein. The *central vein* (**C6**) located in the central sulcus may also form anastomoses with the middle cerebral vein. The *inferior anastomotic vein* (Labbé's vein, Browning 's vein) (**AC7**) connects the superficial middle cerebral vein with the transverse sinus.

Carotid angiogram (venous phase). A diagram of the venous phase of a carotid angiogram is shown in C (for arterial phase, see p. 272). A radiograph taken only seconds after injection of the contrast medium shows its drainage via the venous vascular tree. Superficial and deep veins are seen in a single plane.

Deep cerebral veins (S. 278):

BC8	*Great cerebral vein* (great vein of Galen).
C9	Internal cerebral vein.
C10	Thalamostriate vein (terminal vein).
C11	Vein of the septum pellucidum
C12	Interventricular foramen (foramen of Monro).
C13	Basal vein (Rosenthal's vein).
BC14	Straight sinus.
BC15	Inferior sagittal sinus.
BC16	Confluence of sinuses (p. 104, A19).

(For the large venous channels, the *sinuses of dura mater*, see vol. 2.)

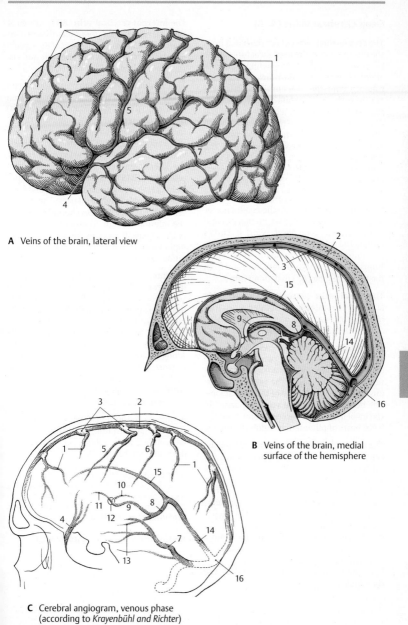

A Veins of the brain, lateral view

B Veins of the brain, medial surface of the hemisphere

C Cerebral angiogram, venous phase (according to *Krayenbühl and Richter*)

Deep Cerebral Veins (A, B)

The **deep cerebral veins** collect the blood from the diencephalon, the deep structures of the hemispheres, and the deep white matter. In addition, there are thin *transcerebral veins* running along the fibers of the corona radiata from the outer white matter and from the cortex. They connect the superficial drainage areas with the deep ones. The deep cranial veins empty their blood into the *great cerebral vein* (great vein of Galen). The drainage system of the deep veins is therefore also known as the system of the great cerebral vein.

The **great cerebral vein** (**AB1**) is a short vascular trunk formed by the confluence of four veins, namely, the two *internal cerebral veins* and the two *basal veins*. It curves around the splenium of the corpus callosum and empties into the straight sinus. Veins from the surface of the cerebellum and from the occipital lobe (**B2**) may drain into it.

The **basal vein** (Rosenthal's vein) (**AB3**) arises at the anterior perforated substance (**A4**) by junction of the *anterior cerebral vein* and the *deep middle cerebral vein*.

The **anterior cerebral vein** (**A5**) receives its blood from the anterior two-thirds of the corpus callosum and the adjacent convolutions. It extends around the genu of the corpus callosum to the base of the frontal lobe. The **deep middle cerebral vein** (**A6**) arises in the insular area and receives veins from the basal parts of putamen and globus pallidus.

The basal vein crosses the optic tract and ascends in the cisterna ambiens around the cerebral peduncle (**A7**) to below the splenium, where it empties into the great cerebral vein. Along its course it receives numerous venous tributaries, namely, veins from the optic chiasm and the hypothalamus, the *interpeduncular vein* (**A8**), the *inferior choroid vein* (**A9**) from the choroid plexus (**A10**) of the inferior horn, and veins from the internal segment of the globus pallidus and from the basal parts of the thalamus.

The **internal cerebral vein** (**AB11**) arises at the interventricular foramen (foramen of Monro) by junction of the *vein of the septum pellucidum*, the *thalamostriate vein,* and the *superior choroid vein*.

The **thalamostriate vein** (*terminal vein*) (**B12**) runs in the terminal sulcus between thalamus (**B13**) and caudate nucleus (**B14**) in rostral direction to the interventricular foramen. It receives venous tributaries from the caudate nucleus, from the adjacent white matter, and from the lateral corner of the lateral ventricle. The **vein of the septum pellucidum** (**B15**) receives venous branches from the septum pellucidum (**B16**) and from the deep frontal white matter. The **choroid vein** (**B17**) runs with the choroid plexus to the inferior horn. In addition to the vessels of the plexus, it receives the veins from the hippocampus and from the deep temporal white matter.

The internal cerebral vein extends from the interventricular foramen across the medial surface of the thalamus at the margin of the roof of the diencephalon to the region of the pineal gland, where it unites with the contralateral internal cerebral vein and the basal veins to form the great cerebral vein. Along its course it receives tributaries from the fornix (**B18**), from the dorsal parts of the thalamus, from the pineal gland (epiphysis) (**B19**) and, variably, from the deep white matter of the occipital lobe.

In summary, the dorsal parts of thalamus, pallidum, and striatum drain into the internal cerebral vein, while the ventral parts drain into the basal vein.

■ **Clinical Note:** Obstruction of a cerebral vein causes congestion and hemorrhage in the affected region. In case of birth trauma, rupture of the thalamostriate vein in newborns may lead to hemorrhage into the ventricles.

A20 Superficial middle cerebral vein.

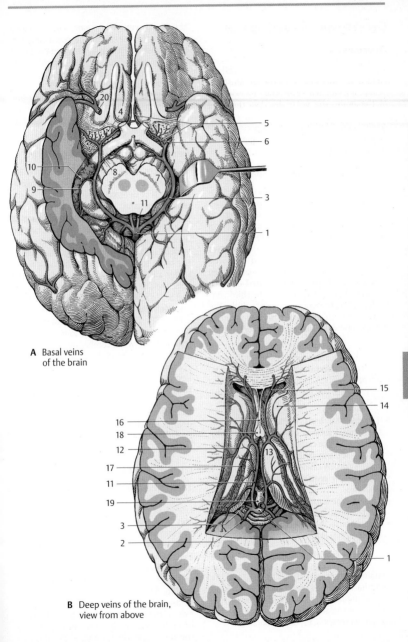

A Basal veins
of the brain

B Deep veins of the brain,
view from above

Cerebrospinal Fluid Spaces

Overview

The central nervous system (CNS) is completely surrounded by the *cerebrospinal fluid* (*CSF*), which also fills the inner cavities of the brain, the ventricles. We therefore distinguish between *internal* and *external cerebrospinal fluid spaces*. They communicate with each other in the region of the fourth ventricle.

Internal Cerebrospinal Fluid Spaces (A–C)

The ventricular system consists of four ventricles, namely, the two *lateral ventricles* (**A1**) of the telencephalon, the *third ventricle* (**A–C2**) of the diencephalon, and the *fourth ventricle* (**A–C3**) of the rhombencephalon (pons and medulla oblongata). The two lateral ventricles are connected with the third ventricle by the **interventricular foramen** (*foramen of Monro*) (**AC4**) located on each side in front of the thalamus. The third ventricle, in turn, communicates with the fourth ventricle through a narrow channel, the **cerebral aqueduct** (*aqueduct of Sylvius*) (**A–C5**).

Owing to the rotation of the hemispheres (p. 208) the **lateral ventricle** has a semicircular configuration with a caudally directed spur. We distinguish the following parts:

- The *anterior horn*, **frontal horn** (**BC6**), in the frontal lobe is bordered laterally by the head of the caudate nucleus, medially by the septum pellucidum, and dorsally by the corpus callosum
- The narrow **central part** (**BC7**) above the thalamus
- The *inferior horn*, **temporal horn** (**BC8**), in the temporal lobe
- The *posterior horn*, **occipital horn** (**BC9**), in the occipital lobe

The lateral wall of the **third ventricle** is formed by the thalamus with the interthalamic adhesion (**C10**) (p. 10, C18) and the hypothalamus. The *optic recess* (**C11**) and the *infundibular recess* (**C12**) project rostrally, and the *suprapineal recess* (**C13**) and the *pineal recess* (**C14**) do so caudally.

The **fourth ventricle** creates a tent-shaped space above the rhomboid fossa between cerebellum and medulla oblongata; on both sides, it sends out a long *lateral recess* (**BC15**). At the end of each lateral recess is the *lateral aperture of the fourth ventricle* (foramen of Luschka, foramen of Key and Retzius). At the attachment of the inferior medullary velum lies the *median aperture* (foramen of Magendie) (p. 282, D14).

External Cerebrospinal Fluid Spaces (A)

The external cerebrospinal fluid space lies between the two layers of the leptomeninges. It is delimited on the inside by the pia mater and on the outside by the arachnoidea mater (subarachnoid space or cavity; p. 288, A13). The space is narrow over the convexity of the hemispheres and widens only in several areas at the base of the brain to form *cisterns*. Whereas the pia mater adheres closely to the surface of the CNS, the arachnoidea spans across sulci and fossae; in regions of deep indentations this arrangement creates larger spaces filled with CSF, the **subarachnoid cisterns**. The largest space is the *cerebellomedullary cistern* (**A16**) between cerebellum and medulla oblongata. In the corner formed by diencephalic floor, cerebral peduncles and pons lies the *interpeduncular cistern* (**A17**), and the *chiasmatic cistern* (**A18**) lies in front of it near the optic chiasm. The cerebellar surface, quadrigeminal plate, and pineal gland delimit the *cisterna ambiens* (**A19**), which is traversed by a loose network of connective tissue.

Circulation of Cerebrospinal Fluid (A)

The cerebrospinal fluid is produced by the *choroid plexus* (p. 282). It flows from the lateral ventricles into the third ventricle, and from there through the aqueduct into the fourth ventricle. Here it passes through the median and lateral apertures into the external cerebrospinal fluid space. Drainage of CSF into the venous circulation takes place partly in the *arachnoid granulations* (p. 288, A15) that protrude into the venous sinuses or lateral lacunae, and partly at the *exits of the spinal nerves* where the fluid enters into the dense venous plexuses and into the nerve sheaths (drainage into the lymphatic circulation).

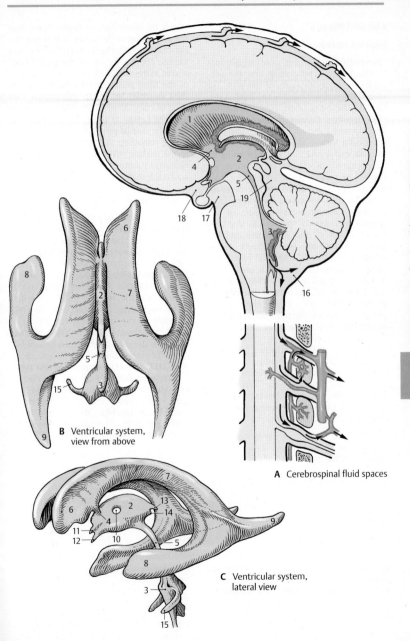

B Ventricular system, view from above

A Cerebrospinal fluid spaces

C Ventricular system, lateral view

Choroid Plexus

Lateral Ventricles (A, C)

The **choroid plexus** consists of convolutions of blood vessels protruding into the ventricles from specific parts of the wall. The area of the wall adhering to the medial surface of the hemisphere (*choroid lamina*) (**A1**) becomes thinner during embryonic development and is pushed into the ventricular lumen by the vascular loops of the pia mater (**A2**). At the beginning of development, all vascular convolutions are covered by a thin layer of hemispheric wall. The latter finally turns into a layer of epithelial cells, the **plexus epithelium**. Thus, the adult choroid plexus consists of two components, namely, the vascularized connective tissue of the pia mater and the plexus epithelium (transformed hemispheric wall). Once the choroid plexus has invaginated into the ventricular cavity, it remains connected with the external pia mater only through a narrow channel, the **choroid fissure** (**A3**). When the choroid plexus is removed, the thinned parts of the hemisphere wall tear away at this location. The line of attachment is called **taenia** (band). One such line is attached to the fornix (p. 230, A6) and to the fimbria of the hippocampus (p. 230, ACD5); it is known as the *taenia fornicis* (taenia of the fornix) (**C4**). The other line runs along the lamina affixa (p. 170, D15, E16) and is known as the *taenia choroidea* (choroid line) (**C5**).

Owing to the rotation of the hemispheres (p. 208), the choroid plexus forms a semicircle along the medial wall of the ventricle, reaching from the interventricular foramen (foramen of Monro) via the central part (**C6**) to the inferior horn (**C7**). The anterior horn (**C8**) and the posterior horn (**C9**) do not contain a plexus.

Tela choroidea (A – C)

When the hemispheres overgrow the diencephalon, the leptomeninges of both parts of the brain come to lie on top of each other to form a *duplication* (**A10**), the **tela choroidea** (**B**), a connective-tissue plate spanning between the telencephalic hemispheres and the diencephalon. At its lateral margins, the pia mater forms the vascular convolution for the choroid plexus of the lateral ventricles. In the middle, the tela covers the roof of the third ventricle (**tela choroidea of the third ventricle**) (**BC11**). In this region, two rows of convoluted blood vessels protrude into the lumen of the third ventricle and form the **choroid plexus of the third ventricle**. When removing the ventricular roof, the *taenia thalami* (**C12**) remains as line of attachment running along the medullary stria across the thalamus.

Fourth Ventricle (D, E)

Above the fourth ventricle, the **tela choroidea of the fourth ventricle** is also formed as a *duplication of the pia mater* by apposition of the lower surface of the cerebellum to the surface of the rhombencephalon (**E**). The roof of the rhombencephalon becomes reduced to a thin epithelial layer and is pushed into the ventricle by the vascular loops originating from the tela. The tela choroidea of the fourth ventricle consist of pia mater only, as the arachnoidea does not adhere to the surface of the cerebellum but spans the cerebellomedullary cistern. At the *obex* (**D13**), the attachment of the tela above a narrow medullary fold, lies the *median aperture* (foramen of Magendie) (**D14**). On both sides open the *lateral apertures* (foramina of Luschka), through which the lateral end of the choroid plexus protrudes (*Bochdalek's flower basket*) (**D15**).

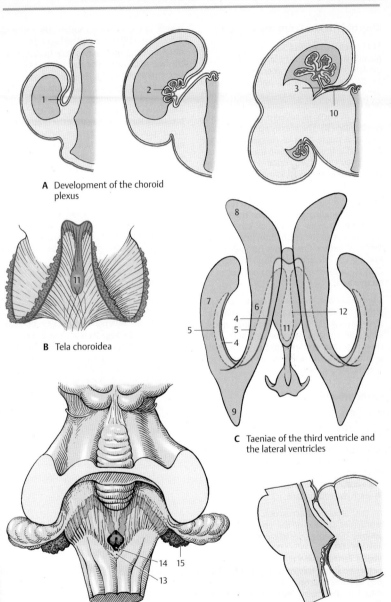

A Development of the choroid plexus

B Tela choroidea

C Taeniae of the third ventricle and the lateral ventricles

D Plexus of the fourth ventricle, view from above

E Plexus of the fourth ventricle, lateral view

Choroid Plexus (continued) (A, B)

The treelike ramification of the choroid plexus (**A**) creates a very large surface. Each branch contains one or more vessels, such as arteries, capillaries, and thin-walled venous caverns. The vessels are surrounded by a loose meshwork of collagen fibers (**B1**) which, in turn, is covered by the plexus epithelium (**B**). The plexus epithelium consists of a single layer of cuboidal cells that carry a fine brush border on their apical surface. Their cytoplasm contains vacuoles and coarse granules as well as inclusions of lipid and glycogen.

The choroid plexus is the *site of cerebrospinal fluid production*. Fluid is transferred from the vascular system of the choroid plexus through the epithelium into the ventricles. Whether this occurs through secretion by the plexus epithelium or through dialysis (a form of filtration) has yet to be established.

Like the pia-arachnoid and dura, the choroid plexus is richly innervated (the meninges are innervated by the trigeminal nerve, vagus nerve, and autonomic fibers). The choroid plexus and meninges are therefore sensitive to pain, whereas the brain tissue itself is largely insensitive.

Ependyma (C, D)

The walls of the ventricular system are lined by a single cell layer, the **ependyma** (**C**). Each ependymal cell has a basal process, the ependymal fiber, which extends into the brain. The cell surface facing the ventricular lumen often carries several cilia, with the *basal bodies*, or *kinetosomes* (**C2**), lined up beneath the cell surface.

In the electron-microscopic image, the ventricular surface of the ependymal cells exhibits numerous vesicle-containing protrusions (**D3**). The cilia (**D4**) contain *microtubules* in the characteristic 9 + 2 arrangement: two single microtubules in the center (**D5**) and nine microtubule doublets (**D6**) arranged around them. The basal body of each cilia is surrounded by a dense zone (**D7**) into

which numerous short rootlets (**D8**) radiate. A *basal foot* (**D9**) is located on one side of the basal body; it may play a role in directing the beat of the cilia. The ependymal cells are interconnected along their lateral surfaces by *zonulae adherentes* (adherent junctions) (**D10**) and by *zonulae occludentes* (tight junctions) (**D11**); the latter seal the cerebrospinal fluid space against the brain. Neuronal processes (**D12**) run between the ependymal cells. The layer underneath the ependyma consists of radially or horizontally running glial fibers (**C13**) and contains only few cells. Below it lies the *subependymal cell layer* (**C14**). It contains undifferentiated cells in addition to astrocytes. According to recent studies, not only glial cells but also neurons are generated here throughout life. Intensive investigations are under way to test whether neuronal stem cells of the subependymal zone can be used for neuronal replacement in various forms of neuronal degeneration.

The structure of the ventricular wall varies widely in different regions. The ependymal cover or the subependymal layer of glial fibers may be completely absent in certain areas. The subependymal glial cell layer is most prominent above the head of the caudate nucleus and at the base of the anterior horn but is absent above the hippocampus.

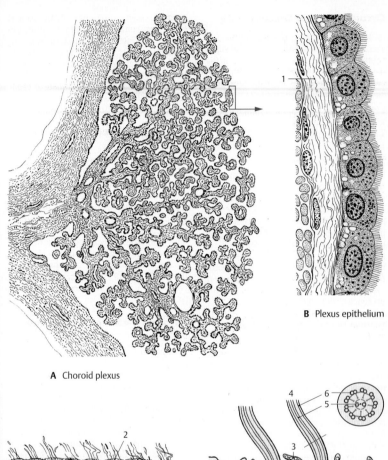

A Choroid plexus

B Plexus epithelium

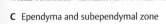

C Ependyma and subependymal zone

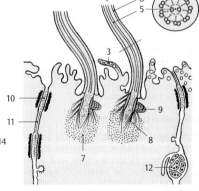

D Ependymal cell, electron-microscopic diagram (according to *Brightman and Palay*)

Circumventricular Organs (A – D)

In lower vertebrates, the ependyma has secretory functions and probably also receptor functions. This has resulted in the development of special structures that can be demonstrated also in mammals. Such circumventricular organs include:

- The *vascular organ of the terminal lamina*
- The *subfornical organ*
- The *area postrema*
- The *paraphysis*
- The *subcommissural organ*

These organs are regressed in humans, and some of them (paraphysis and subcommissural organ) appear only temporarily during embryonic development. Their functions are unknown. It is assumed that they play a role in the regulation of CSF pressure and composition and that they are somehow related to the neuroendocrine system of the hypothalamus. What is striking is their *location at narrow passages of the ventricular system*, their high *vascularization,* and the presence of *fluid-filled spaces.*

Vascular Organ of the Terminal Lamina (A, D)

This organ (**A, D1**) lies in the terminal lamina which spans as rostral end of the third ventricle between the anterior commissure and the optic chiasm. We distinguish between a highly vascularized *outer zone* beneath the pia mater and an *inner zone* consisting mainly of glia. The vessels form a dense plexus with sinuslike dilatations. In the inner zone run nerve fibers from the supraoptic nucleus that contain Gomori-positive material (Herring bodies, p. 204, B5). It also receives peptidergic fibers from the hypothalamus.

Subfornical Organ (B, D)

The **subfornical organ** (**B, D2**) lies as a small, pinhead-sized nodule between the two interventricular foramina in the roof of the third ventricle at the rostral end of the tela choroidea. In addition to glial cells and isolated neurons, it contains large round parenchymal cells, the neuronal character of which is disputed.

Electron microscopy has demonstrated *ependymal canaliculi* penetrating from the surface and interconnecting with wide intercellular clefts. Vascular loops penetrate from the tela choroidea into the inside of the subfornical organ. Peptidergic nerve fibers (somatostatin, luliberin) terminate at capillaries and in the area of the ependymal canaliculi.

Area postrema (C, D)

The area postrema (**C, D3**) consists of two symmetrical narrow structures at the floor of the rhomboid fossa; they lie at the funnel-shaped entrance of the central canal. The loose tissue of this region contains numerous small cavities. It consists of glia and parenchymal cells, which are generally regarded as neuronal cells. The tissue contains numerous convoluted capillaries that appear highly fenestrated in the electron-microscopic image. The area postrema is therefore one of the few brain regions where the *blood–brain barrier* is permeable (p. 44, C–E).

Paraphysis and Subcommissural Organ (D)

Both structures appear in humans only temporarily during embryonic development (*transitory structures*). The **paraphysis** is a small saclike evagination of the roof of the third ventricle caudally to the interventricular foramina. The **subcommissural organ** (**D4**) consists of a complex of cylindrical ependymal cells underneath the epithalamic commissure. These cells produce a secretion that does not dissolve in the CSF but condenses to form a long, thin filament, Reissner's fiber. In animals in which the central canal is not obliterated, Reissner's fiber extends into the lower spinal cord.

D5 Choroid plexus.
D6 Interventricular foramen (foramen of Monro).

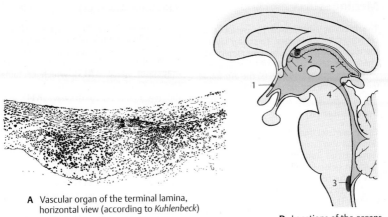

D Locations of the organs

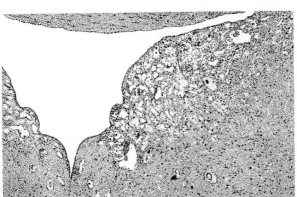

A Vascular organ of the terminal lamina, horizontal view (according to *Kuhlenbeck*)

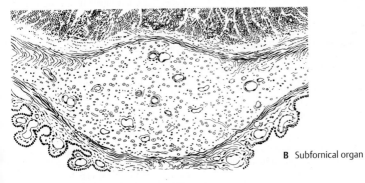

B Subfornical organ

C Area postrema

Meninges

The brain is surrounded by mesodermal coverings, the **meninges**. The outer layer is the tough *pachymeninx*, or *dura mater* (**A1**). The inner layer is the soft *leptomeninx* which consists of two sheets, the *arachnoid mater* (**A2**) and the *pia mater* (**A3**).

Dura Mater (B)

The dura mater lines the inner surface of the skull and also forms the periosteum. Sturdy septa extend from it deep into the cranial cavity. A sickle-shaped fold of the dura, the **falx of the cerebrum** (**B4**), suspends vertically between the two cerebral hemispheres. It is attached rostrally to the crista galli and extends over the frontal crest to the internal occipital protuberance, where it turns into the **tentorium of the cerebellum** (**B5**) spanning both sides. The falx divides the superior part of the cranial cavity in such a way that each hemisphere is supported in its own space. The tentorium of the cerebellum stretches like a tent across the cerebellum, lying in the posterior cranial fossa. It is attached along the transverse sulcus of the occipital bone and the upper margin of the petrous bone, leaving rostrally a wide opening for the passage of the brain stem (**B6**). At the lower surface of the tentorium and along the occipital crest, the falx of the cerebellum projects into the posterior cranial fossa.

The large venous channels, the *sinuses of the dura mater*, are embedded between the two sheets of the dura mater (see vol. 2). The diagram shows cross sections of the *superior sagittal sinus* (**B8**) and the *transverse sinus* (**B9**).

Certain structures are encapsulated by dural pockets and thus separated from the rest of the inner cavity. The *sellar diaphragm* (**B10**) spans the sella turcica and contains an aperture for passage of the hypophysial stalk, the *diaphragmatic hiatus* (**B11**). The trigeminal ganglion (semilunar ganglion, Gasser's ganglion) on the anterior surface of the petrous bone is enclosed by a dural pocket, the *trigeminal cave* (Meckel's space).

Arachnoid Mater (A)

The *arachnoid mater* (**A2**) adjoins closely the inner surface of the dura mater and is separated from it only by a capillary cleft, the **subdural space** (**A12**). It encloses the **subarachnoid space** (**A13**), which contains cerebrospinal fluid and is connected with the pia mater by trabeculae (**A14**) and septae that form a dense meshwork and create a system of communicating chambers.

Peduncled mushroomlike vegetations of the arachnoid protrude into the large sinuses, the **arachnoid granulations** (*meningeal granules*, or *pacchionian bodies*) (**A15**). They consist of an arachnoid meshwork and are covered by mesothelium. The dura mater, which encloses them, is reduced to a membrane. These arachnoid villi are most abundant in the vicinity of the superior sagittal sinus (**A16**) and at the lateral lacunae (**A17**), and less frequent at the exits of spinal nerves. The CSF is absorbed into the venous blood in the area of the granulations. In older people, the granulations may also penetrate into the bone (formation of granular foveolae) (**A18**) and invaginate into the diploic veins.

Pia Mater (A)

The pia mater (**A3**) is the meningeal covering that *contains the blood vessels*. It borders directly on the brain and forms the mesodermal side of the pia–glia barrier. From here vessels enter the brain, and they are surrounded by the pia mater for some distance (*pial funnel*).

A19 Scalp.
A20 Skull.
A21 Diploë (see vol. 1).

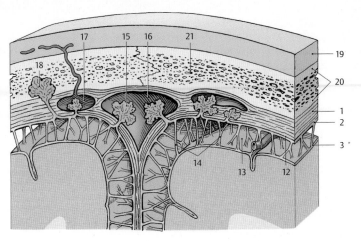

A Meninges and subarachnoid space

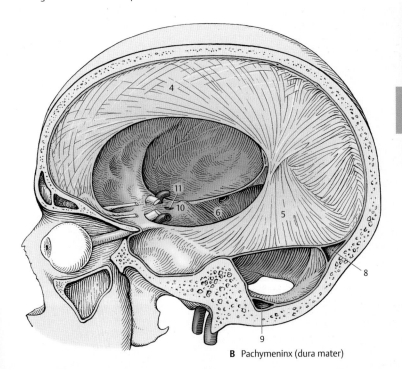

B Pachymeninx (dura mater)

Autonomic Nervous System

Overview

The **vegetative** or **autonomic nervous system** supplies the internal organs and their coverings. Almost all tissues of the body are permeated by a plexus of very delicate nerve fibers. We distinguish between **afferent** (*viscerosensory*) and **efferent** (*visceromotor* and *secretory*) fibers. The neurons with sensory fibers lie in the *spinal ganglia*. The neurons giving rise to efferent fibers form cell clusters scattered throughout the body; these clusters are surrounded by a connective-tissue sheath and are known as *autonomic ganglia*.

The main function of the autonomic nervous system is to *stabilize the internal environment* of the organism and to *regulate the function of the organs* in accordance with the changing requirements of the surroundings. This regulation is achieved by interaction of two antagonistic parts of the autonomic system, the **sympathetic nervous system** (**A1**) (yellow) and the **parasympathetic nervous system** (**A2**) (green). The sympathetic nervous system is stimulated by increased physical activity, resulting in elevated blood pressure, accelerated heart rate and respiratory rate, dilated pupils, raised hair, and increased perspiration. At the same time, the peristaltic activity of the gastrointestinal tract is suppressed and secretion by intestinal glands is reduced. When the parasympathetic system predominates, it increases peristaltic activity and intestinal secretion, stimulates defecation and urination, and reduces the heart rate and respiratory rate, while the pupils constrict. The sympathetic nervous system is responsible for *increased performance* under stress and in states of emergency, while the parasympathetic nervous system promotes metabolism, *regeneration*, and the buildup of body reserves.

The division of the autonomic nervous system into a sympathetic part and a parasympathetic part refers to *visceromotor* and *secretory* fibers. Such a distinction is not possible in case of viscerosensory fibers.

Central Autonomic System

We distinguish between a peripheral and a central autonomic nervous system. The central cell groups of the sympathetic and parasympathetic nervous systems lie in different regions. Parasympathetic neurons form nuclei in the brain stem (see p. 106):

- The *Edinger-Westphal nucleus* (**A3**)
- The *salivatory nuclei* (**A4**)
- The *dorsal nucleus of the vagus nerve* (**A5**)

The sacral spinal cord also contains parasympathetic neurons (**A6**). The sympathetic neurons occupy the lateral horn in the thoracic and upper lumbar segments of the spinal cord (**A7**) (p. 297, A1). Localization of the parasympathetic nuclei is therefore *craniosacral*, while that of the sympathetic nuclei is *thoracolumbar*.

The highest integration organ of the autonomic nervous system is the **hypothalamus**. It also regulates the endocrine glands through its connection to the hypophysis, and it coordinates the autonomic nervous system and the endocrine system. Cell groups in the *reticular formation* of the brain stem also participate in the central regulation of organ functions (heart rate, respiratory rate, blood pressure, p. 146).

A8 Sympathetic trunk.
A9 Superior cervical ganglion.
A10 Stellate ganglion (cervicothoracic ganglion).
A11 Celiac ganglion.
A12 Superior mesenteric ganglion.
A13 Inferior mesenteric ganglion.
A14 Hypogastric plexus.
A15 Greater splanchnic nerve.

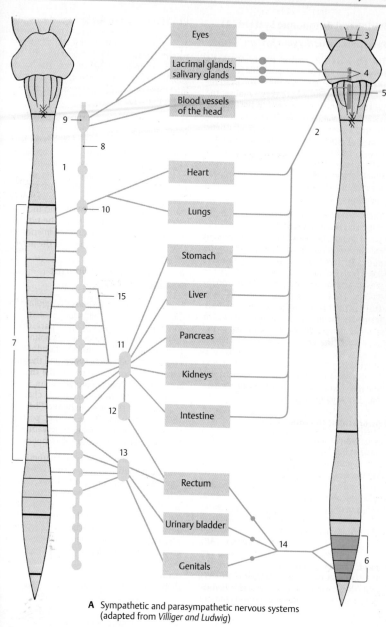

A Sympathetic and parasympathetic nervous systems
(adapted from *Villiger and Ludwig*)

Peripheral Autonomic System (A)

Parasympathetic Nervous System

The fibers of the central parasympathetic neurons run within various cranial nerves to the parasympathetic ganglia in the head region (pp. 128, 130) where they synapse; the postganglionic fibers extend to the effector organs. The **vagus nerve** (**A1**), which is the principal nerve of the parasympathetic nervous system, descends together with the large cerebral vessels (neurovascular trunk of the neck); after passing through the superior thoracic aperture, it divides into plexuses in the regions of the thoracic and abdominal viscera (p. 116).

The cells lying in the intermediolateral nucleus and intermediomedial nucleus of the sacral spinal cord send their axons through the third and fourth sacral root (**A2**) to the pudendal nerve; from here the fibers pass as *pelvic nerves* into the *inferior hypogastric plexus* and to the pelvic organs (urinary bladder [**A3**], rectum, and genitals). Synapses with postganglionic neurons are formed in the inferior hypogastric plexus or in small ganglia of the various organ plexuses. As is the case with the sympathetic nervous system, the peripheral supply is provided by two neurons: the first neuron (preganglionic neuron) in the spinal cord, and the second neuron (postganglionic neuron) in the ganglia (p. 297, B, C).

Sympathetic Nervous System

The sympathetic neurons in the thoracic and lumbar lateral horn send their axons via the *communicating branches* (**A4**) to the **sympathetic trunk** (**A5**) (p. 296, D). The latter consists of a chain of sympathetic ganglia which lie on each side of the vertebral column, in front of the transverse processes of each vertebra, and extend from the base of the skull to the coccyx. They are interconnected by *interganglionic branches* (**A6**). There are three ganglia in the **cervical segment**, namely, the *superior cervical ganglion*, the variable *middle cervical ganglion* (**A7**), and the *stellate ganglion (cervicothoracic ganglion)* (**A8**). The **thoracic segment** contains

10 – 11 ganglia, the **lumbar segment** usually four, and the **sacral segment** also four ganglia. The chain is completed by the small *unpaired ganglion* (**A9**) which lies in the middle in front of the coccyx. The sacral ganglia receive their preganglionic fibers via interganglionic branches from spinal cord levels T12 – L2.

From the thoracic and lumbar sympathetic trunk ganglia, nerves extend to ganglia that lie within dense nervous plexuses on both sides of the abdominal aorta. The upper group of ganglia are the *celiac ganglia* (**A10**) to which the *greater splanchnic nerve* (**A11**) extends from the fifth to the ninth sympathetic trunk ganglia. Below it lies the *superior mesenteric ganglion* (**A12**) and the *inferior mesenteric ganglion* (**A13**). The *superior hypogastric plexus* (**A14**) and the *inferior hypogastric plexus* (**A15**) expand in the pelvis.

Adrenergic and Cholinergic Systems

The transmission of impulses is mediated in the sympathetic nervous system by norepinephrine, and in the parasympathetic nervous system by acetylcholine. The *sympathetic nervous system* is therefore also known as the **adrenergic system** and the *parasympathetic nervous system* as the **cholinergic system**. All preganglionic fibers of the sympathetic nervous system are cholinergic, and only the postganglionic fibers are noradrenergic (p. 297, C). The postganglionic sympathetic fibers innervating the sweat glands of the skin are also cholinergic.

The antagonism between sympathetic and parasympathetic nervous systems is clearly apparent for some organs (heart, lungs). Other organs are regulated by the increased or decreased tone of just one system. Thus, the adrenal glands and the uterus are supplied only by sympathetic fibers (the adrenal gland as paraganglion is supplied by preganglionic fibers, see vol. 2). The function of the urinary bladder is regulated by parasympathetic fibers; the role of sympathetic fibers in this control is disputed.

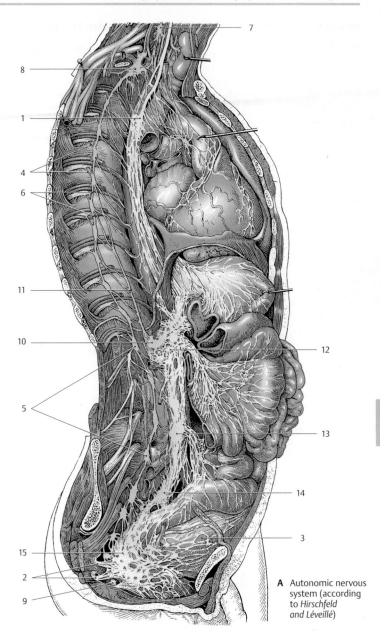

A Autonomic nervous system (according to *Hirschfeld* and *Léveillé*)

Neuronal Circuit (A–C)

The sympathetic neurons in the **intermediomedial nucleus** and in the **intermediolateral nucleus** (lateral horn) (**A1**) of the thoracic spinal cord send their axons through the anterior root (**A2**) into the spinal nerve. They run through the **white communicating branch** (**A3**) to reach the **sympathetic trunk ganglion** (**A4**) as preganglionic fibers. Here, some of them terminate at neurons from where postganglionic fibers return into the spinal nerve via the **gray communicating branch** (**A5**). The preganglionic fibers entering the ganglion are myelinated so that the connecting branch appears white (white communicating branch). The postganglionic fibers exiting the ganglion are unmyelinated so that the connecting branch appears gray (gray communicating branch).

Other postganglionic fibers (**A6**) extend from the sympathetic trunk via autonomic nerves to the organs. Some preganglionic fibers (**A7**) pass through the ganglion without synapsing and terminate in the **prevertebral ganglia** (**A8**) which lie on both sides of the aorta. Numerous small and very small **terminal ganglia** (**A9**) lie in the regions of internal organs. They are part of the neuroplexuses spreading through each organ; they are found in the sheaths (*extramural ganglia*) as well as inside the organs (*intramural ganglia*). Whereas preganglionic and postganglionic fibers of the parasympathetic nervous system are both cholinergic (**B**), the cholinergic preganglionic fibers of the sympathetic nervous system synapse in the ganglia with noradrenergic neurons (**C**).

According to their localization, we distinguish three different types of ganglia in which preganglionic fibers synapse with postganglionic neurons:

- The sympathetic trunk ganglia
- The prevertebral ganglia
- The terminal ganglia

The sympathetic trunk ganglia and the prevertebral ganglia are sympathetic ganglia, whereas the terminal ganglia are mostly, but not exclusively, parasympathetic ganglia.

Sympathetic Trunk

Cervical and Upper Thoracic Segments (D)

The cervical ganglia are reduced to three; the uppermost, the **superior cervical ganglion** (**D10**), lies below the base of the skull near the nodose ganglion (inferior ganglion of the vagus nerve). It receives fibers from the upper thoracic segment via the interganglionic branches. Its postganglionic fibers form plexuses around the internal carotid artery and external carotid artery. Branches extend from the *internal carotid plexus* to the meninges, to the eyes, and to the glands of the head region. The superior tarsal muscle of the upper eyelid and the ophthalmic muscles at the posterior wall of the orbit are innervated by sympathetic fibers. Injury to the superior cervical ganglion therefore leads to drooping of the upper eyelid (*ptosis*) and to a backward displacement of the eyeball (*enophthalmos*).

The *middle cervical ganglion* (**D11**) may be absent, and the *inferior cervical ganglion* has in most cases fused with the first thoracic ganglion to form the **stellate ganglion** (**D12**). Its postganglionic fibers form plexuses around the subclavian artery and around the vertebral artery. Fiber bundles connecting the stellate ganglion with the middle cervical ganglion extend across the subclavian artery and form the *subclavian ansa* (**D13**). Nerves from the cervical ganglia (**D14**) and nerves from the upper thoracic ganglia (**D15**) extend to the heart and to the hila of the lungs, where they participate together with the parasympathetic fibers of the vagus nerve in the formation of the *cardiac plexus* (**D16**) and the *pulmonary plexus* (**D17**). The branches of the fifth to ninth sympathetic trunk ganglia join to form the *greater splanchnic nerve* (**D18**) which extends to the celiac ganglia.

Autonomic Nervous System

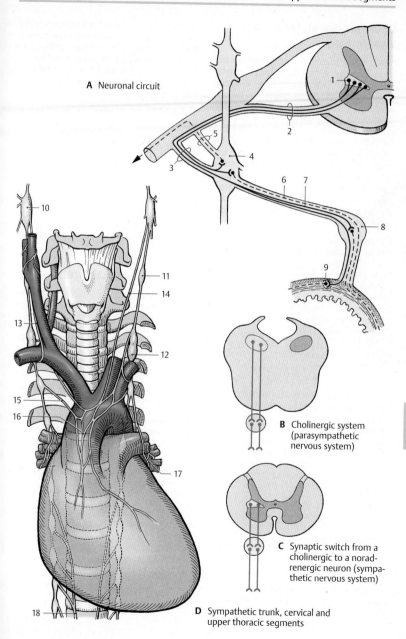

A Neuronal circuit

B Cholinergic system (parasympathetic nervous system)

C Synaptic switch from a cholinergic to a noradrenergic neuron (sympathetic nervous system)

D Sympathetic trunk, cervical and upper thoracic segments

Autonomic Nervous System

Lower Thoracic and Abdominal Segments (A)

Branches originating from the thoracic and upper lumbar sympathetic trunk ganglia extend to the prevertebral ganglia of the *abdominal aortic plexus*. There are several groups of ganglia. At the exit of the celiac trunk lie the **celiac ganglia** (**A1**) where the *greater splanchnic nerve* (**A2**) (T5–T9) and the *lesser splanchnic nerve* (**A3**) (T9–T11) terminate. Their postganglionic fibers extend with the branches of the aorta to stomach, duodenum, liver, pancreas, spleen, and adrenal gland (gastric plexuses, hepatic plexus, splenic plexus, pancreatic plexus, suprarenal plexus). Preganglionic fibers run to the adrenal medulla (see vol. 2).

The postganglionic fibers from the **superior mesenteric ganglion** (**A4**), together with the branches of the celiac ganglion, supply the small intestine, the ascending colon, and the transverse colon. The fibers from the **inferior mesenteric ganglion** (**A5**) supply the descending colon, the sigmoid colon, and the rectum. The preganglionic fibers (lumbar splanchnic nerves) of both ganglia originate from levels T11–L2. Some branches extend to the renal plexus, which also contains fibers from the celiac ganglia and the superior hypogastric plexus.

Parasympathetic fibers also participate in the formation of the visceral plexuses. Stimulation of parasympathetic fibers in the digestive tract leads to increased peristalsis and secretion as well as relaxation of the sphincter muscles, while stimulation of sympathetic fibers causes reduced peristalsic and secretion as well as contraction of the sphincter muscles.

The pelvic organs are supplied by the *superior hypogastric plexus* (**A6**) and the *inferior hypogastric plexus*. Both plexuses receive preganglionic sympathetic fibers from the lower thoracic and upper lumbar spinal cord, and parasympathetic fibers from the sacral spinal cord.

The **urinary bladder** is predominantly innervated by the parasympathetic fibers of the *visceral plexus* that supply the muscles for bladder contraction (*detrusor muscle*). The sympathetic fibers terminate at the smooth muscles of the orifice of the urethra and both ureteral orifices. Regulation of bladder tone and urination takes place via spinal reflexes which, in turn, are controlled by the hypothalamus and by cortical areas.

The **genitals** are supplied by the *prostatic plexus* in the male and by the *uterovaginal plexus* in the female. Stimulation of parasympathetic fibers dilates the vessels of the cavernous bodies and thus triggers erection in the male (*nervi erigentes*). Stimulation of sympathetic fibers leads to vasoconstriction and to ejaculation. The uterine muscles are innervated by sympathetic and parasympathetic fibers. Their functional significance is not clear because even a denervated uterus is fully functional during pregnancy and parturition.

Innervation of the Skin (B)

The sympathetic fibers returning from the sympathetic trunk ganglia into the spinal nerves (p. 297, A5) run in the peripheral nerves to the skin, where they innervate vessels, sweat glands, and erector pili muscles (vasomotor, sudomotor, pilomotor functions) in the corresponding dermatomes (p. 66). Segmental loss of these functions is of diagnostic significance in injuries to the spinal cord.

Clinical Note: In specific skin regions, **Head's zones** (*zones of hyperalgesia*) (**B**), disease of an organ may give rise to pain or hypersensitivity, with each organ being represented in a specific zone: diaphragm (**B7**) (C4), heart (**B8**) (T3/4), esophagus (**B9**) (T4/5), stomach (**B10**) (T8), liver and gallbladder (**B11**) (T8 – 11), small intestine (**B12**) (T10), large intestine (**B13**) (T11), urinary bladder (**B14**) (T11 – L1), kidneys and testes (**B15**) (T10 – L1). The Head's zones are of practical importance in the diagnosis.

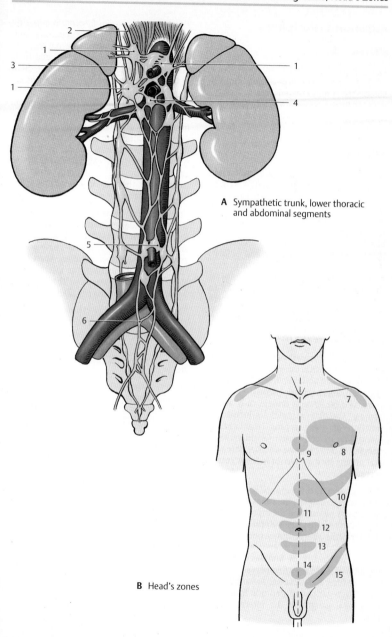

A Sympathetic trunk, lower thoracic and abdominal segments

B Head's zones

Autonomic Periphery

Efferent Fibers

The preganglionic fibers are myelinated, while the postganglionic fibers are unmyelinated. The unmyelinated fibers are surrounded by the cytoplasm of Schwann cells, and a single Schwann cell envelopes several axons (p. 38, A8).

Afferent Fibers

The viscerosensory fibers are myelinated. They are regarded as being neither sympathetic nor parasympathetic in nature. In general, they accompany the sympathetic nerves and enter the spinal cord through the posterior roots. The fibers from the heart run through the upper thoracic roots, those from the stomach, liver, and gallbladder run through the middle thoracic roots, and those from the colon and appendix run through the lower thoracic roots. The respective dermatomes of these roots correspond roughly to the various Head's zones.

Intramural Plexus (A–E)

The autonomic nerves enter the internal organs together with the vessels and form a fine network of noradrenergic (**A**) or cholinergic (**B**) fibers (**enteric plexus**). The fibers terminate on smooth-muscle cells and on glands. Vascular muscles influence the function of many organs (regulation of blood flow by contraction or dilatation of vessels). It is not known whether parenchymal organs such as liver or kidneys contain secretory fibers.

The digestive tract is supplied by two different plexuses, namely, the *submucous plexus* (*Meissner's plexus*) and the *myenteric plexus* (*Auerbach's plexus*). The **submucous plexus** (**C1**, **D**) forms a three-dimensional network throughout the entire submucosa. It is an irregular meshwork of bundles containing medium-sized to very fine nerve fibers that become finer and narrower toward the mucosa. Aggregations of neurons form small intramural ganglia at the intersection of fibers.

The **intramural ganglia** (**E**) contain primarily multipolar, rarely unipolar, neurons with granular Nissl substance. The neurons are surrounded by flat sheath cells, and their numerous long dendrites are thin and often indistinguishable from axons. The axons are extremely thin, unmyelinated or poorly myelinated, and often arise from a dendrite rather than the perikaryon. Between the neurons lies a dense network of fibers in which it is difficult to differentiate between dendrites, terminating axons, and fibers passing through the ganglion. Sympathetic and parasympathetic neurons have the same shape; they can only be distinguished histochemically.

The **myenteric plexus** (**C2**) is embedded in a narrow space between the transverse and longitudinal muscles of the intestine. It consists of a relatively regular network of coarse and fine fiber bundles. Apart from the intramural ganglia at the intersections, there are numerous neurons that are often arranged in rows along the fiber bundles.

The enormous number of neurons scattered throughout the tissue almost amounts to an independent nervous organ. This explains the local autonomy of the gastrointestinal tract, which remains functional even after denervation.

As derivatives of the neural crest (p. 62, C2), the *paraganglia* and the *adrenal medulla* are regarded as part of the autonomic nervous system (see vol. 2).

A, B Vas deferens in the guinea pig (after J. Winkler)

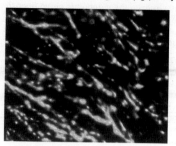

A Sympathetic fibers, fluorescence microscopy

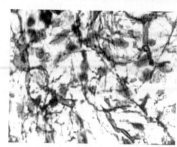

B Parasympathetic fibers, acetylcholine esterase histochemistry

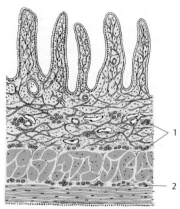

C Intestinal wall, diagram

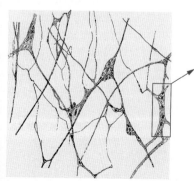

D Submucous plexus

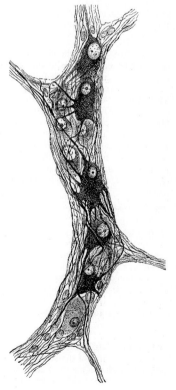

E Intramural ganglion

Autonomic Neurons

Structure (A, B, D)

The autonomic nervous system consists of numerous individual elements, the autonomic neurons (**A**) (*neuron doctrine*). The *theory of neuronal continuity*, which had long been postulated for the terminal ramifications of the autonomic nervous system, has been disproved by electron-microscopic studies. According to the continuity theory, the terminal ramifications of an intramural plexus were thought to form a continuous network (*terminal reticulum*), in which the processes of different neurons would fuse with each other and with the innervated muscle cells and glandular cells. The network was thought to represent a syncytium where nerve fibers share a common cytoplasm. The electron-microscopic image does not show such continuity.

However, the postganglionic neurons do have specific features. The bundles of nerve fibers exhibit numerous *axo-axonal synapses* (**D1**), not only among sympathetic and parasympathetic fibers themselves but also between sympathetic and parasympathetic fibers. In the area of terminal ramification, specific structures corresponding to the motor end plates (p. 313, B and C) at striated muscles are absent. The only remarkable features are *varicose swellings* (varicosities) (**A–D2**) along the terminal branches of the axons.

The axon swellings may lead to indentations on the smooth-muscle cells or may even invaginate the cells. In general, however, they lie between the muscle cells without direct membrane contact (as they would in a synapse, see p. 27, D). The swellings contain clear and granular vesicles (**C3**) similar to those in presynaptic terminal boutons (p. 27, C). The granular vesicles have been shown to contain norepinephrine, the neurotransmitter of the sympathetic nervous system. The *sheath of Schwann cells* (**B4**) surrounding the terminal branches is absent in the area of the swelling, and the adjacent wall segment of the smooth-muscle cell (**BC5**) lacks a basement membrane (**B6**). This is the site of signal transmission from the autonomic nerve fiber to the smooth-muscle cell.

Signal Transmission (C)

The vesicles contained in the axon swellings release their content into the intercellular space (**C7**). The neurotransmitter molecules diffuse into the intercellular space, thereby transmitting the signal to a large number of smooth-muscle cells. Signal transmission also spreads via membrane contacts between the muscle cells. The smooth-muscle cells are interconnected by *gap junctions* (**B8**) which function as electrical synapses (see p. 26). There is no basement membrane at gap junctions.

Efferent autonomic nerve fibers innervate the smooth muscles and glandular cells (secretory fibers). Innervation of glandular cells (blue cells in **C**) is essentially the same as that of muscle cells. The neurotransmitters released from the vesicles in the axon swellings (green dots in **C**) activate G protein-coupled receptors (**C9**) on the surface of the innervated cell. G protein mediates the opening of ion channels (**C10**), thus triggering an intracellular cascade of signals. This results in the contraction of smooth-muscle cells or in the synthesis and release of glandular secretion by the glandular cells.

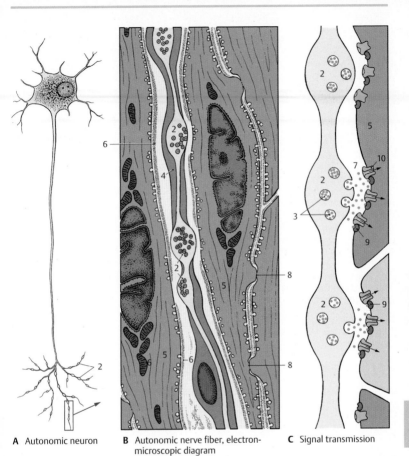

A Autonomic neuron

B Autonomic nerve fiber, electron-microscopic diagram

C Signal transmission

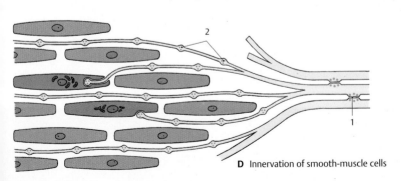

D Innervation of smooth-muscle cells

Functional Systems

Brain Function

The central nervous system enables the organism to adapt to the environment and to survive. It receives stimuli from outside and inside the body through the sensory organs; it then filters the stimuli and processes them into information. In accordance with this information, it sends impulses to the periphery of the body so that the organism can react in a meaningful way to the constantly changing conditions (p. 2, EF). The functional systems described in this chapter are by no means independent, isolated systems. The highly simplified and schematic description is supposed to provide an approximate illustration of the extremely complex interactions among billions of nerve cells.

For centuries there has been the simple mechanistic idea that sensations reach the brain and the brain then triggers motor reactions. According to **Descartes**, optic stimuli are transmitted from the eyes to the pineal gland (epiphysis), which then sends impulses that travel to the muscles (**A**). He viewed the pineal gland as the seat of the soul. **Franz Gall** was the first to postulate the importance of the cerebral convolutions and the cerebral cortex for brain function. He localized the "organs of the soul" on the surface of the hemispheres, and he believed that mental faculties could be determined to various degrees by the topographical features on the surface of the skull (*phrenology*) (**B**). He used Roman numerals for localizing these features: I, reproductive instinct; II, love of offspring; III, friendship; IV, bravery; V, instinct to eat meat; VI, intelligence; VII, greed and cleptomania; VIII, pride and arrogance; IX, vanity and ambition; and so on.

On the basis of brain injuries, **Kleist** arrived at localizations of higher cerebral functions (**C**). He assumed that positive capacities, or "functions," would correspond to deficits in recognition and thinking, motivation and action, among others, which were expressed as negative pathological findings. However, this is not so. While it is possible to localize symptoms of deficit, this cannot be done for capacities (von Monakow). Critics of this *theory of localization and centers* spoke of "brain mythology." Ultimately, specific capacities cannot be attributed to specific brain regions because immense numbers of other neuronal groups are always participating in the stimulation, inhibition, or modulation of a capacity. The so-called "centers" can be regarded, at best, as important *relay stations* for a particular capacity. Nor is the central nervous system a rigid apparatus; rather it exhibits a considerable degree of **plasticity**. Especially in the infant brain, other centers can take over and perform functions in place of injured parts of the brain. Plasticity of our principal organ is also a prerequisite for the capacity to learn (language, writing, physical skills).

Information processing in the telencephalon is known as **integration**. It refers to the combination and interconnection of sensations, including stored experience, to form a higher and complex functional unit. In this way, the functions of the organism are guided by means of the meticulous mutual coordination between groups of neurons. During human evolution, the integrative processes of regulating and coordinating elementary biological tasks have developed into conscious recognition, thought, and action. Cybernetics and computer technology have provided us with models for the function of the brain (cognitive science). According to these models, various "functions" are based on the continuously changing stimuli within interconnected neuronal circuits.

Despite our highly developed instrument technology, there is ultimately only one instrument available for brain research: the brain itself. In other words, a human organ is involved in an attempt to investigate its own structure and function.

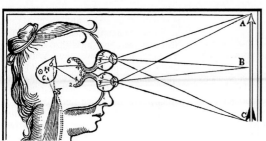

A Transmission of visual stimuli from the eyes to the pineal gland (according to *René Descartes 1662*)

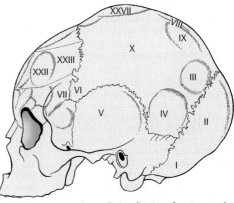

B Localization of centers on the skull (according to *Gall 1810*)

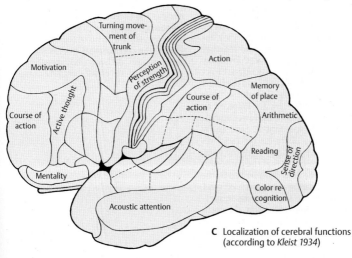

C Localization of cerebral functions (according to *Kleist 1934*)

Motor Systems

Corticospinal Tract (A, B)

The **corticospinal tract** (pyramidal tract) and the *corticonuclear fibers* (p. 58, A; p. 140, A) are regarded as **pathways of voluntary movements**. It is through them that the cortex controls the subcortical motor centers. The cortex can have a reducing and inhibiting effect, but it also produces a continuous tonic stimulus that promotes quick, sudden movements. The mechanical and stereotyped motions controlled by the subcortical motor centers must be modified by the influence of pyramidal impulses in order to produce specific, fine-tuned movements.

The fibers of the corticospinal tract originate in the precentral areas 4 (**A1**) and 6 (**A2**), in regions of the parietal lobe (areas 3, 1, and 2), and in the second sensorimotor area (area 40) (see pp. 246, 248, 250). Roughly two-thirds stem from the precentral area, and one-third from the parietal lobe. Only about 60% of the fibers are myelinated; the other 40% are unmyelinated. The thick fibers of Betz's giant pyramidal cells (p. 246, A) in area 4 account for only 2 – 3% of the myelinated fibers. All other fibers stem from smaller pyramidal cells.

The fibers of the pyramidal tract pass through the internal capsule (p. 258, A2, B). At the transition to the midbrain, they approach the base of the brain and together with the corticopontine tracts form the cerebral peduncles. The corticospinal tract fibers occupy the central part, and the fibers from the parietal cortex occupy the most lateral position (**B3**). They are followed by the corticospinal tracts for the lower limb (L, S), trunk (T), and upper limb (C), and finally the corticonuclear fibers for the facial region (**B4**). When passing through the pons, the fiber tracts rotate so that the corticonuclear fibers now lie dorsally, followed by the bundles terminating in the cervical, thoracic, lumbar, and sacral regions, respectively. In the medulla oblongata, the corticonuclear fibers terminate on the cranial nerve nuclei (p. 140, A). Most of the fibers

(70 – 90%) cross to the opposite side in the pyramidal decussation (**AB5**) (p. 59, A1) and form the **lateral corticospinal tract** (**AB6**). The fibers for the upper limb cross to the fibers for the lower limb. Within the lateral corticospinal tract, the fibers for the upper limb lie medially, while the long fibers for the lower limb lie laterally (p. 58A). The uncrossed fibers continue in the **anterior corticospinal tract** (**AB7**) and cross to the opposite side only at the level of their termination above the white commissure (p. 50, A14). The anterior tract shows variable degrees of development; it may be asymmetric or even completely absent. It reaches only into the cervical or thoracic spinal cord.

The majority of corticospinal tract fibers terminate on interneurons in the intermediate zone between anterior and posterior horns (p. 52, A6). Only a small portion reaches the motor neurons of the anterior horn, predominantly those supplying the distal segments of the limbs, which are under the special control of the corticospinal tract. Impulses from the corticospinal tract activate neurons that innervate the *flexor muscles* but inhibit neurons that innervate the *extensor muscles*.

Fibers originating from the parietal lobe terminate in the dorsal column nuclei (*gracile nucleus* and *cuneate nucleus*) and in the *substantia gelatinosa* of the posterior horn. They regulate the input of sensory impulses. Hence, the corticospinal tract is not a uniform motor pathway but contains descending systems of different functions.

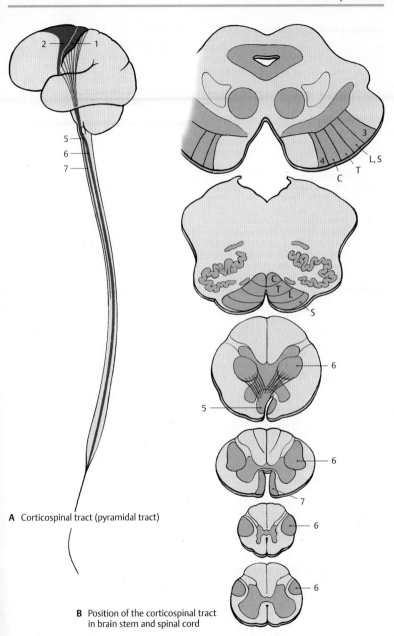

A Corticospinal tract (pyramidal tract)

B Position of the corticospinal tract
in brain stem and spinal cord

Extrapyramidal Motor System (A)

Apart from the precentral area and corticospinal (pyramidal) tract, numerous other cortical areas and pathways control motor activity. They are collectively known as the **extrapyramidal motor system**. It is phylogenetically older than the corticospinal tract and, unlike it, consists of chains of multisynaptic neurons. Originally, the term referred to a group of nuclei characterized by its high iron content: *neostriatum* (putamen [**A1**] and caudate nucleus [**A2**]), *pallidum* (**A3**), *subthalamic nucleus* (**A4**), *red nucleus* (**A5**), and *substantia nigra* (**A6**) (**extrapyramidal motor system** in **the narrower sense**). This group of nuclei is connected to other centers that are important for motor activity; however, these are integration centers rather than motor nuclei: *cerebellum* (**A7**), *thalamic nuclei* (**A8**), *reticular formation*, *vestibular nuclei* (**A9**), and some cortical areas. They are collectively known as the **extrapyramidal system in the broader sense**.

Function

When one limb is moved voluntarily, muscle groups of the other limbs and of the trunk are simultaneously activated so that balance and posture are maintained under the changed static conditions and the movement can be completed smoothly. The accompanying muscle activities, which are often nothing more than the increased tension or relaxation of certain muscle groups, are carried out involuntarily and are not experienced consciously. Without them, however, coordinated movement would be impossible. Such **unconscious motions** include associated movements (synkinesis) (such as arm movements while walking) as well as many movements that have been practiced for a long time and thus occur automatically. They are all under the control of the extrapyramidal system; this may be compared to a servomechanism that supports all voluntary movements in an autonomic way and without reaching the level of consciousness.

Pathways

Afferent pathways reach the system through the cerebellum. Cerebellar tracts terminate in the red nucleus (*cerebellorubral tract*) (**A10**) and in the centromedian nucleus of thalamus (**A11**) from which fibers extend to the striatum. *Cortical fibers* run to the striatum (**A12**), to the red nucleus (**A13**), and to the substantia nigra (**A14**). *Vestibular fibers* terminate in the interstitial nucleus of Cajal (**A15**).

The **efferent pathway** of the system is the *central tegmental tract* (**A16**) (p. 144, A). Other descending pathways are:

- The *reticulospinal tract* (**A17**)
- The *rubrospinal tract*
- The *vestibulospinal tract* (**A18**)
- The *interstitiospinal tract* (**A19**)

The extrapyramidal centers are interconnected by numerous **neuronal circuits** to ensure mutual control and adjustment. There are two-way connections between pallidum and subthalamic nucleus, and between striatum and substantia nigra (**A20**). A large neuronal circuit runs from the cerebellum through the centromedian nucleus of thalamus to the striatum and from there via pallidum, red nucleus, and olive (**A21**) back to the cerebellum. Other functional circuits are formed by cortical fibers to the striatum, with a recurrent circuit passing through pallidum, ventral anterior nucleus and ventral lateral nucleus of thalamus, and back to the cortex.

The frontal and occipital visual fields (p. 248, C; p. 254) together with the regions of parietal and temporal lobes, from where complex massive movements can be elicited by a strong electrical current, are known as *extrapyramidal cortical fields*. However, the inclusion of cortical fields in the extrapyramidal system is controversial, even though numerous corticostriatal connections have been demonstrated (**A12**).

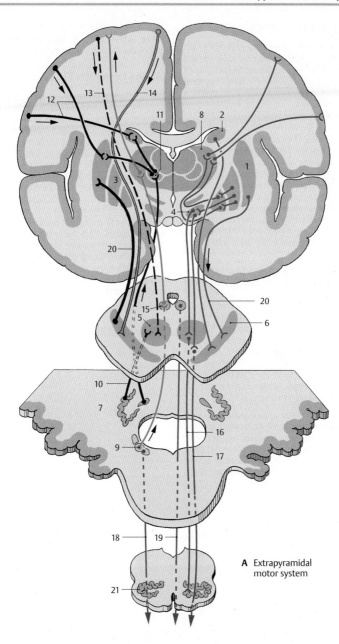

A Extrapyramidal motor system

Motor End Plate (A – C)

The axons of the motor neurons (**A1**) arborize in the muscles so that each muscle fiber (**AB2**) is reached by an axonal branch (**A3**). The number of muscle fibers supplied by one axon varies considerably. While a single axon may innervate two to three muscle fibers in the muscles of eyes and fingers, it may supply 50 – 60 muscle fibers in other muscles. The anterior horn cell and its axon (α-motoneuron) together with the group of muscle fibers it supplies is called a **motor unit**. When the neuron is stimulated, the muscle fibers contract in unison. The terminal branches of the axon lose their myelin sheaths before terminating and form tangled ramifications. In the terminal region, the surface of the muscle fiber forms a flat eminence (hence the term *end plate*) (**A4**).

The area of axonal arborization (**A5**) contains a number of cell nuclei. The nuclei lying on top of the axonal ramifications belong to Schwann cells that envelop the axon terminals (*teloglia*) (**B6**). The nuclei lying beneath the ramifications (**B7**) are muscle fiber nuclei in the region of the end plate. At the junction between axoplasm and sarcoplasm, the axon terminals are surrounded by a palisade layer (**B8**) which consists of infoldings of the sarcolemma, as shown by electron microscopy.

The axons terminate with boutonlike swellings (**B9**) that dip into the surface of the end plate. These grooves are lined by the membrane of the sarcoplasm (sarcolemma) and a basement membrane. The heavily folded sarcolemma of the grooves (**subneural clefts**) (**C10**) greatly enlarges the surface area of the muscle fiber.

The motor end plate is a specialized synapse. Its presynaptic membrane is the axolemma (**C11**), and its postsynaptic membrane is the folded sarcolemma (**C12**). The substance transmitting nerve impulses to the muscle fiber is **acetylcholine**. It is contained in clear synaptic vesicles (**BC13**). Upon stimulation of the axon, the neurotransmitter is released into the synaptic cleft, resulting in receptor-mediated (nicotinic acetylcholine receptors) depolarization of the membrane of the muscle fiber.

Tendon Organ (D – F)

The **Golgi tendon organ** lies at the junction between tendon and muscle. It consists of a group of collagen fibers (**D14**), which is surrounded by a thin sheath of connective tissue and is innervated by a myelinated nerve fiber (**D15**). The nerve fiber loses its myelin sheath after passing through the connective tissue capsule and divides into a number of branches that coil around the collagen fibers. The loosely arranged collagen fibers (**E**) are thought to tighten upon tension (**F**) and thus exert pressure onto the nerve endings. The resulting impulse is transmitted by the nerve fiber via the posterior root to the spinal cord. Here it has an inhibitory effect on the motor neurons, thus preventing the muscle from excessive stretching or contraction.

B16 Basement membrane of the axon terminal.

B17 Basement membrane of the muscle fiber.

C18 Synaptic cleft with common basement membrane of axon terminal and muscle fiber.

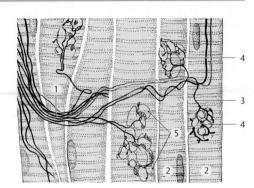

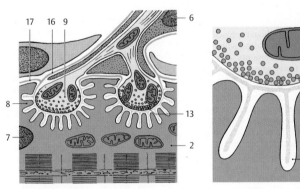

A Motor end plates, overview

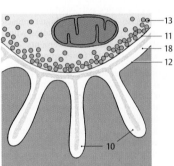

B End plate, electron-microscopic diagram (according to *Couteaux*)

C Magnified detail of B (according to *Robertson*)

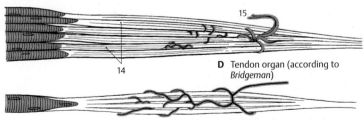

D Tendon organ (according to *Bridgeman*)

E Tendon organ with muscle relaxed (according to *Bridgeman*)

F Tendon organ with muscle contracted (according to *Bridgeman*)

Functional Systems

Functional Systems *(sidebar)*

Muscle Spindle (A – F)

A muscle spindle, or neuromuscular spindle, consists of 5 – 10 thin striated muscle fibers (**intrafusal muscle fibers**) (**A1**) which are surrounded by a fluid-filled connective-tissue capsule (**A2**). The fibers of the up to 10 mm long spindles are arranged in parallel with the other fibers of the muscle (*extrafusal muscle fibers*) and attach either to the tendons of the muscle or to the connective-tissue poles of the capsule. As the intrafusal fibers lie in the same longitudinal orientation as the extrafusal fibers, stretching and shortening of the muscle affects them in the same way. The number of spindles within individual muscles is quite variable. Muscles participating in delicate and precise movements (finger muscles) possess a large number of spindles, whereas muscles for simple movements (trunk muscles) contain far fewer spindles.

The central equatorial part (**A3**) of an intrafusal fiber contains several cell nuclei but no myofibrils; this part of the spindle is not contractile. Only the two segments (**A4**) that contain striated myofibrils are contractile. A thick sensory nerve fiber (**A5**) terminates at the central part; its terminal branches wind around the muscle fibers like spirals and form the **annulospiral endings** (**AC6**; **B**). A delicate sensory fiber (**A7**) attaches in an umbelliform fashion (**flower-spray ending**) (**A8**, **D**) at one side, or at both sides, of the annulospiral ending.

Both contractile polar segments are innervated by thin fusimotor fibers (γ-**fibers**) (**A9**). Their small motor end plates have only poorly developed subneural clefts; like the extrafusal muscle fibers, they are *epilemmal*. The sensory annulospiral endings lie below the basement membrane of the muscle fiber (**C10**) and, hence, are *hypolemmal*. The γ-fibers stem from small motor neurons in the anterior horn (γ-motoneurons); impulses from these neurons cause contraction of the polar segments of the intrafusal fiber. This results in stretching of the equatorial segment and does not only stimulate the annulospiral ending but alters the sensitivity of the spindle as well.

The muscle spindle is a **stretch receptor**, which is stimulated when the muscle is stretched but becomes inactive when the muscle is contracted. Upon stretching the muscle the impulse frequency increases with the change in muscle length. This way, the spindles transmit information on the prevailing length of the muscle. The impulses are transmitted not only via the spinocerebellar tracts to the cerebellum but also via reflex collaterals directly to the large anterior horn cells (α-motoneurons). Stimulation of the latter neurons during sudden stretching results in immediate muscle contraction (*stretch reflex*, p. 50, F).

The muscle spindle contains two different types of intrafusal fibers: the **nuclear chain fibers** (**EF11**) and the **nuclear bag fibers** (**EF12**). Both types of fiber are innervated by annulospiral endings. Flower-spray endings are predominantly found at nuclear chain fibers. The thicker nuclear bag fibers respond to the ongoing stretching of the muscle, whereas the continuous state of muscle stretching is registered by the thinner nuclear chain fibers. Thus, muscle spindles transmit to the cerebellum not only information on the length of the muscle but also on the speed of muscle stretching.

Apart from tendon organs and muscle spindles, there are sensory end organs in joint capsules and ligaments (**tension receptors**) that constantly send information on movement and posture of trunk and limbs to the cerebellum (anterior and posterior spinocerebellar tracts).

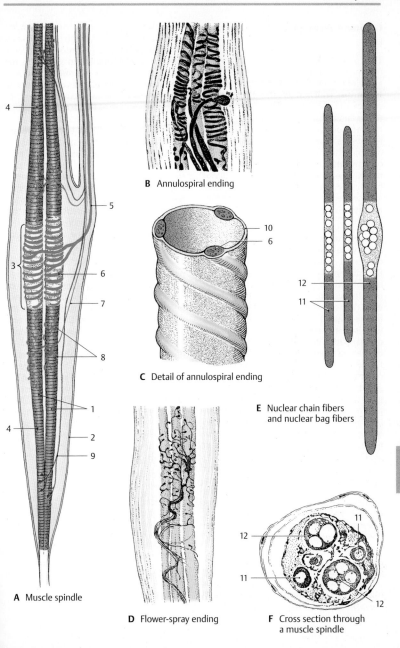

B Annulospiral ending

C Detail of annulospiral ending

D Flower-spray ending

E Nuclear chain fibers and nuclear bag fibers

A Muscle spindle

F Cross section through a muscle spindle

Functional Systems

Common Terminal Motor Pathway (A)

The common terminal pathway of all centers involved in motor activity is the large **anterior horn cell** (**A1**) and its axon (α-**motoneuron**), which innervates the voluntary skeletal muscles. Most of the tracts running to the anterior horn do not end directly on the anterior horn cells but terminate on interneurons. The latter influence the motor neurons either directly or act by inhibiting or activating the reflexes between muscle receptors and motor neurons. The anterior horn is therefore not simply a relay station as described earlier (p. 50) but a complex integration apparatus that regulates motor activity.

The **central regions** that influence motor activity via descending pathways are interconnected in many ways. The most important afferent pathways stem from the cerebellum, which receives the impulses of muscle receptors via the **spinocerebellar tracts** (**A2**) and the stimuli of the cortex via the **corticopontine tracts** (**A3**). The cerebellar impulses are transmitted via the parvocellular part of the *dentate nucleus* (**A4**) and the *ventral lateral nucleus of thalamus* (**A5**) to the *precentral cortex* (area 4) (**A6**). The *corticospinal (pyramidal) tract* (**A7**) descends from area 4 to the anterior horn and gives off collaterals in the *pons* (**A8**) that return to the cerebellum. Additional cerebellar impulses are transmitted via the *emboliform nucleus* (**A9**) and the *centromedian nucleus of the thalamus* (**A10**) to the *striatum* (**A11**) and via the magnocellular part of the *dentate nucleus* (**A12**) to the *red nucleus* (**A13**). From here fibers run in the *central tegmental tract* (**A14**) via the *olive* (**A15**) back to the cerebellum and in the *rubroreticulospinal tract* (**A16**) to the anterior horn. Fibers from the *globose nucleus* (**A17**) run to the *interstitial nucleus of Cajal* (**A18**) and from there in the *interstitiospinal fasciculus* (**A19**) to the anterior horn. Finally, **cerebellofugal fibers** are relayed in the *vestibular nuclei* (**A20**) and in the *reticular formation* (**A21**) to the *vestibulospinal tract* (**A22**) and the *reticulospinal tract* (**A23**), respectively.

The descending pathways can be divided into two groups according to their effect on the muscles: one group stimulates the *flexor muscles*, and another group stimulates the *extensor muscles*. The **corticospinal tract** and the **rubroreticulospinal tract** activate mainly the neurons of the flexor muscles and inhibit the neurons of the extensor muscles. This corresponds to the functional importance of the corticospinal tract for delicate and precise movements, especially those of hand and finger muscles where flexor muscles play an important role. In contrast, the fibers of the **vestibulospinal tract** and the fibers from the **pontine reticular formation** inhibit the flexors and activate the extensors. They belong to a phylogenetically old motor system that is directed against the effect of gravity and, thus, is of special importance for body posture and balance.

The **peripheral fibers** that run through the posterior root into the anterior horn originate from the muscle receptors. The afferent fibers of the annulospiral endings (**A24**) terminate with their collaterals directly on the α-motoneurons, while the fibers of the tendon organs (**A25**) terminate on interneurons. Many descending pathways influence the α-neurons via the spinal reflex apparatus. They terminate on the large α-neurons and on the small γ-neurons (**A26**). Since the γ-neurons have a lower threshold of stimulation, they are stimulated first, which results in the activation of muscle spindles. The latter send their impulses to the α-neurons. Thus, the γ-neurons and muscle spindles have a *starter function* for voluntary movements.

A27 Accessory olive.
A28 Skeletal muscles.
A29 Muscle spindle.

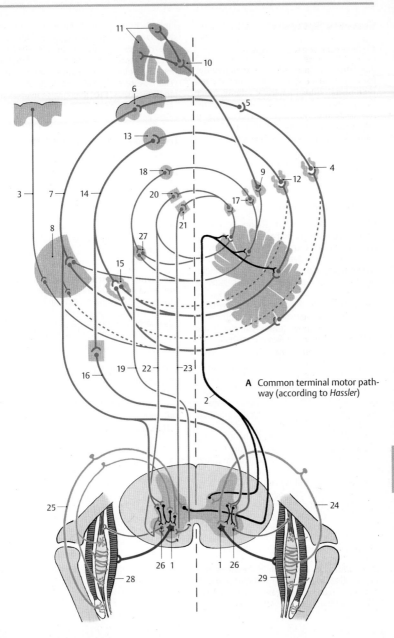

A Common terminal motor pathway (according to *Hassler*)

Sensory Systems

Cutaneous Sensory Organs

The skin is endowed with a large number of end organs that differ in their structures and their sensitivities to specific stimuli. Since the assignment of different sensory qualities to specific end organs is a matter of controversy, the nerve endings are classified according to morphological aspects:

- *Free nerve endings*
- *Encapsulated nerve endings*
- *Transitional forms* between these two types

Free Nerve Endings (A – C, F)

Free nerve endings are found in almost every tissue of the body. In the skin, they reach into the lower layers of the epidermis (stratum germinativum). At the end of their terminal branches, the axons send nodular or fingerlike evaginations through gaps in the Schwann cell sheath. These evaginations are covered only by the basement membrane and represent the receptor segments of free nerve endings to which the *sensations of pain and cold* are attributed.

Delicate nerve fibers encircle the **hair follicles**, and their terminal segments ascend or descend parallel to the hair shaft (**A**). The receptor terminals lose their myelin sheath and are enclosed between two Schwann cells (**B1**) (*sandwich arrangement*), which leave a cleft along the entire terminal segment (**B2**). Through this cleft the axon terminal reaches the surface where it is covered only by the basement membrane. The nerve endings (**C3**) are radially arranged around the hair follicle (**C4**) in such a way that the sensory clefts face the follicle. Every movement of the hair causes mechanical stimulation of the nerve endings (red arrows in C), which is perceived as *touch*.

Also associated with hair follicles are **Merkel's touch cells** (**F5**). These are large, clear epithelial cells that lie between the basal cells (**F6**) of the outer root sheath and send out fingerlike processes (**F7**) into their surroundings. Deformation of these cells through movement of the hair results in stimulation of the associated nerve fiber (**F8**). The nerve fiber loses its myelin sheath as it penetrates the basement membrane (**F9**) and sends branches to several tactile cells. The terminal segment widens into a *tactile meniscus* (**F10**) and forms synapselike membrane contacts (**F11**) with the Merkel cell.

Encapsulated Nerve Endings (D, E)

Meissner's tactile corpuscles (**D**, **E**) lie in the papillae of the dermis. They are most densely located in the palmar and plantar surfaces of hands and feet. Their densities at the distal ends, or digital pulps (especially at the digital pulp of the index finger), are far higher than at the rest of the surface. This pattern of distribution is an indication of how important these corpuscles are for delicate *touch sensation*.

The tactile corpuscles (**E**) are ovoid structures that consist of stacks of flattened cells (probably Schwann cells) and are surrounded by a thin capsule. One or more axons (**E12**) enter the corpuscle from below, lose their myelin sheath, and run in convoluted spirals between the stacked cells. The club-shaped swellings (**E13**) of the axons represent the receptor terminals. Collagen fiber bundles (**E14**) radiate into the capsule of the tactile corpuscle. They are continuous with the tonofibrils in the epidermal cells and transmit any mechanical deformation of the skin surface to the receptor terminals.

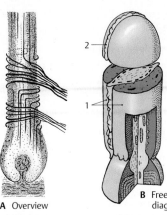

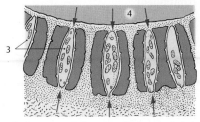

A–C Free nerve endings associated with hair follicles

C Arrangement of nerve endings around a hair follicle (according to *Andres and von Düring*)

A Overview

B Free nerve ending, electron-microscopic diagram (according to *Andres*)

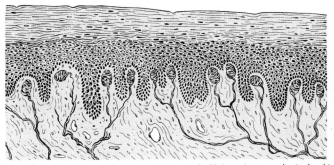

D Meissner's corpuscles in the skin

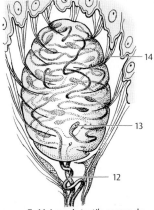

E Meissner's tactile corpuscle

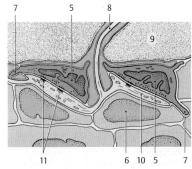

F Merkel's touch cells, electron-microscopic diagram (according to *Andres and von Düring*)

Cutaneous Sensory Organs (continued)

Encapsulated Nerve Endings (continued) (A–C)

Pacini's corpuscles, or *lamellar corpuscles* (**A–C**), are relatively large, up to 4 mm long bodies located below the skin in the subcutaneous tissue. They are also found in the periosteum, in the vicinity of joints, and at the surface of tendons and fasciae. They consist of a large number of concentric lamellae which are arranged in three layers:

- The capsule
- The *lamellar outer core*
- The *inner core*

The **capsule** (**A1**) is formed by a few densely packed lamellae that are reinforced by connective-tissue fibers. The **lamellar outer core** (**AC2**) consists of closed annular protoplasmic lamellae (**B3**) that are separated from each other by noncommunicating, fluid-filled spaces. Capsule and lamellar outer core are considered to be differentiation products of the perineurium. The tightly packed layers of the fluid-free **inner core** (**A–C4**) are formed by Schwann cells. The inner core consists of two symmetrical stacks of semiannular lamellae (**B5**) that are separated by a radial cleft. The protoplasmic lamellae are arranged alternately in such a way that two juxtaposed lamellae stem from two different cells. The nerve fiber (**AC6**) enters the lamellar corpuscle at the lower pole and runs in the center of the inner core to the end of the core. Upon entering the inner core, the axon loses its myelin sheath (**A7**), and it terminates with a few club-shaped swellings. The unmyelinated segment of the axon (**A–C8**) in the inner core represents the receptor portion. Perineurium (**A9**) (see p. 41, CD9).

Pacini's corpuscles are not only pressure receptors but, above all, *vibration receptors*. Electrical recordings from isolated corpuscles have demonstrated that they are stimulated by both deformation and decompression, but not by continuous pressure. In experiments, these highly sensitive vibration receptors registered, for example, the vibration of the floor caused by someone passing by.

Transitional Forms (D, E)

There are numerous different transitional forms between free and encapsulated nerve endings. In all of them the nerve fibers ramify into terminal complexes in which the delicate axonal branches form glomeruli or treelike ramifications and terminate with club-shaped swellings. These formations are enclosed in a more or less prominent connective tissue capsule. They include the ovoid *bulboid corpuscles* (*Krause's end bulbs*) in the papillae of the dermis, the round *Golgi–Mazzoni corpuscles*, and the elongated *Ruffini's corpuscles* (**D**) in the subcutaneous tissue. Such structures are found not only in the skin but also in the mucosa, in joint capsules, in the coverings of inner organs, and in the adventitial coat of large arteries. The most diverse forms are found in the genitals, especially in glans (**E**) and clitoris.

The assignment of the different sensory qualities to specific end organs is a matter of controversy. It is, however, recognized that the corpuscular end organs are important mechanoreceptors and that free nerve endings are pain receptors. How the sensation of heat and cold is brought about has not yet been elucidated. However, the different structures of the receptors suggest that there is selectivity with respect to the type of stimulus processed.

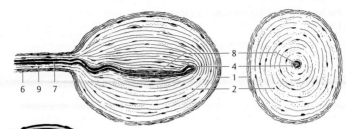

A Pacini's corpuscle, longitudinal section and cross section

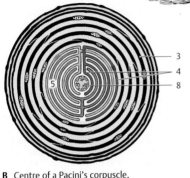

B Centre of a Pacini's corpuscle, electron-microscopic diagram (according to *Quilliam*)

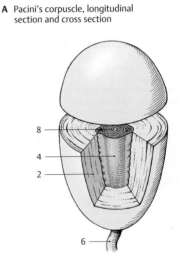

C Pacini's corpuscle (according to *Munger*)

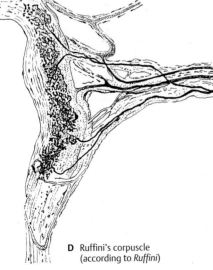

D Ruffini's corpuscle (according to *Ruffini*)

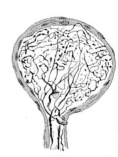

E End organ from the skin of the glans penis (according to *Dogiel*)

Pathway of the Epicritic Sensibility (A – C)

The nerve fibers transmitting impulses for the **senses of touch**, **vibration**, and **joints** originate from neurons of the spinal ganglia (**A1**), while the fibers for face and sinuses stem from neurons of the trigeminal ganglion (Gasser's ganglion, semilunar ganglion) (**A2**) (1st neuron). Touch stimuli are transmitted by two types of fibers; *thick, well-myelinated nerve fibers* terminate at the corpuscular end organs, while *thin nerve fibers* terminate at hair follicles. The centripetal axons of the neurons enter the spinal cord via the posterior root, with the thick myelinated fibers running through the medial portion of the root (p. 63, F8). They merge with the posterior funiculi (**AB3**) in such a way that the newly entering fibers border laterally; as a result, the sacral and lumbar fibers lie medially and the thoracic and cervical fibers laterally. The sacral, lumbar, and thoracic bundles form the *gracile fasciculus* (Goll's tract) (**B4**), while the cervical fibers form the *cuneate fasciculus* (Burdach's tract) (**B5**).

The primary fibers (*gracile funiculus* and *cuneate funiculus*) terminate in a corresponding arrangement on the neurons of the dorsal column nuclei (**A6**) (2 nd neuron), *gracile nucleus* (**B7**), and *cuneate nucleus* (**B8**), which therefore exhibit the same somatotopic arrangement as the posterior funiculi. Each neuron in the dorsal column nuclei receives its impulses from a specific type of receptor. The cutaneous supply area of a neuron is small in the distal segments of the limbs (hand, finger) but larger in the proximal segments. As electrophysiological studies have demonstrated, the neurons receiving impulses from specific receptors show a somatotopic arrangement as well; close to the surface of the nuclei lie the neurons for the hair follicle receptors, in the middle those for the touch organs, and still deeper those for the vibration receptors.

Corticofugal fibers from the central region (precentral gyrus and postcentral gyrus) run via the corticospinal (pyramidal) tract to the dorsal column nuclei; fibers from the lower limb area of the central region terminate in the gracile nucleus, while fibers from the upper limb area terminate in the cuneate nucleus. The corticofugal fibers have a postsynaptic or presynaptic inhibitory effect on neurons of the dorsal column nuclei and therefore attenuate the incoming afferent impulses. Thus, the cortex is able to regulate in these relay nuclei the input of impulses coming from the periphery.

The secondary fibers ascending from the dorsal column nuclei (2 nd neuron) form the *medial lemniscus* (**B9**). In the *decussation of the medial lemnisci* (**B10**) the fibers cross to the opposite sides, with the fibers from the gracile nucleus lying ventrally and those from the cuneate nucleus lying dorsally. Later, the gracile fibers take a lateral position (**B11**) and the cuneate fibers take a medial position (**B12**). The secondary fibers from the trigeminal nuclei (**B13**), *trigeminal lemniscus*, join the *medial lemniscus* at the level of the pons and become located dorsomedially to it in the midbrain (**B14**).

The medial lemniscus extends to the lateral part of the *ventral posterior nucleus of thalamus* (**AC15**); the fibers of the gracile nucleus terminate laterally, while those of the cuneate nucleus terminate medially. The trigeminal fibers (**C16**) terminate in the medial part of the ventral posterior nucleus. This results in a somatotopic organization of the nucleus. The arrangement of fibers is preserved in the projections of the thalamocortical fibers (*3rd neuron*) to the cortex of the postcentral gyrus (**A17**) and forms the basis for the somatotopic organization of the postcentral area (p. 250).

Hence, the pathway of the epicritic sensibility consists of three neurons relayed in tandem, with a demonstrated somatotopic organization in each relay station and at the terminal station.

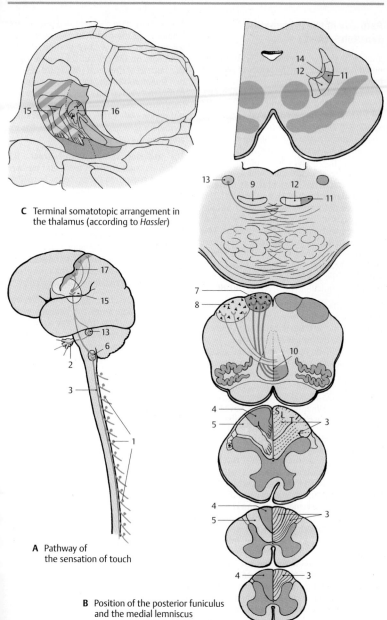

C Terminal somatotopic arrangement in the thalamus (according to *Hassler*)

A Pathway of the sensation of touch

B Position of the posterior funiculus and the medial lemniscus

Pathway of the Protopathic Sensibility (A – C)

The thin, poorly myelinated or unmyelinated nerve fibers for the **senses of pain and temperature** originate from the small neurons of the spinal ganglia (**A1**) (1st neuron). Their centripetal axons enter the spinal cord through the lateral part of the posterior root (p. 63, F7). They bifurcate in *Lissauer's tract* and terminate in the dorsal border region of the substantia gelatinosa and in the posterior horn (p. 57, A2). The secondary fibers cross to the opposite side and ascend in the anterolateral funiculus as **lateral spinothalamic tract** (**B2**) (2nd neuron). The tract does not form a discrete fiber bundle but consists of loosely arranged fibers that are mixed with fibers of other systems. The fibers entering at various root levels join ventromedially. Thus, the sacral fibers lie at the surface, and the cervical fibers that joined last lie in the inner part of the anterolateral funiculus (p. 57, A1; p. 140, B8).

The input of impulses is controlled by descending fibers that originate in the central region, in the anterior lobe of the cerebellum, and in the reticular formation. These fibers terminate in the substantia gelatinosa, a relay station in which the peripheral impulses are modulated by the excitatory or inhibitory influences of higher centers. Numerous axo-axonal synapses, which are typical for presynaptic inhibition, have been demonstrated in the substantia gelatinosa.

In the medulla oblongata, the lateral spinothalamic tract (*spinal lemniscus*) is located at its lateral margin above the olive and gives off numerous collaterals to the reticular formation. Here, too, a considerable portion of the fibers (*spinoreticular tract*) terminate. The reticular formation is part of the ascending activation system (p. 146), the stimulation of which puts the organism into a state of alertness. Hence, the impulses transmitted via the **pain pathway** not only cause a conscious sensation but also increase the attention via the reticular formation. By contrast, the pathway of the epicritic sensibility runs through the brain stem without giving off any collaterals.

The spinothalamic fibers join the *medial lemniscus* in the midbrain and take a dorsolateral position. A large portion of them terminate on the cells of the *ventral posterior nucleus of thalamus* (**AC3**) (3rd neuron) in somatotopic organization, predominantly in a ventral parvocellular region. Tertiary fibers extend from here to the postcentral region (**A4**). Other spinothalamic fibers terminate in other thalamic nuclei, for example, in the intralaminar nuclei.

The **anterior spinothalamic tract** (**B5**) transmits **crude senses of touch** and **pressure**. Its fibers cross from the posterior horn (2 nd neuron) to the contralateral anterior funiculus (p. 57, A3). The position of the tract in the medulla oblongata is a matter of controversy. It is though to lie either medially to the medial lemniscus (**B6**) or laterally to the olive (**B7**). In the pons and midbrain, the fibers join the medial lemniscus (**B8**) and terminate on the cells of the *ventral posterior nucleus of thalamus* (3rd neuron).

Pain and temperature fibers for the *face and sinuses* originate from the neurons of the trigeminal ganglion (**A9**), the centripetal axons of which terminate in the *spinal nucleus of the trigeminal nerve* (**AB10**) (p. 125, BC4). In the spinal tract of the trigeminal nerve, the pain-transmitting fibers are thought to lie laterally and those transmitting temperature further medially. The secondary trigeminal fibers (**B11**) (*trigeminal lemniscus*) join the medial lemniscus.

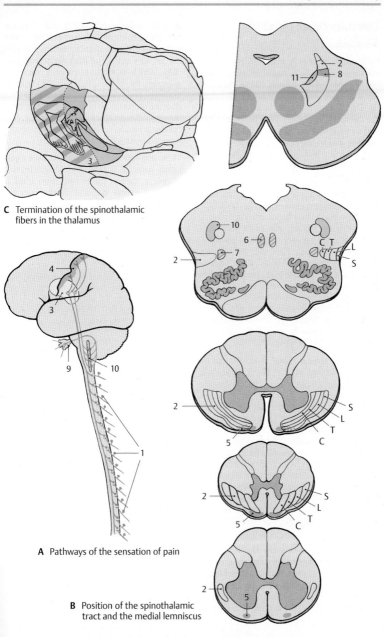

C Termination of the spinothalamic fibers in the thalamus

A Pathways of the sensation of pain

B Position of the spinothalamic tract and the medial lemniscus

Gustatory Organ (A–E)

Taste Buds

Different taste sensations are registered by the taste buds which, together with the olfactory epithelium, constitute the *chemoreceptors*. **Taste buds**, or *gustatory receptors* (**B–D1**; **E**), are found in large numbers in the lateral walls of the *vallate papillae* (**AB2**), and in moderate numbers in the *fungiform papillae* and in the *foliate papillae*. In addition, there are isolated taste buds in the soft palate, in the posterior pharyngeal wall, and in the epiglottis.

A taste bud consists of up to 50 spindle-shaped, modified epithelial cells. Each barrel-shaped taste bud has a small opening at the epithelial surface, the *taste pore* (**C3**), into which the sensory cells send short cytoplasmic processes. New sensory cells are continuously generated from epithelial cells at the base of the taste bud. The conversion of epithelial cells into sensory cells takes place at the periphery, while the fully developed sensory cells lie in the center. Taste buds can multiply by fission, and twin buds or multiple buds are found occasionally (**C4**).

The base of the taste bud is innervated by thin, myelinated nerve fibers (**D5**) that branch and supply the adjacent epithelium as well. We therefore distinguish between *extragemmal* and *intragemmal* nerve fibers. Upon entering the taste bud, the intragemmal nerve fibers (usually 2 – 3) lose their Schwann cell sheath and arborize. According to electron microscopic studies, the delicate nerve endings lie between the sensory cells and invaginate these so deeply that they appear to be intracellular. When a taste nerve is severed, the taste buds degenerate. Regeneration of the nerve induces the formation of new taste buds in the epithelium of the tongue.

Electron micrographs show the long neck (**E6**) of the sensory cell extending into the taste pore and ending with a dense border of microvilli (**E7**). This is the receptor portion of the cell. The microvilli considerably enlarge the cell surface coming into contact with the substances to be tasted. The taste bud contains three different cell types that represent different developmental stages of the sensory cells: the dark *supporting cells* (**E8**), which contain numerous secretory granules, the clear *taste cells* (**E9**) without granules, and the small *basal cells* (**E10**). The floor of the taste pore as well as the clefts between the receptor portions of the cells contain a mucous substance that is probably secreted by the taste cells.

There are different theories as to how a gustatory stimulus is produced. Molecules that trigger a specific taste sensation bind to specific receptor sites on the sensory cell membrane. These sites recognize the configuration of the molecule and cause a change in membrane potential, which finally leads to excitation of the associated axon terminal.

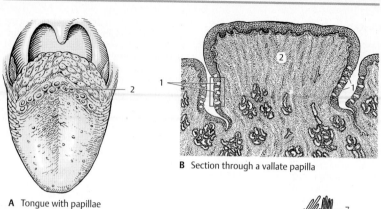

A Tongue with papillae

B Section through a vallate papilla

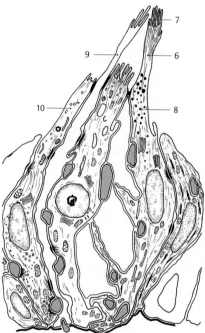

C Overview, detail of **B**

E Electron-microscopic diagram
(according to *Popoff*)

C–E Taste buds

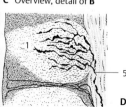

D Nerve endings

Gustatory Organ (continued)

Taste Buds (continued) (A)

The four basic taste sensations (sweet, sour, salty, and bitter) are not equally well perceived by all taste buds. Some buds react only to sweet or only to sour, while others register two or three qualities without exhibiting morphological differences. The perception of individual qualities on the surface of the tongue is differentially distributed. *Sour* is perceived especially at the lateral margins of the tongue (**A1**), *salty* at the margins and the tip (**A2**), *bitter* at the base of the tongue (**A3**), and *sweet* at the tip of the tongue (**A4**).

Taste Fibers (B, C)

The taste fibers are assigned to three cranial nerves, namely, the **facial nerve** (*intermediate nerve*) (**B5**), the **glossopharyngeal nerve** (**B6**), and the **vagus nerve** (**B7**). The fibers originate from pseudounipolar neurons in the cranial nerve ganglia, namely, the *geniculate ganglion* (**B8**), the *petrosal ganglion* (inferior ganglion of the glossopharyngeal nerve) (**B9**), and the *nodose ganglion* (inferior ganglion of the vagus nerve) (**B10**). Taste fibers of the facial nerve run in the chorda tympani to the lingual nerve (**B11**) (p. 131, AB9) and supply the receptors of the fungiform papillae on the anterior two-thirds of the tongue. Taste fibers of the glossopharyngeal nerve run in the lingual branches of the glossopharyngeal nerve to the posterior third of the tongue and supply the receptors of the vallate papillae. In the tonsillar branches of the glossopharyngeal nerve, taste fibers run to the soft palate. Taste fibers of the vagus nerve reach the epiglottis and the epipharynx through the pharyngeal branches of the vagus nerve.

The centripetal axons of the neurons enter the medulla oblongata and form the solitary tract. The *primary taste fibers* terminate in the **solitary nuclear complex** (**BC12**) approximately at the level of the nerve entry. The solitary nuclear complex widens in this region and contains a cell column also referred to as the *gustatory nucleus*.

The *secondary taste fibers* originate from the solitary nuclear complex. Their exact course in the brain stem is not known. It is assumed that the majority of fibers cross to the opposite side as arcuate fibers and join the **medial lemniscus** (**C13**). They are thought to occupy the most medially located part in the lemniscus. The secondary taste fibers terminate in the medial part of the **ventral posterior nucleus of thalamus** (**C14**). From here, *tertiary taste fibers* project to a cortical area on the ventral aspect of the parietal operculum (**C15**), below the postcentral region. The terminals in thalamus and cerebral cortex have been confirmed by experiments in monkeys. In humans, a loss of taste perception on the contralateral half of the tongue has been observed when these regions are destroyed by disease.

Some of the secondary taste fibers run to the hypothalamus. They are thought to branch off from the medial lemniscus in the midbrain and to run through the mamillary peduncle to the mamillary body. Other fibers are relayed in the ventral tegmental nucleus and are thought to reach the hypothalamus through the posterior longitudinal fasciculus (Schütz's bundle) (p. 144B).

Collaterals extend from the neurons of the solitary nuclear complex to the parasympathetic salivatory nuclei. This way, taste sensation can cause reflex *secretion of saliva*. Collaterals running to the posterior nucleus of the vagus nerve form the connection by which taste stimuli produce reflex *secretion of gastric juice*.

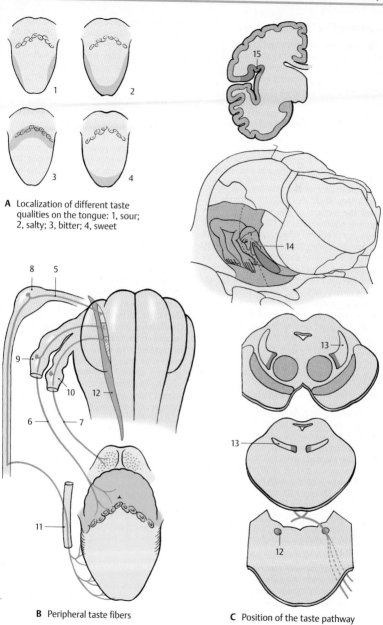

A Localization of different taste qualities on the tongue: 1, sour; 2, salty; 3, bitter; 4, sweet

B Peripheral taste fibers

C Position of the taste pathway

Olfactory Organ (A–D)

In humans, the **olfactory epithelium** occupies a small region in both nasal cavities (**olfactory region**) (**A1**) at the upper margin of the superior nasal concha and on the opposite surface of the nasal septum. The multilayered sensory epithelium is composed of **supporting cells** (**C2**) and **receptor cells** (**C3**) that are characterized by pale, deep-lying cell nuclei. The olfactory region also contains small mucous glands, the *olfactory glands* (*Bowman's glands*). Their secretion covers the olfactory mucosa as a thin film.

The apical part of the sensory cell tapers into a thin shaft that slightly extends beyond the surface of the epithelium. This knoblike *olfactory vesicle* (**C4**) is occupied by a number of *olfactory hairs*. At the basal end, the ovoid cell body forms a fine process that, together with several other processes, is enveloped by Schwann cells. The bundled processes (*fila olfactoria*) represent the **olfactory nerves** (**AC5**) and extend through the openings of the cribriform plate to the **olfactory bulb** (**A6**) (p. 228, A). In the olfactory bulb, the processes terminate in the *olfactory glomeruli*, where they form synaptic contacts with the dendrites of *mitral cells*. The epithelial sensory cells are bipolar neurons; the short dendrite represents the receptor part and the long axon runs as centripetal fiber to the olfactory bulb.

Electron micrographs of olfactory cells (in the cat) show that the apical shaft of the cell (**D7**) terminates with a knob (**D8**) from which numerous long *olfactory cilia* (**D9**) extend. The terminal parts of the sensory cilia lie in the mucous layer (**D10**) that seals the entire surface of the olfactory epithelium against the air space. Shaft and knob contain microtubules, numerous mitochondria (**D11**), and some lysosomes (**D12**). The knob extends above the surface of the supporting cells, which exhibit a dense border of microvilli (**D13**).

How the sensory cells receive the different odors is currently the topic of intensive research. The odoriferous substances must be water-soluble in order to dissolve in the superficial mucous layer and reach the sensory cilia where they bind to specific membrane receptors. In sufficiently high concentrations, they induce depolarization of the membrane, which is conducted as an action potential in the cell's axon. As in the case of taste, it is assumed that there are a few basic qualities of odor and that one sensory cell registers only one particular basic quality via specific receptors. Since substances belonging to a group of odors have roughly the same molecular size, it seems possible that the membrane of an olfactory cilium reacts to only one particular molecular size. Indeed, recent studies suggest that each sensory cell expresses only one receptor type. In the mouse, the sensory cells of one receptor type project to only two of the 1800 glomeruli in the olfactory bulb (central olfactory pathway, p. 228, AB).

Apart from the olfactory nerves, two other paired nerves run from the nasal cavity to the brain, namely, the terminal nerve and the vomeronasal nerve. The **terminal nerve** (**B14**) consists of a bundle of delicate nerve fibers that extends from the nasal septum through the cribriform plate to the terminal lamina and enters below the anterior commissure into the brain. The bundle contains numerous neurons and is regarded as an autonomic nerve. The **vomeronasal nerve** (**B15**), which runs from the *vomeronasal organ* (*Jacobson's organ*) to the accessory olfactory bulb, is well developed in lower vertebrates but can be demonstrated in humans only during embryonic development. In reptiles, the vomeronasal organ is a sensory epithelium in a pocket of the mucosa of the nasal septum; it is thought to play an important role in tracking down prey.

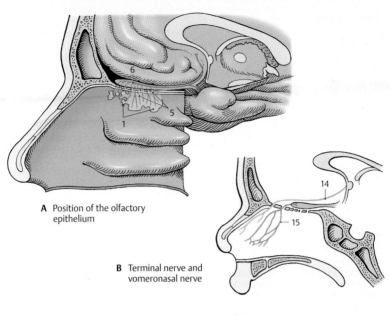

A Position of the olfactory epithelium

B Terminal nerve and vomeronasal nerve

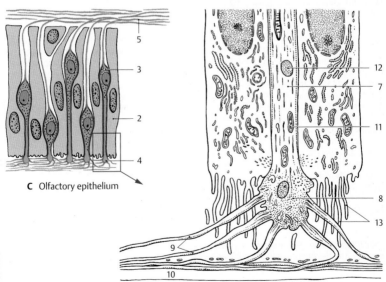

C Olfactory epithelium

D An olfactory cell between two supporting cells, electron-microscopic diagram (according to *Andres*)

Limbic System

Overview

The phylogenetically old parts of the telencephalon, together with its border zones and its connections to subcortical centers, are collectively known as the limbic system. It is not a circumscribed, topically organized system of pathways but rather a collection of functionally closely related nuclei and cortical areas. The system has also been called the *visceral* or *emotional brain*, a catchphrase highlighting its functional significance. As the concept of the limbic system is based on functional connections, its basic anatomical structures are only vaguely defined.

Subdivision (A)

The cortical regions belonging to the limbic system form a C-shaped complex on the medial aspect of the hemisphere that consists of the *parahippocampal gyrus* (**A1**), the *cingulate gyrus* (**A2**), and the *subcallosal area* (parolfactory area) (**A3**). The cingulate gyrus, which is also called *limbic gyrus*, gave the system its name. An inner and an outer arch can be distinguished at the medial aspect of the hemisphere. The **outer arch** (parahippocampal gyrus, cingulate gyrus, and parolfactory area) is formed by the border zones of the archicortex (*periarchicortex*) and by the indusium griseum of the corpus callosum (*rudimentary archicortex*). The **inner arch** consists of archicortical and paleocortical regions, namely, the *hippocampal formation* (**A4–C5**), the *septal area* (**A6**), the *diagonal band of Broca*, and the *paraterminal gyrus* (**A7**). An important component is also the *amygdaloid body*. Some subcortical nuclei with close fiber connections to the limbic cortex are included in the system, such as the *mamillary body*, the *anterior nuclei of thalamus*, the *habenular nucleus*, and additionally, in the midbrain the *posterior tegmental nucleus*, the *anterior tegmental nucleus*, and the *interpeduncular nucleus*.

Pathways (B, C)

The limbic system is connected with the olfactory centers through several fiber bundles. The fibers of the lateral olfactory stria terminate in the cortical parts of the amygdaloid body (p. 228, B).

The limbic system influences the hypothalamus via three pathways:

- The **fornix**, the precommissural fibers of which terminate in the preoptic area (**B8**) and in the nuclei of the tuber cinereum (**B9**)
- The **terminal stria** (**B10**), which runs from the amygdaloid body (**B11**) to the tuberal nuclei
- The **ventral amygdalofugal fibers** (**B12**)

The connection to the tegmental nuclei of the midbrain is established through the descending bundle of the habenular nucleus (*habenulotegmental tract* and *habenulopeduncular tract*) and through the pathways of the mamillary body (*mamillary peduncle* and *mamillotegmental fasciculus*). The efferent mamillotegmental fasciculus (**C13**) and the afferent peduncle of the mamillary body (**C14**) form a neuronal circuit.

Within the limbic system runs a multiple pathway, the **neuronal circuit of Papez**. The efferent fibers of the hippocampus reach the mamillary body (**C15**) through the fornix (**C5**). Here the impulses are relayed to the bundle of Vicq d'Azyr (**C16**), which extends to the anterior nuclei of thalamus (**C17**). The latter project to the cortex of the cingulate gyrus (**C18**), from where the cingulate fiber bundles (**C19**) return to the hippocampus.

Connections of the neocortex to the limbic system exist especially via the entorhinal area in the parahippocampal gyrus (**A1**), which projects to the hippocampus (perforant pathway, see p. 234, A4).

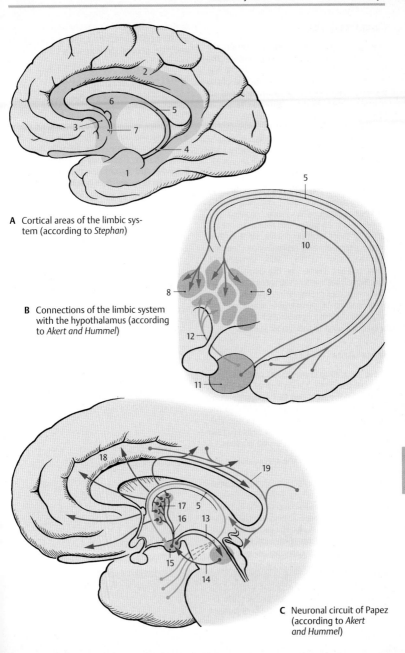

A Cortical areas of the limbic system (according to *Stephan*)

B Connections of the limbic system with the hypothalamus (according to *Akert and Hummel*)

C Neuronal circuit of Papez (according to *Akert and Hummel*)

Functional Systems

Cingulate Gyrus

The cingulate gyrus is connected with the olfactory cortex, the hypothalamus, the frontal cortex, the caudal portion of the orbital cortex, and the rostral portion of the insular cortex. Electrical stimulation of its rostral region in humans leads to changes in blood pressure, pulse rate, and respiration rate. Changes in temperature, erection of the hair, dilatation of pupils, increased salivation, and altered gastric motility have been observed in stimulation or lesion experiments in monkeys.

The cortex of the cingulate gyrus influences the hypothalamus and the autonomic nervous system. The limbic system obviously plays an important role in the regulation of basic vital processes, such as *food intake, digestion,* and *reproduction* (see also hypothalamus, p. 198, and amygdaloid body, p. 226). These are the primary vital functions that serve self-preservation as well as species preservation and are always accompanied by *pleasurable sensations* or *reluctance*. Hence, emotional states have been attributed to the limbic system.

Septal Area (A–C)

There are strong connections to the hippocampus, the central structure of the limbic system. Cholinergic and GABAergic neurons of the medial septal nucleus project to the hippocampus and the dentate gyrus; collaterals of the CA3 pyramidal cells (p. 234) project back to the lateral septal nucleus.

As with stimulation of the amygdaloid body, electrical stimulation of the septal area (**BC1**) triggers oral reactions (licking, chewing, retching), excretory reactions (defecation, urination), and sexual reactions (erection). The septal area, especially the *diagonal band of Broca*, is also the preferred localization for self-stimulation experiments in the rat (**A, B**). Rats carrying an implanted electrode that facilitates stimulation of this area continuously stimulate themselves by pressing a button. The urge for stimulation is so strong that the animals prefer stimulation to feeding, even after periods of hunger

or thirst. Stimulation near the anterior commissure (**C2**) produces a euphoric reaction and the general feeling of well-being in humans.

Klüver–Bucy syndrome (D). This syndrome was observed following bilateral removal of the temporal lobe (**D3**) in the monkey. The lobectomy affected the neocortex, hippocampus, and amygdaloid body, and resulted in a polysymptomatic behavioral syndrome. The animals became tame and trusting; they lost their natural behavior and shyness even toward dangerous objects or animals (such as snakes). In addition, they completely lost any sexual inhibition (*hypersexuality*). Familiar objects were no longer recognized visually (*visual agnosia*); instead, they were put in the mouth (*oral tendency*) and examined repeatedly this way as if they were completely unknown each time they were encountered. The disturbed memory indicates a special role of the hippocampus in processes of learning and memory. Information from the neocortex, such as visual impressions, is transmitted to the hippocampus via the entorhinal area and examined for its novelty value.

There is no doubt that the limbic system includes very divergent and complex functions. It is likely that increases in knowledge will soon lead us to abandon the summary and vague concept of *one* limbic system.

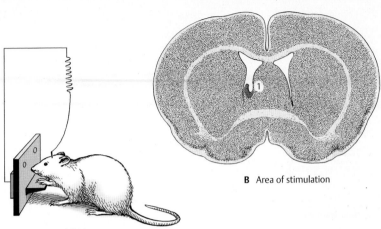

B Area of stimulation

A Experimental design

A, B Self-stimulation experiment in the rat (according to *Olds*)

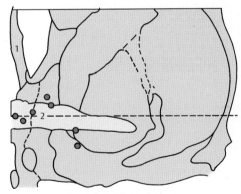

C Foci of stimulation for euphoric reactions in humans (according to *Schaltenbrand, Spuler, Wahren, and Wilhelm*)

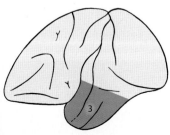

D Removal of the temporal lobe, Klüver–Bucy syndrome (according to *Klüver and Bucy*)

Functional Systems

Sensory Organs

The Eye

Structure

Eyelids, Lacrimal Apparatus, and Orbital Cavity

Eyelids (A–C)

The *eyeball* is embedded in the *orbit* and is covered by the eyelids. The **upper eyelid** (**A1**) and the **lower eyelid** (**A2**) demarcate the **palpebral fissure**. The latter ends in the *medial angle of the eye* (**A3**) with a recess enclosing the *lacrimal caruncle* (**A4**).

In Oriental People, the upper eyelid continues medially onto the side of the nose as a vertical fold of skin, the *palpebronasal fold*. The fold can also be observed as a transitory formation in infants and is known as *epicanthus*.

The eyelids are reinforced by firm plates of connective tissue consisting of collagenous fibers, *tarsus superior* (**B5**) and *tarsus inferior* (**B6**); the tarsal plates, or palpebral cartilages, are attached to the margin of the orbit by the *lateral palpebral ligament* (**B7**) and the *medial palpebral ligament*. The tarsal plates contain elongated *tarsal glands*, the meibomian glands (**C8**), which spread over the entire height of the eyelids. Their secretion prevents the tears from flowing over the edge of the eyelids. They open at the *posterior margin* of the eyelid. Several rows of eyelashes, the *cilia* (**C10**), emerge from the *anterior margin* (**AC9**). The inner wall of the eyelids is lined by the **conjunctiva** (**C11**), which extends to the anterior aspect of the eyeball at the *conjunctival fornix* (**C12**). The smooth *superior tarsal muscle* (**C13**) and *inferior tarsal muscle* (**C14**) (innervated by the sympathetic nervous system), which control the size of the palpebral fissure, attach at the tarsus. The eyelids are closed by the *orbicular muscle of the eye* (**C15**) (facial nerve, p. 122). The upper lid is lifted by the *levator muscle of the upper eyelid* (**BC16**) (oculomotor nerve), which attaches at the upper margin of the optic canal. Its superficial tendinous membrane (**C17**) penetrates into the subcutaneous connective tissue of the upper lid, while the deep tendinous membrane (**C18**) attaches at the upper margin of the tarsus.

Lacrimal Apparatus (B)

The **lacrimal gland** (**B19**) lies above the lateral angle of the eye; it is divided by the tendon of the levator muscle of the upper eyelid into an *orbital part* (**B20**) and a *palpebral part* (**B21**). Its excretory ducts at the conjunctival fornix secrete the lacrimal fluid (tears), which keeps the anterior aspect of the eyeball continuously moist and collects in the *lacrimal lake* of the medial angle of the eye. Here at the inner surface of each eyelid is a small opening, the *lacrimal point* (**B22**), which leads to the *lacrimal canaliculus* (**B23**). The canaliculi ascend and descend, respectively, and then turn at a right angle to join and open into the **lacrimal sac** (*tear sac*) (**B24**), from where the **nasolacrimal duct** (*tear duct*) (**B25**) leads to the inferior meatus of the nose. Blinking not only results in uniform moistening of the eyeball surface but also has a sucking effect on the flow of tears by expanding and constricting the nasolacrimal duct.

Orbit (C)

The orbital cavity (eye socket) is lined by the periosteum (*periorbita*) (**C26**) and is filled with fatty tissue, the *orbital fat body* (**C27**), in which the eyeball (**C28**), the optic nerve (**C29**), and the eye muscles (**C30**) are embedded. At the anterior margin of the orbit, the fatty tissue is demarcated by the orbital septum (**BC31**). The fatty tissue is separated from the eyeball by a connective tissue capsule, the *bulbar sheath* (**C32**), which encloses the *sclera* (**C33**).

C34 *Choroidea*.
C35 Osseous wall of the orbit.

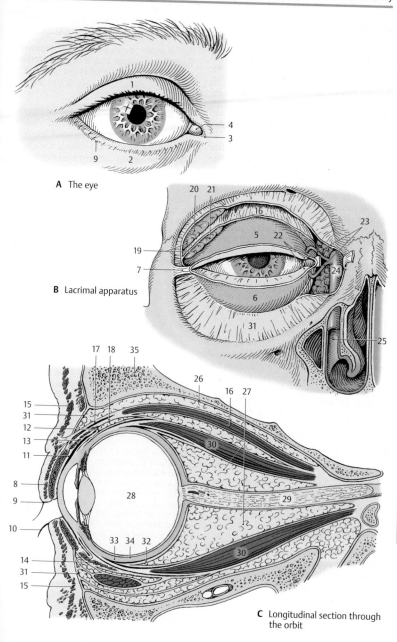

A The eye

B Lacrimal apparatus

C Longitudinal section through the orbit

Muscles of the Eyeball (A–E)

The eyeball is attached by membranes to the capsule of the orbital fat body, and it can move in all directions. Movements are achieved by six *extra-ocular muscles*, namely four rectus muscles and two oblique muscles. The tendons of origin of the rectus muscles form a funnel-shaped ring around the optic canal, the *common annular tendon* (**AB1**). The **superior rectus muscle** (**A–C2**) (oculomotor nerve) runs above the eyeball in a slightly oblique, outward direction. The **inferior rectus muscle** (**A–C3**) (oculomotor nerve) runs beneath the eyeball in the same direction. At the nasal aspect of the eyeball lies the **medial rectus muscle** (**AC4**) (oculomotor nerve), and at the temporal aspect lies the **lateral rectus muscle** (**A–C5**) (abducens nerve). At a distance of 0.5–1 cm from the margin of the cornea, the flat tendons of the rectus muscles attach to the sclera of the eyeball. The **superior oblique muscle** (**AC6**) (trochlear nerve) originates medially at the lesser wing of the sphenoid bone and extends almost to the margin of the orbit. Here, its tendon passes through the **trochlea** (**A7**), a wide loop consisting of fibrous cartilage and lined with a synovial sheath. The tendon then turns at an acute angle in posterolateral direction and attaches underneath the superior rectus muscle on the temporal side of the upper eyeball. The **inferior oblique muscle** (**BC8**) (oculomotor nerve) originates medially at the infraorbital margin and runs to the temporal side of the eyeball. *Levator muscle of upper eyelid* (**B9**).

Movements of the eyeball:

- Rotation around the **vertical axis** toward the nose (*adduction*) and toward the temple (*abduction*)
- Rotation around the **horizontal axis** upward (*elevation*) and downward (*depression*)
- Rotation around the **sagittal axis** with rolling of the upper half of the eyeball toward the nose (inward rotation, or *intorsion*) and toward the temple (outward rotation, or *extorsion*)

The *medial rectus muscle* (**C4**) causes **adduction**; the *lateral rectus muscle* (**C5**) **abduction**.

The *superior rectus muscle* (**C2**) raises the eyeball and causes also a slight adduction and **intorsion**; the *inferior rectus muscle* (**C3**) lowers the eyeball and causes also a slight adduction and **extorsion**.

The *superior oblique muscle* (**C6**) rotates the upper pole of the eyeball inward slightly depresses and abducts the eyeball; the *inferior oblique muscle* (**C8**) rotates the upper pole of the eyeball outward and slightly elevate and abducts the eyeball.

This functional description applies only when looking straight ahead (medial gaze) and when both eyeballs have parallel axes of vision. During ocular movements and simultaneous reactions of convergence (p. 358, C) and divergence, the functions of individual muscles changes, e.g., the two medial rectus muscles are synergists during convergence and antagonists during lateral gaze. The *change in function* is determined by the deviation of the visual axis from the anatomical axis of the orbit. When the two axes overlap during abduction of the eyeball by 23°, the *superior rectus muscle* (**D10**, looking straight ahead) and the *inferior rectus muscle* lose their accessory functions; the first turns into a true levator of the eyeball (**D11**) and the latter into a true depressor of the eyeball. During maximal adduction of the eyeball up to 50°, the *superior oblique muscle* (**E12**, looking straight ahead) turns into a true **depressor of the eyeball** (**E13**) and the *inferior oblique muscle* into a true **levator of the eyeball**. All extra-ocular muscles are involved with tension and relaxation during every eye movement, and the position of the eyeball determines the function of each muscle.

Precision and speed of muscle function rely on structural characteristics. Apart from the intrafusal fibers of muscle spindles, numerous extrafusal fibers are supplied with sensory anulospiral nerve endings. The motor units are extremely small; about six ocular muscle fibers are supplied by one nerve fiber. For comparison, one nerve fiber innervates 100–300 muscle fibers in the muscle of the fingers, in other muscles often more than 1500 fibers.

■ **Clinical Note:** Paralysis of individual eye muscles causes double vision, when objects in the surroundings are perceived twice. The relative position of the two images—side by side or shifted obliquely above or obliquely below one another—indicates which muscle is paralyzed.

The Eye

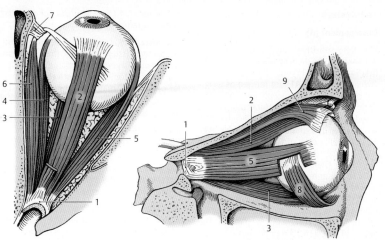

A Extraocular muscles, view from above (right eye)

B Extraocular muscles, lateral view

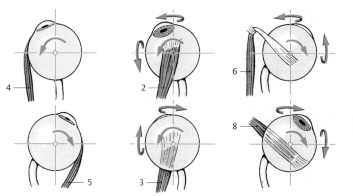

C Functions of the extraocular muscles of the right eye

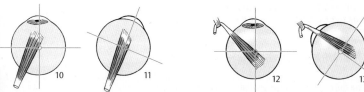

D Superior rectus muscle of the right eye, gaze straight ahead (left) and abducted by 23° (right)

E Superior oblique muscle of the right eye, gaze straight ahead (left) and adducted by 50° (right)

The Eye

The Eyeball

Development (A)

The light-sensitive part of the eye is a derivative of the diencephalon. At end of the first month of embryonic development, the two *optic vesicles* (**A1**) are formed as evaginations of the prosencephalon (**A2**). The optic vesicles then induce thickenings in the ectoderm of the head, the *lens placodes* (**A3**), which later invaginate as *lens vesicles* (**A4**). The epithelial cells of the posterior vesicle wall elongate into *lens fibers* (**A5**), which later form the main part of the lens. The cells of the anterior vesicle wall persist as lens epithelium. The anterior and posterior walls of the optic vesicle approximate each other to form the *optic cup* (**A6**). The lumen of the vesicle, originally a part of the ventricular system, the *optic ventricle* (**A7**), becomes a narrow cavity. The optic cup consists of an inner layer, the *neural layer* (**A8**), and an outer layer, the *pigmented layer* (**A9**); both are layers of the retina. The *hyaloid artery* (**A10**) first extends to the lens but later regresses.

Structure (B)

The anterior aspect of the **eyeball** consists of the transparent **cornea** (**B11**). Behind it lies the **lens of the eye** (**B12**), overlaid by the **iris** (**B13**) with its central opening, the *pupil*. The **optic nerve** (**B14**) exits at the posterior wall of the eyeball, slightly medial to the optic axis. There are three cavities in the eye:

- The **anterior chamber** (**B15**), which is bordered by cornea, iris, and lens
- The **posterior chamber**, which forms a ring around the lens (**B16**)
- The interior of the eye, which contains the **vitreous body** (**B17**)

The vitreous body is a transparent gel consisting mostly of water. The two eye chambers contain a clear fluid, the aqueous humor.

The wall of the eyeball consists of three layers:

- The fibrous tunic of the eyeball, or *sclera*
- The vascular tunic of the eyeball, or *uvea*
- The internal (sensory) tunic of the eyeball, or *retina*

The **sclera** (**B18**) is a thick, stretch-resistant connective-tissue capsule that consists mainly of collagenous fibers and some elastic fibers; in conjunction with the intraocular pressure, it maintains the shape of the eyeball.

The vascular **uvea** forms the iris and the **ciliary body** (**B19**) in the anterior part of the eyeball, and the **choroidea** (**B20**) in the posterior part.

The posterior part of the **retina** (*optic part*) (**B21**) contains the light-sensitive receptor cells as well as pigmented epithelium, while the anterior part (*blind part*) (**B22**) contains only pigmented epithelium. The border between the two retinal parts is known as the **ora serrata** (**B23**).

The eyeball has an anterior pole (**B24**) and a posterior pole (**B25**); the *equator of the eyeball* (**B26**) runs between them. Some blood vessels and muscles follow the *meridians of the eyeball* (**B27**) which run from pole to pole.

The eyeball can be divided into an anterior part and a posterior part, which fulfill different functions. The anterior part contains the *image-forming apparatus*, the refracting lens system. The posterior part contains the *photoreceptive surface*, the retina. Hence, the eye can be compared to a camera that possesses a lens system with an aperture in the front—the iris of the eye—and a light-sensitive film at the back—the retina.

B28 Subarachnoidal cavity (p. 351, D9).

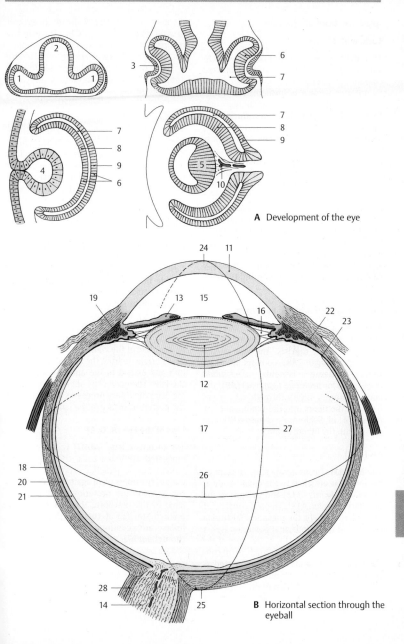

A Development of the eye

B Horizontal section through the eyeball

Anterior Part of the Eye

Cornea (A, B)

The **cornea** (**A1**, **B**) is positioned like a watch-glass on the eyeball. Because of its marked curvature, it has the effect of a focusing lens. Its anterior surface is formed by a multi-layered, nonkeratinizing squamous epithelium (**B2**) that rests upon a thick basement membrane, called the *anterior limiting lamina* (Bowman's membrane) (**B3**). Below this lies the *stroma of the cornea* (substantia propria) (**B4**); its straight collagen fibers form lamellae that lie parallel to the surface of the cornea. At the posterior surface lies another basement membrane, the *posterior limiting lamina* (Descemet's membrane) (**B5**) and a single-layered endothelium (**B6**). The cornea contains unmyelinated nerve fibers but no blood vessels. Its transparency is due to a specific fluid content and the state of swelling of its components. Any change in the turgor causes turbidity of the cornea.

Anterior Chamber of the Eye (A)

The anterior chamber of the eye (**A7**) contains the aqueous humor generated by the blood vessels of the iris. The wall of the *iridocorneal angle* (**A8**) consists of loose connective-tissue strands (*reticulum trabeculare*, or *pectinate ligament*) (**A9**); the aqueous humor filters through the spaces between the fibers into the *venous sinus of the sclera*, or Schlemm's canal (**A10**), and passes into the bloodstream.

Iris (A, C)

The **iris** (**A11**) forms an aperture in front of the lens. It is attached to the ciliary body at the *root of the iris* (**A12**) and extends to the *pupillary margin* (**A13**). The iris consists of two layers, namely, the mesodermal *stroma* (**AC14**) and the ectodermal posterior aspect of the iris, known as the *iridial part of the retina* (**AC15**). The stroma is made up of connective-tissue strands and is pigmented. A high pigment content results in brown eyes, while a low pigment content gives the eyes a green or blue appearance. Numerous blood vessels branch radially off the *greater arterial circle of the iris* (**AC16**). The ectodermal part of the iris, a derivative of the optic cup (p. 342, A6), gives rise to two smooth muscles, namely, the *sphincter muscle of the pupil* (**AC17**) and the thin *dilator muscle of the pupil* (**AC18**).

Ciliary Body (A, D)

The lens is suspended at the circular **ciliary body** (**A19**, **D**). The muscles of the ciliary body control the curvature of the lens (p. 358, B) and, hence, the visual acuity for close and distant vision. It consists of a radially folded surface, the *ciliary disk* (**D20**), from where approximately 80 *ciliary processes* (**D21**) protrude (*ciliary crown*). The anterior part of the disk is occupied by the ciliary muscle; its muscle fibers, the *meridional fibers* (**A22**), span between the ora serrata and the reticulum trabeculare as well as Descemet's membrane. From here, radial muscle fibers extend inward and bend to take a circular course (*circular fibers*) (**A23**). The posterior surface of the ciliary body is covered by the *ciliary part of the retina* (**A24**). Delicate fibers, the *zonular fibers* (**AD25**), extend from the ciliary body to the lens and form the *ciliary zone* (**D26**). Many fibers originate from the area of the ora serrata and extend to the anterior surface of the lens. They cross the shorter fibers arising from the ciliary processes and ending at the posterior surface of the lens.

Lens of the Eye (A, D, E)

The biconvex **lens** (**AD27**) is made up by elongated epithelial cells, the lens fibers, which show a lamellar arrangement. The ends of the fibers push against each other at the anterior and posterior surfaces of the lens; in the newborn, they form a three-pointed star (**E**) at their border. The lens fibers are continuously being formed throughout life.

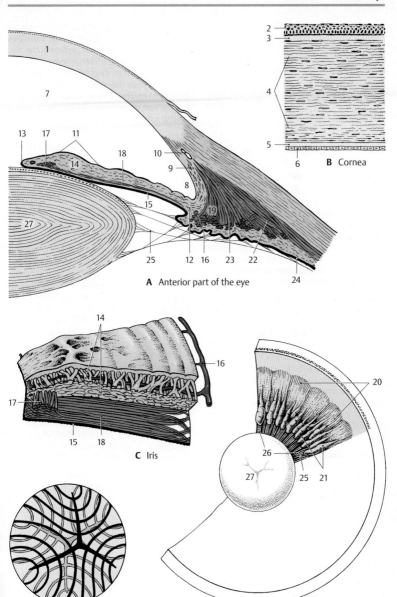

A Anterior part of the eye

B Cornea

C Iris

E Fibers in the lens of a newborn

D Ciliary body and lens, posterior view

The Eye

Vascular Supply (A)

The eye has two different vascular systems: the **ciliary arteries** and the **central retinal artery**. All these vessels arise from the *ophthalmic artery* (p. 273, C16). The posterior ciliary arteries are the branches supplying the vascular tunic of the eyeball, or uvea, which forms the iris (**A1**), the ciliary body (**A2**), and the choroidea proper of the posterior wall of the eyeball (**A3**). This vascular system not only supplies the blood, it is also essential for maintaining the intraocular pressure and the tension of the eyeball.

Long posterior ciliary arteries (A4). These arteries penetrate the sclera near the exit of the optic nerve. One of them runs in the temporal, the other one in the nasal wall of the eyeball to the ciliary body and to the iris. At the root of the iris they form the *greater arterial circle of the iris* (**A5**), from where vessels radiate to the *lesser arterial circle of the iris* (**A6**) near the pupil.

Short posterior ciliary arteries (A7). These form the vascular plexus of the choroidea, which extends from the posterior wall of the eyeball to the *ora serrata* (**A8**). The inner choroid layer consists of especially wide capillaries, the *choriocapillary layer*, and borders on the pigmented epithelium of the retina. Whereas the choriocapillary lamina is firmly attached to the pigmented epithelium, the outer aspect of the choroidea is separated from the sclera by the *perichoroidal space* and, thus, can be displaced.

Anterior ciliary arteries (A9). These run from the rectus muscles to the sclera where they branch in the episcleral tissue and in the conjunctiva. In the conjunctiva they form the *marginal loops* (**A10**) around the margin of the cornea.

The veins unite to form the four *posterior ciliary veins*, or **vorticose veins** (**A11**), which obliquely penetrate the sclera at the posterior wall of the eyeball.

Central retinal artery (A12). This artery enters the optic nerve approximately 1 cm behind the eyeball and extends in the middle of the nerve to the papilla of the optic nerve (see below). It then divides into branches which run along the inner surface of the retina within the layer of nerve fibers. The retinal vessels are *end arteries*. Their capillaries reach as far as the inner nuclear layer (see p. 349, A12). The venules unite to form the *central vein of the retina* (**A13**), which takes a course similar to that of the central artery.

The visual cells are nourished from both sides of the retina: from the outside by the capillary system of the short posterior ciliary arteries, and from the inside by the central arteries.

Fundus of the Eye (B)

The posterior pole of the eyeball, the **fundus**, can be examined through the pupil with an ophthalmoscope. It is reddish in color. In the nasal half lies the **papilla of the optic nerve** (*blind spot*) (**B14**), where all nerve fibers of the retina combine to leave the eye as the optic nerve. The papilla is a whitish disk with a central shallow depression, the *excavation of the optic disk* (**AB15**) (p. 351, D). In the papilla, the central artery divides into several branches, and the veins unite to form the central vein. The arteries are relatively light in color and are thin, while the veins are darker and slightly thicker. The vessels run radially in the nasal direction, while they arch in the temporal direction. Numerous vessels run to the **macula** (*yellow spot*) (**B16**), the *area of highest visual acuity*. Its traversely oval, slightly yellowish surface contains a small depression in the center, the **central fovea** (**AB17**).

A18 Optical part of the retina.
A19 Ciliary part of the retina.
A20 Iridial part of the retina.

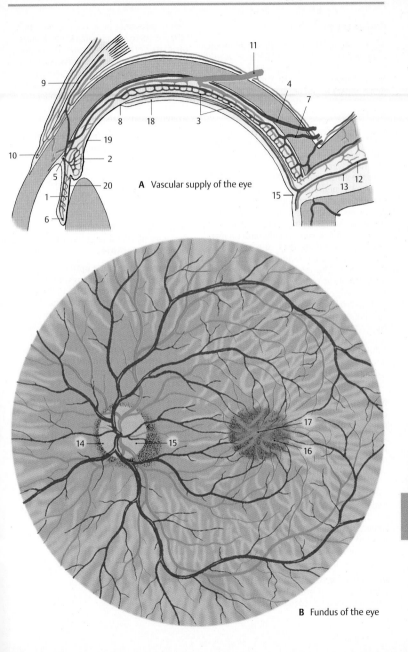

A Vascular supply of the eye

B Fundus of the eye

Retina (A, B)

The **retina** consists of two layers, namely, the outer *pigmented layer* and the inner *neural layer* (see p. 342, A8, A9). The two layers firmly adhere to each other only in the regions of papilla and ora serrata.

Neural Layer

The neural layer contains three layers of cells. Adjacent to the pigmented epithelium (**AB1**) lies the *neuroepithelial layer of the retina*, the layer of photoreceptors. Next follows the *ganglionic layer of the retina*, a layer of bipolar neurons. Last comes the *ganglionic layer of the optic nerve*, a layer of multipolar neurons; their axons form the optic nerve. Thus, the sensory cells of the retina do not face the incoming light with their receptor parts; rather, they turn away from the light and are covered by neurons and nerve fibers. This reverse order is known as *inversion* of the retina. The inner surface of the retina is separated from the vitreous body by a basement membrane, the *internal limiting membrane* (**B2**). A glial membrane, the *external limiting membrane* (**B3**), separates the receptor parts of the sensory cells from the rest of the neural epithelium. Between both membranes expand elongated *supporting glial cells of Müller* (**B4**) with their leaf-like processes.

Neuroepithelial layer (AB5). The **neural epithelium** contains two different types of photoreceptor cells, the retinal **rods** (**B6**) and **cones** (**B7**). The rod cells are for light–dark perception in dim light (night vision), while the cone cells are responsible for color perception in bright light (color vision) and visual acuity (*duplicity theory*). The cell nuclei of the photoreceptors form the *external nuclear layer* (**A8**). The rods and cones rest on the pigmented epithelium and do not function without contacting with the pigmented epithelium. The photoreceptors represent the 1st neuron of the visual pathway.

In lower vertebrates, the pigmented epithelium has a brush border of microvilli (**B9**) that reach between the rods and cones.

When exposed to light, the pigment moves into the microvilli and surrounds the receptor parts of the sensory cells. The pigment returns to the cell body in darkness. In some species, the sensory cells can elongate and press against the pigmented epithelium when exposed to light. Such *retinomotoric movement* has not been clearly demonstrated in mammals.

Ganglionic layer of the retina (AB10). This layer consists of **bipolar neurons** (**B11**) (2nd neuron). Their dendrites extend to the sensory cells and contact the synaptic terminals of the photoreceptors. Their axons have contact with the large neurons of the ganglionic layer of the optic nerve. The cell nuclei of the bipolar neurons form the *inner nuclear layer* (**AB12**). Their synapses lie in the *outer plexiform layer* (**AB13**) and *inner plexiform layer* (**AB14**).

Ganglionic layer of the optic nerve (AB15). The uppermost cell layer consists of a row of large **multipolar neurons** (**B16**) (3rd neuron); their short dendrites form synapses in the inner plexiform layer with the axons of the bipolar neurons. Their axons extend as unmyelinated fibers in the layer of nerve fibers (**AB17**) to the papilla of the optic nerve (blind spot). They form the optic nerve and terminate in the lateral geniculate body (p. 256).

Association cells. There are transverse connections in the individual layers. In the outer plexiform layer they are formed by the axons of the *horizontal cells* (**B18**) and in the inner plexiform layer by the dendrites of the *amacrine cells* (**B19**).

A20 Choriocapillary layer.
A21 Direction of incoming light.

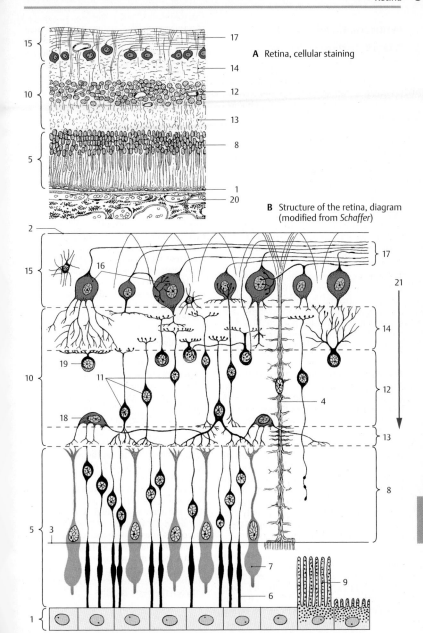

A Retina, cellular staining

B Structure of the retina, diagram
(modified from *Schaffer*)

Retina (continued) (A – C)

Regional Structures of the Retina

The retina is divided into three parts:

- The **optic part of the retina** (p. 347, A18) covers the fundus of the eye and contains the light-sensitive sensory cells (photoreceptors).
- The **ciliary part of the retina** (p. 344, A24; p. 347, A19) lies below the ciliary body and is separated from the optic part by the *ora serrata* (p. 343, B23).
- The **iridial part of the retina** (p. 344, A15; p. 347, A20) covers the posterior surface of the iris.

Neither the ciliary part nor the iridial part contain sensory cells; they consist only of a double-layered pigmented epithelium. They are therefore collectively known as the *blind part of retina*.

The human retina contains approximately 120 million rod cells and 6 million cone cells. Their distributions vary regionally. The rod-to-cone ratio in the vicinity of the central fovea is 2 : 1. The portion of cones decreases toward the periphery, and the lateral region of the retina is dominated by rods.

Macula and central fovea. The **macula lutea** does not contain any rods but only cones, and the unusually long and narrow shape of its cones (**AB1**) differs from the usual structure of cones. The ganglionic layer of the retina and the ganglionic layer of the optic nerve are relatively thick in the macula, but they diminish in the region of the central fovea, leaving only a thin tissue layer on top of the cone cells. Thus, the incoming light has immediate access to the photoreceptors in the fovea. Macula and fovea are the *areas of highest visual acuity*. This is not only due to the absence of the upper retinal layers but also to special neuronal circuits.

Each cone in the central fovea is connected with a single bipolar neuron (**B2**). Each bipolar neuron in the region around the fovea (*perifovea*) synapses with six cone cells. Such convergence (p. 34) is even more pronounced at the *periphery* of the retina

(**C**). Here a single optic nerve neuron may receive input from over 500 receptors. A total of only one million of optic nerve fibers correspond to 130 million photoreceptors.

A circumscribed fiber bundle, the *papillomacular bundle*, extends from the neurons of the macula to the papilla.

Neuronal Circuits (B, C)

The retina is an extremely complex relay system. The eye is not just a sensory organ transmitting light impulses. Rather, these impulses already undergo processing in the retina. Electrophysiological studies have shown that groups of sensory cells are combined to *receptive fields* and react as functional units. A receptive field is formed by the dendritic tree of a large optic nerve ganglion cell and the dendrites of the amacrine interneurons in the inner plexiform layer. Stimulation of a receptive field inhibits neurons in the vicinity. This inhibition is made possible by the transverse connections of the horizontal interneurons in the outer plexiform layer.

Optic Nerve (D)

The nerve fibers of the retina extend in bundles to the *papilla of the optic nerve* (**D3**) where they unite to form the optic nerve before exiting the eyeball. Sclera and choroidea are very thin at the site of penetration, and the sclera is perforated (*lamina cribrosa*) (**D4**). Once the extremely delicate nerve fibers have passed the sclera, they become enveloped by myelin sheaths. The optic nerve is actually a fiber tract of the CNS and contains astrocytes and oligodendrocytes; hence, its nerve fibers do not have Schwann cell sheaths. As part of the brain, the optic nerve is surrounded by mengines. The *dural sheath* (**D5**) and the *arachnoid sheath* (**D6**) merge with the sclera (**D7**). Between *arachnoid sheath* and *pial sheath* (**D8**) lies a CSF-filled space (**D9**) which makes a shift between nerve and sheath possible. A number of septae from the pia mater extend between the nerve bundles.

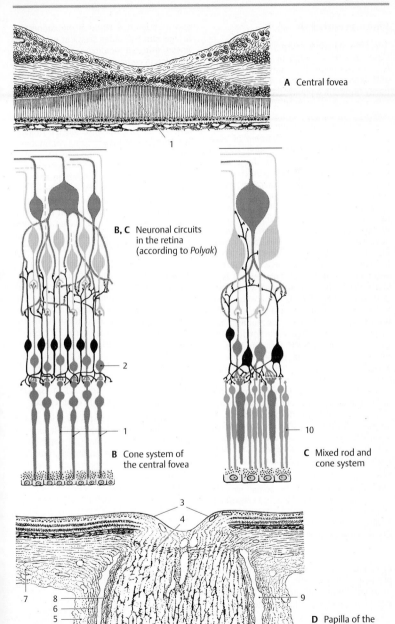

A Central fovea

B, C Neuronal circuits in the retina (according to *Polyak*)

B Cone system of the central fovea

C Mixed rod and cone system

D Papilla of the optic nerve

The Eye

Photoreceptors (A – D)

The light-sensitive sensory cells have the same structural design in all vertebrates. Next to the pigmented epithelium lies the **outer segment** of the photoreceptor cell; it is partly buried in a pigmented epithelial cell. The outer segment of rod cells is a cylinder (**ABC1**) containing several hundreds of stacked, disk-shaped membrane pouches of uniform size. The outer segment of cone cells (**B2, D**) is of a conical shape, and the proximal membrane folds are larger than the distal ones. A thin, eccentric cytoplasmic bridge, the *connecting cilium* (**ACD3**), links the outer segment to the **inner segment** (**AB4**). The bridge contains a modified cilium with 9 pairs of microtubules but without the central pair characteristic of other cilia (see p. 285, D5, D6). The connecting cilium is relatively long in some species, leaving a distinct space between the two segments (**A**). In humans, however, it is so short that both segments touch each other without leaving a visible space between them (**B**). The inner segment contains numerous Golgi stacks, ribosomes, and longitudinally arranged mitochondria. The cell body then tapers to form an axonlike process (**AB5**) containing neurofilaments and microtubules. The cell nucleus (**AB6**) lies either at the transition from the inner segment to the axon or within the axon. The cell terminates with an end bulb, the **synaptic terminal** (**A7**). In addition to the usual synapses, the terminal develops *invaginated synapses* (**A8**) in which the presynaptic membrane becomes invaginated and surrounds the postsynaptic complex on all sides.

The outer segment is the actual *receptor part* of the cell where the light is absorbed. The stacked membranes are formed by infoldings of the plasma membrane (**C9**) in the proximal part of the outer segment. In the rod cells, they detach from the outer membrane and form isolated disks in the distal part of the segment (**C10**). The *visual pigment* of the rod cells, **rhodopsin**, is bound to the membrane of the disks. The formation of visual pigment in the inner segment and its migration through the connecting bridge into the outer segment can be followed using autoradiography by labeling the protein component of rhodopsin with a radioactive amino acid. Once the labeled substance has passed the bridge, it forms a band that migrates to the outer end and then disappears (in rats within 10 days). The migrating band represents a membrane disk that has incorporated labeled rhodopsin. Thus, new disks are being continuously formed in the rod cells, migrate to the distal end, and are shed there. Fragments of the shed disks have been found in the pigmented epithelial cells. There is no new formation of disks in the outer segments of cone cells (**D**). The infoldings of the membrane are permanent and, in contrast to rod cells, there is no detachment of membrane invaginations from the plasma membrane.

Only rod cells contain rhodopsin. The absorption of light changes the molecular structure of rhodopsin, causing it to break down into its protein and pigment components. From these components, rhodopsin is continuously resynthesized in the rods (*rhodopsin–retinin cycle*). It absorbs light of all wavelengths and, thus, is not involved in color vision. Rod cells are **light-dark receptors**. The three different types of cone cells each contain a different pigment that absorbs only light of a specific wavelength. Cone cells are **color receptors**.

There are some animal species in which the retina contains only cones, while it contains only rods in other species (cat, cattle). Animals with a retina containing only rods cannot distinguish between colors. The bull, which is said to react to red, is actually color blind.

The Eye

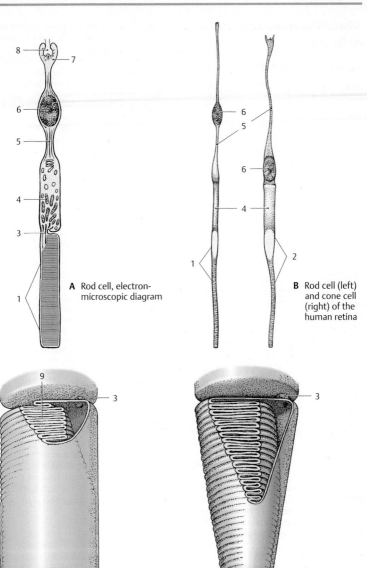

A Rod cell, electron-microscopic diagram

B Rod cell (left) and cone cell (right) of the human retina

C Outer segment of a rod cell

D Outer segment of a cone cell

Visual Pathway and Ocular Reflexes

Visual Pathway (A, B)

The visual pathway consists of four neurons connected in tandem:

- 1st neuron, the **photoreceptors**
- 2nd neuron, the **bipolar neurons** of the retina, which transmit the impulses from the rods and cones to the large ganglion cells of the retina
- 3rd neuron, the **large ganglion cells**, the axons of which combine to form the optic nerve and extend to the primary visual centers (*lateral geniculate nucleus*)
- 4th neuron, the **geniculate cells**, the axons of which project as the optic radiation to the visual cortex (*striate area*)

The **optic nerve** (**A1**) enters the cranial cavity through the optic canal. At the base of the diencephalon, together with the contralateral optic nerve, it forms the **optic chiasm** (**A2**). The fiber bundle starting from the chiasm is known as the **optic tract** (**A3**). The two tracts run around the cerebral peduncles to the two lateral geniculate bodies (**A4**). Before reaching these, each tract divides into a *lateral root* (**A5**) and a *medial root* (**A6**). Whereas most of the fibers run through the lateral root to the lateral geniculate body, the medial fibers continue below the medial geniculate body (**A7**) to the superior colliculi. They contain visual reflex pathways (p. 358, A16). The optic nerve fibers are thought to give off collaterals to the pulvinar of the thalamus (**A8**) prior to terminating in the lateral geniculate body. The optic radiation (*radiation of Gratiolet*) (**B9**) begins at the **lateral geniculate body** and extends as a broad fiber plate to the calcarine sulcus at the medial aspect of the occipital lobe and, while doing so, forms the outward-arching *temporal genu* (**B10**) (p. 258, C16). Numerous fibers bend rostrally (*occipital genu*) (**B11**) in the occipital lobe to reach the anterior areas of the visual cortex.

The optic nerve fibers originating from the nasal halves (**B12**) of the retina cross in the optic chiasm. The fibers from the temporal halves (**B13**) do not cross but continue on the ipsilateral side. Hence, the right tract contains the fibers from the temporal half of the right eye and from the nasal half of the left eye. The left tract contains fibers from the temporal half of the left eye and from the nasal half of the right eye. In a cross section of the tract, the crossed fibers lie mostly ventrolaterally, and the uncrossed fibers dorsomedially; in between, the fibers are mixed.

The crossed and uncrossed fibers of the optic tract extend to different cell layers of the lateral geniculate body (p. 256, B). The number of geniculate cells, approximately one million, corresponds to the number of optic nerve fibers. However, the optic nerve fibers usually end on five to six cells located in different cell layers. Corticofugal fibers of the occipital cortex also end in the lateral geniculate body. They probably control the input of impulses, as suggested by the presence of axo-axonal synapses characteristic for presynaptic inhibition.

The axons of the geniculate cells form the optic radiation. Its fibers are arranged according to the different regions of the retina (p. 356). The fibers for the lower half of the retina, especially those for the periphery of the retina, arch most rostrally in the temporal genu. The fibers for the upper half of the retina and for the central region of the macula arch only slightly in the temporal lobe.

In the **striate area** (**B14**) of the right hemisphere terminate the fibers for the right halves of the retinae; hence, it receives sensory input from the left halves of the visual fields. In the striate area of the left hemisphere there terminate the fibers of the left halves of the retinae with input from the right halves of the visual fields. The right hand and the right visual field are therefore both represented in the left hemisphere, which dominates in right-handed persons (p. 262).

B15 Visual fields.

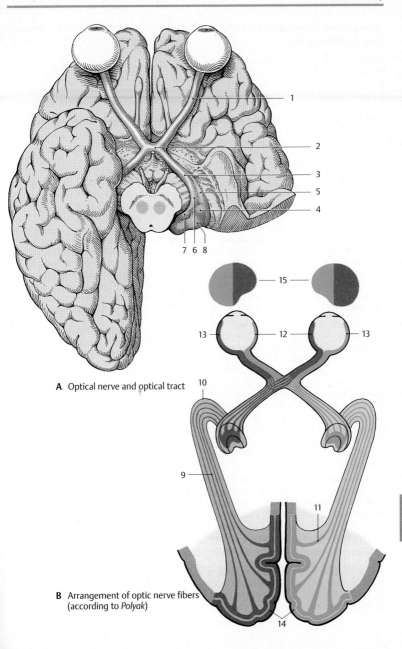

A Optical nerve and optical tract

B Arrangement of optic nerve fibers
(according to *Polyak*)

The Eye

Topographic Organization of the Visual Pathway (A)

The fibers from individual regions of the retina occupy specific positions in the different parts of the optic system. To illustrate this in a simple way, the retina is subdivided into four quadrants that all share the center, namely, the macula with the central fovea (area of highest visual acuity). The fibers of the fovea show a regular point-to-point connection between fovea, lateral geniculate body, and striate area.

The halves of the visual fields of each eye (*visual hemifields*) (**A1**) are projected onto the respective contralateral halves of the respective **retina** (*hemiretinas*) (**A2**). Immediately after the exit of the **optic nerve** from the eyeball (**A3**), the macular fibers lie on the lateral side of the nerve, with the fibers from the nasal half of the macula lying in the center, surrounded by fibers of the temporal half of the macula. Further along the optic nerve, the macular bundle comes to occupy the center (**A4**).

The fibers of the nasal hemiretinas (continuous lines) cross to the opposite side in the **optic chiasm** (**A5**). While doing so they take a strange course. The *medial fibers* cross, then run a short distance into the contralateral optic nerve, and finally turn in a right angle into the contralateral optic tract. The *lateral fibers* run a short distance into the ipsilateral optic tract and then turn abruptly into the contralateral tract. The fibers of the temporal hemiretinas (broken lines) do not cross but continue in the ipsilateral tract.

The **optic tract** (**A6**) thus contains the fibers of the corresponding halves of both retinas: the left tract contains the fibers of the left hemiretinas, the right tract contains the fibers of the right hemiretinas. The fibers of the two superior retinal quadrants lie ventromedially, those of the two inferior quadrants lie dorsolaterally, while the fibers of the macula take a central position. Prior to radiating into the lateral geniculate body (**A7**), the fibers rearrange so that the macular fibers form a central wedge, the fibers

from the upper retinal quadrants come to lie medially, and the fibers from the lower retinal quadrants lie laterally.

The fibers in the **lateral geniculate body** (**A8**) terminate in the same arrangement. The central wedge of the terminal macular fibers makes up almost half of the geniculate body. The fibers from the periphery of the retina terminate in the most anterior and ventral regions of the lateral geniculate body. The terminals of the ipsilateral and contralateral fibers in the geniculate layers are shown schematically in light gray and dark gray (**A9**) (see also p. 257, A). The geniculate cells of the central wedge project to the posterior region of the **striate area** (**A10**). The area of highest acuity, which in the human retina measures slightly more than 2 mm in diameter, is represented by the largest portion of the visual cortex. Rostrally to it lie the much smaller areas for the remaining parts of the retina. The upper quadrants of each retina are represented in the upper lip of the calcarine sulcus, and the lower quadrants in the lower lip.

▦ **Clinical Note:** Corresponding to the arrangement of fibers, injury to the visual pathway in specific segments results in various patterns of loss of vision. It should be taken into consideration that the lower halves of each retina register the input from the upper halves of the visual fields, while the upper halves of each retina register the input from the lower halves of the visual fields. The same is true for the left and right halves of each retina. Injury to the optic tract, lateral geniculate body, or visual cortex on the left side affects the left halves of each retina and the right halves of each visual field. The result is homonymous hemianopia on the right side. In case of bitemporal heteronymous hemianopia, injury to the crossing nasal fibers of both retinae (e.g., in case of tumors of the hypophysis near the optic chiasm) results in bilateral loss of the temporal halves of the visual fields. Damage of both visual cortices causes visual agnosia.

A11 Blind spot (papilla of the optic nerve).

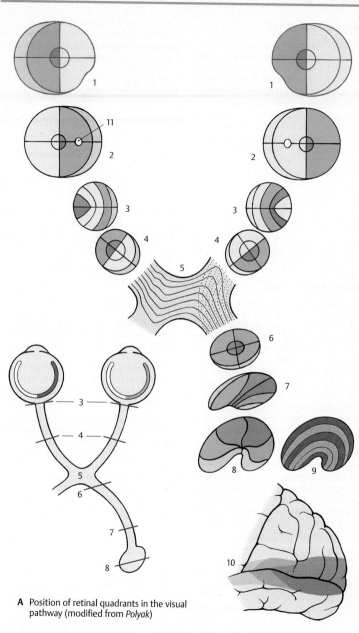

A Position of retinal quadrants in the visual pathway (modified from *Polyak*)

Ocular Reflexes (A – C)

During the visual process the eye must continuously compensate for changes from light to dark and from near to far. Both the aperture and lens system must continuously adapt to the prevailing conditions. Whereas **light–dark adaptation** is achieved by dilatation or contraction of the pupil, **near-far adaptation** requires a change in the curvature of the lens (*accommodation*), a change in the lines of sight (*convergence*), and a change in pupillary width. During *negative accommodation* (adjustment for long distances), the surface of the lens is only slightly curved, the lines of sight run parallel, and the pupils are dilated. During *positive accommodation* (adjustment for short distances), the surface of the lens is distinctly curved, the lines of sight cross at a distance that corresponds to the fixated object, and the pupils are constricted.

Pupillary Light Reflex (A)

When light falls onto the retina, the pupil constricts. The *afferent loop* of this reflex arch is formed by optic nerve fibers (**A1**) extending to the pretectal nucleus (**A2**). The latter is connected with the rostral part of the Edinger–Westphal nucleus (accessory oculomotor nucleus) (**A3**), the fibers of which (**A4**) extend as the *efferent limb* of the reflex arch to the *ciliary ganglion* (**A5**) (p. 128). The postganglionic fibers (**A6**) innervate the sphincter muscle of the pupil (**A7**). Both pretectal nuclei are connected via the epithalamic commissure (**A8**). In addition, the optic nerve fibers of each side terminate in the two pretectal nuclei. This explains the bilaterality of the light reflex; when light falls only onto one eye, the pupil of the other eye constricts as well (*consensual pupillary response*). Ciliospinal center (**A9**); sympathetic fibers (**A10**) for the dilator of the pupil (**A11**).

Accommodation (B)

The accommodation apparatus consists of the lens, its suspension mechanism (ciliary zone), the ciliary body, and the choroidea. These parts form a tense, elastic system which spans the entire eyeball and maintains the flat, slightly curved form of the lens (**B12**) (**negative accommodation**). During **positive accommodation** the circular ciliary muscle (**B13**) contracts. The meridional muscle fibers pull the origins of the long zonular fibers forward, and the circular fibers move the ciliary processes closer to the margin of the lens. This relaxes the zonular fibers (**B14**), and the lens capsule, thus causing the lens to round off (**B15**).

The fiber tracts of the **accommodation reflex** are less well known. As fixation of an object is the prerequisite of accommodation, the optic nerve is the *afferent loop*. The reflex arch runs propably via the visual cortex (striate area) to the pretectal nuclei, possibly also via the superior colliculi (**A16**). The *efferent loop* begins in the caudal part of the Edinger–Westphal nucleus. Its fibers synapse in the ciliary ganglion with postganglionic fibers that innervate the ciliary muscle.

Convergence (A, C)

When an object approaching from a far distance is fixated with both eyes, the medial rectus muscles (**C17**) increasingly adduct both eyeballs, and the lines of sight which initially run parallel (broken arrows) begin to cross each other. The fixated object remains in the intersection of the lines of sight while being projected onto each macula.

The **visual fixation reflex** probably runs via the visual pathway to the occipital cortex and via the corticofugal fibers (**A18**) to the superior colliculi, to the pretectal region, and to the nuclei of the ocular muscles (**A19**). The occipital cortex is therefore regarded as a reflex center (*occipital eye fields*).

The Eye

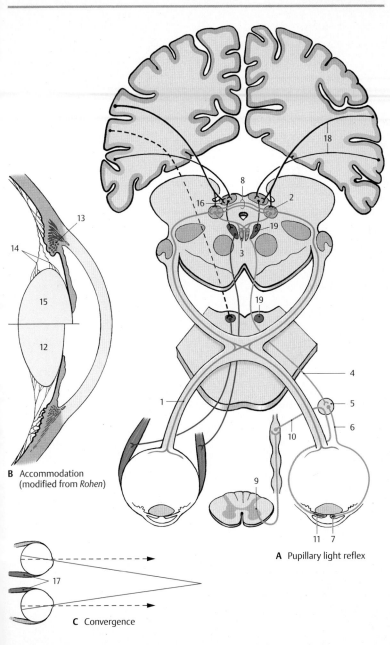

B Accommodation (modified from *Rohen*)

A Pupillary light reflex

C Convergence

The Eye

The Ear

Structure

Overview (A, D)

The ear contains two sensory organs with different functions; morphologically they form a single complex, the inner ear, or *vestibulocochlear organ*. One part of it, the *cochlea*, is the **organ of hearing**, or *spiral organ* (organ of Corti). The other part consists of *saccule*, *utricle*, and *semicircular ducts*; it registers changes in body position, especially those of the head, and represents the **organ of balance**, or *vestibular apparatus*. The ear is divided into three parts: the *external ear*, the *middle ear* (p. 364f), and the *internal ear* (p. 368ff).

The **external ear** includes the auricle (**A**, **D1**) and the external acoustic meatus (**D2**).

The **middle ear** consists of the *tympanic cavity* (**D3**), the *mastoid cells* (air cells) (p. 364, A6), and the *auditory tube* (eustachian tube) (**D4**). The tympanic cavity with the auditory ossicles is a narrow space filled with air. It lies behind the eardrum and extends as the *epitympanic recess* (**D5**) above the external acoustic meatus. The tympanic cavity merges anteriorly into the auditory tube (*tympanic opening*) (**D6**). The tube extends obliquely downward and forward and opens in front of the posterior pharyngeal wall into the pharyngeal cavity (*pharyngeal opening*) (**D7**). The auditory tube is lined with a ciliated epithelium and consists of an osseous and a cartilaginous section that join at the *isthmus of the tube* (**D8**). The *tubal cartilage* (**D24**) leaves open a cleft that is lined by connective tissue (*membranous lamina*). The tendon of the *tensor tympani muscle* (**D9**) attaches to the base of the manubrium of the malleus (p. 365, A25). The tympanic cavity communicates with the pharyngeal cavity through the auditory tube, thus permitting exchange of air and the equilibration of pressure in the middle ear. The opening of the auditory tube, however, is normally closed and opens only when the pharyngeal muscles contract (swallowing).

The **internal ear** consists of the bony labyrinth (**D10**), which contains the membranous labyrinth and the internal acoustic meatus.

Outer Ear (A–C)

The **auricle**, or *pinna* (**A**, **D1**), with the exception of the lobule, contains a scaffold of elastic cartilage. The shapes of the auricular projections and depressions are different in each person and are genetically determined. The shapes of the following parts are inherited: helix (**A11**), antihelix (**A12**), scapha (**A13**), concha (**A14**), tragus (**A15**), antitragus (**A16**), and triangular fossa (**A17**). In the past, the features of the auricle were of special importance for establishing paternity.

The entrance of the **outer ear canal**, or *external acoustic meatus* (**D2**), is formed by a groovelike continuation of the auricular cartilage (**D25**) and is completed with connective tissue to form a uniform passage. The passage is lined with epidermis, and large *ceruminous glands* lie beneath the epidermis.

The outer ear canal ends with the **eardrum**, or *tympanic membrane* (**B**, **D18**), which is obliquely placed in the meatus. When viewed from the outside, the *mallear stria* (**B19**) can be recognized; it is caused by the attachment of the manubrium of the malleus, which reaches to the *umbo of the tympanic membrane* (**B20**), the innermost point of the funnel-shaped eardrum. Above the upper end of the mallear stria (*mallear prominence*) lies a lax, thin part of the eardrum, the reddish *pars flaccida* (**B21**), which is separated from the firm, gray and shiny *pars tensa* (**B22**) by two mallear folds. The eardrum is covered externally by skin and internally by a mucosa. Between these lies the lamina propria of the pars tensa; it contains radial and nonradial fibers (**C**). The latter are circular, parabolic, and transversal. The *fibrocartilaginous annulus* (**C23**) forms the anchoring tissue of the eardrum.

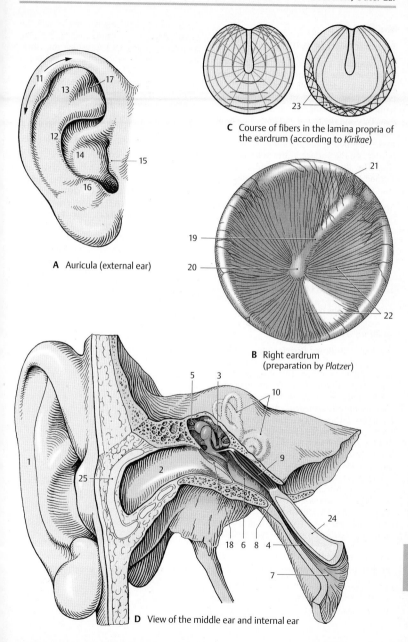

A Auricula (external ear)

C Course of fibers in the lamina propria of the eardrum (according to *Kirikae*)

B Right eardrum (preparation by *Platzer*)

D View of the middle ear and internal ear

The Ear

Middle Ear

Tympanic Cavity (A, B, D)

The **tympanic cavity** is a narrow, vertical space with the eardrum (**AD1**) obliquely placed in its lateral wall. Its medial wall has two openings leading to the internal ear, namely, the *oval window*, or **vestibular window** (**D2**), and the *round window*, or **cochlear window** (**D3**). The roof of the tympanic cavity, the tegmental wall, is relatively thin and borders on the surface of the petrous pyramid. The floor of the tympanic cavity is formed by a thin layer of bone; beneath it runs the jugular vein.

The tympanic cavity continues in anterior direction as the auditory tube (**A4**) (p. 363, D4). Its superior part opens posteriorly into the **mastoid antrum** (**A5**), a round space into which numerous small cavities open, the **mastoid air cells** (**A6**). These air-containing cavities are lined with mucosa and form a system of chambers that penetrates the entire mastoid bone; they may even extend into the petrous bone.

Auditory Ossicles (A, C, D)

The three **auditory ossicles**, or *ear bones*, form together with the eardrum the sound-conducting apparatus. They are called the *hammer*, or **malleus** (**CD7**), the *anvil*, or **incus** (**CD8**), and the *stirrup*, or **stapes** (**CD9**). The *manubrium of the malleus* (**ACD10**) is firmly attached to the eardrum and connected via its neck (**C11**) to the head of the malleus (**C12**). The malleus has a saddle-shaped articular surface that contacts the *body of the incus* (**C13**). Its *lenticular process* (**AC14**), which is attached to the long limb of the incus and projects at a right angle, carries the articular surface for the *head of the stapes* (**C15**). The footplate of the stapes closes the vestibular window; it is attached at the margin by the annular ligament of the stapes (**D16**). Several ligaments (**A17**) connected to the wall of the tympanic cavity keep the ossicles in place.

The auditory ossicles transmit to the inner ear the vibration of the eardrum caused by sound waves. For this purpose, malleus and incus act like an angular lever, and the stapes performs a tilting movement. The footplate of the stapes transmits the vibration to the fluid of the inner ear. The movement of the fluid through the cochlea is simplified in diagram (**D**); in reality, it takes a spiral course inside the cochlea (p. 369, A; p. 371, C). Tension in the system is controlled by two muscles with antagonistic effects, namely, the *tensor tympani muscle* (**A18**) (p. 363, D9) and the *stapedius muscle* (**A19**) (p. 367, C22).

The mucosa lining the tympanic cavity and covering the auditory ossicles forms various folds, such as the anterior (**A20**) and posterior mallear folds (**A21**) which envelope the chorda tympani (**A22**). The folds form several mucosal pouches. The *superior recess of the tympanic membrane*, Prussak's pouch (**D23**), lies between the pars flaccida of the eardrum and the neck of the malleus; it plays an important role in ear infections.

A24 Facial nerve.
A25 Tendon of the tensor tympani muscle.

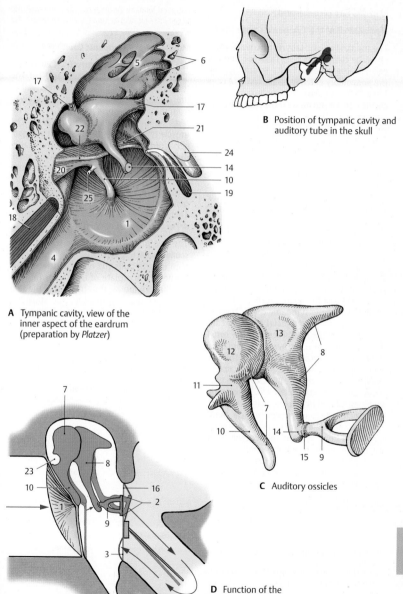

A Tympanic cavity, view of the inner aspect of the eardrum (preparation by *Platzer*)

B Position of tympanic cavity and auditory tube in the skull

C Auditory ossicles

D Function of the auditory ossicles

The Ear

Middle Ear (continued)

Medial Wall of the Tympanic Cavity (A–C)

The medial wall, or **labyrinthic wall**, separates the tympanic cavity from the internal ear. The prominence in its middle region, the **promontory of the tympanic cavity** (**A1**), is caused by the basal convolution of the cochlea. In a bifurcated groove, the *sulcus of the promontory* (**A2**), lies the *tympanic plexus* (**C3**); it is formed by the *tympanic nerve* (**C4**) of the glossopharyngeal nerve and by the sympathetic fibers of the carotid plexus of the internal carotid artery. The promontory is anteriorly delimited by the *tympanic cells* (**A5**). In the medial wall, the *oval window*, **vestibular window** (**A6**), and the *round window*, **cochlear window** (**A7**), open into the inner ear. The *stapes* (**C8**) rests in the vestibular window and closes it with its foot plate. The cochlear window is closed by the *secondary tympanic membrane.* In the posterior wall opening to the *mastoid antrum* (**A9**) run two arched canals, the *facial canal* (**A10**) and the *lateral semicircular canal* (**A11**); both canals cause protrusions on the wall of the tympanic cavity, namely, the *prominence of the facial canal* and the *prominence of the lateral semicircular canal.* A bony protrusion, the *pyramidal eminence* (**A12**), contains an opening at its tip through which the tendon of the *stapedius muscle* (**C13**) enters. In anterior direction, the tympanic cavity leads into the *semicanal of the auditory tube* (**A14**). Above it lies the *semicanal of the tensor tympani muscle* (**A15**). The two semicanals are incompletely separated by a bony septum and together form the *musculotubal canal.* The medial wall at the level of the opening of the tympanic tube (carotid wall) separates the tympanic cavity from the *carotid canal* (**A16**); the bony floor (jugular wall) separates it from the *jugular fossa* (**A17**). Also shown are the *jugular vein* (**B18**) and the *internal carotid artery* (**B19**).

▨▨▨ **Clinical Note:** The osseous roof and floor of the tympanic cavity may be very thin. In case of purulent otitis media, the infection can penetrate either through the roof and progress to meninges and brain (meningitis, cerebral abscess in the temporal lobe), or through the floor and progress to the internal jugular vein (jugular thrombosis).

Muscles of the Tympanic Cavity (C)

The **tensor tympani muscle** (**C20**) originates from the cartilaginous wall of the tympanic tube and from the osseous wall of the canal. Its narrow tendon bends away from the *cochleariform process* (**C21**) and inserts on the manubrium of malleus. The muscle is innervated by the *tensor tympani nerve* of the mandibular nerve. The **stapedius muscle** (**C22**) originates in a small bony canal which usually communicates with the facial canal. Its small tendon passes through the opening of the pyramidal eminence and inserts at the head of the stapes. The muscle is innervated by the *stapedius nerve* of the facial nerve (**C23**).

The two muscles control the **tension of the sound-conducting apparatus**. The tensor tympani muscle pulls the eardrum inward and pushes the footplate of the stapes into the vestibular window, thus increasing the sensitivity of the transmission. The stapedius muscle levers the footplate of the stapes out of the vestibular window and thus dampens the transmission. Hence, the two muscles are antagonists.

▨▨▨ **Clinical Note:** Paralysis of the facial nerve causes loss of function in the stapedius muscle and deficient dampening of sounds; the patients suffer from hyperacusis, an increased sensitivity to sound.

The Ear

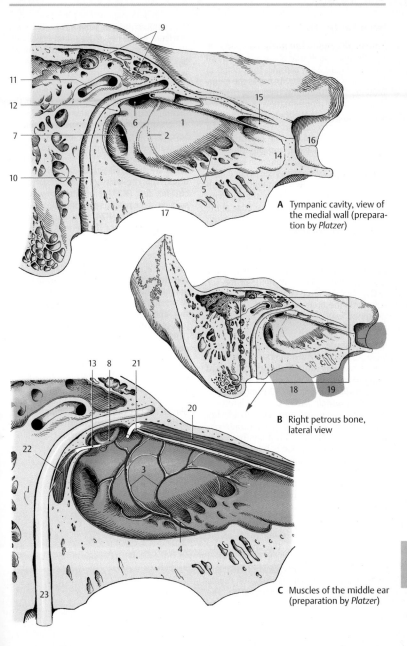

A Tympanic cavity, view of the medial wall (preparation by *Platzer*)

B Right petrous bone, lateral view

C Muscles of the middle ear (preparation by *Platzer*)

Inner Ear (A–D)

The **membranous labyrinth** is a system of vesicles and canals that is surrounded on all sides by a hard, bony capsule. The cavities in the bone have the same shapes as the membranous structures, and their cast (**C**) provides a crude representation of the membranous labyrinth. We therefore distinguish between an osseous (bony) labyrinth and a membranous labyrinth. The **osseous labyrinth** contains a clear, aqueous fluid, the **perilymph** (light greenish-blue), in which the membranous labyrinth is suspended. The perilymphatic space communicates with the subarachnoid space via the *perilymphatic duct* (**A1**) at the posterior edge of the petrous bone. The membranous labyrinth contains the **endolymph** (dark greenish-blue) which is a viscous fluid.

The *vestibular window* (**AC2**) is closed by the stapes and leads into the middle part of the osseous labyrinth, the *vestibule of the ear* (**AC3**). The vestibule communicates anteriorly with the bony *cochlea* (**C4**) and at its posterior wall with the bony *semicircular canals* (**C5**).

The **vestibule** contains two membranous parts, the **saccule** (**AB6**) and the **utricle** (**AB7**). Both structures contain sensory epithelium in a circumscribed part of the wall (blue), the *macula of the saccule* (**AB8**) and the *macula of the utricle* (**AB9**), and are interconnected by the *utriculosaccular duct* (**AB10**). The latter gives off the slender *endolymphatic duct* (**A11**) which runs to the posterior surface of the petrous bone and ends beneath the dura mater as a flattened vesicle, the *endolymphatic sac* (**A12**). The *uniting duct* (**AB13**) forms a connection between the saccule and the membranous cochlear duct.

The **osseous cochlea** (**C4**) has about two and a half turns. The *spiral canal of the cochlea* (**C14**) contains the membranous *cochlear duct* (**AB15**) which starts with a blind end, the *vestibular cecum* (**B16**), and ends in the tip of the cochlea, or *cupula* (**C17**), with the *cupular cecum* (**B18**). The perilymphatic spaces are above and below the cochlear duct, or *scala media*; the *scala vestibuli* (**AB19**) lies above it and opens into the vestibule, and the *scala tympani* (**AB20**) lies beneath it and is closed by the *cochlear window* (**A–C21**).

The three bony *semicircular canals* (**C5**) emanating from the vestibule contain the membranous **semicircular ducts** (**A22**), which are connected to the utricle. They are surrounded by perilymph and attached to the walls of the perilymphatic space by connective-tissue fibers. The three semicircular ducts are arranged perpendicularly to each other. The convexity of the *anterior semicircular duct* (**B23**) is oriented toward the surface of the petrous pyramid, the *posterior semicircular duct* (**B24**) runs parallel to the posterior surface of the petrous bone, and the *lateral semicircular duct* (**B25**) runs horizontally.

Each semicircular duct has a dilatation at its transition to the utricle, the *membranous ampulla* (**B26**), which corresponds to an osseous ampulla in the bony canal. The anterior and the posterior semicircular ducts join to form the *common membranous crus* (**AB27**). Each ampulla contains sensory epithelium, the *ampullary crest*.

The courses taken by the semicircular ducts do not correspond to the axes of the body. The anterior and posterior semicircular ducts diverge from the median and frontal planes by 45°; the lateral semicircular duct is tilted in posterocaudal direction by 30° towards the horizontal plane.

C28 Eardrum.

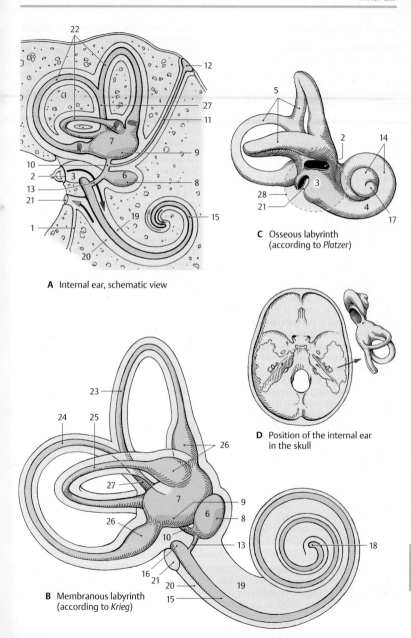

A Internal ear, schematic view

C Osseous labyrinth
(according to *Platzer*)

D Position of the internal ear
in the skull

B Membranous labyrinth
(according to *Krieg*)

The Ear

Inner Ear (continued)

Cochlea (A–C)

The **cochlea** spirals around a conical bony axis, the *central pillar of cochlea*, or **modiolus** (**AC1**), which contains the neurons of the *spiral ganglion* (**AB2**) (p. 377, D9), the nerve fibers originating from them (**AB3**), and the *radix cochlearis* (**A4**) (p. 377, D11) in the center. A double plate of bone, the *osseous spiral lamina* (**A–C5**), protrudes from the modiolus far into the cochlear duct (**A6, B**). It forms a spiral like the cochlea, but it does not reach into the end of the uppermost convolution and terminates in a free, hook-like process, the *hamulus of spiral lamina* (**C7**). The bony spiral lamina is mostly hollow and contains nerve fibers extending to the organ of Corti. Opposite to it at the lateral wall, in the lower half of the basal convolution, lies the *secondary spiral lamina*.

The spiral canal of cochlea contains the membranous **cochlear duct** (*scala media*) (**A–C8**) which is filled with endolymph. Above the duct lies the **scala vestibuli** (**A–C9**) and below it the **scala tympani** (**A–C10**); both these spaces contain perilymph. The lower wall of the cochlear duct is formed by the **basilar membrane** (**B11**) which carries the sensory receptor for hearing, the **organ of Corti** (**B12**). The width of the membrane varies in individual convolutions. The fine fibers of the membrane radiate like a fan prior to attaching to the lateral wall of the cochlear canal, forming the *spiral ligament of cochlea* (**B13**), which looks like a sickle in cross section. Its part above the basilar membrane forms the lateral wall of the cochlear duct; it is known as *vascular stria* (**B14**) because it is rich in endolymph-producing capillaries. The upper wall of the cochlear duct is a thin membrane of double-layered epithelium, **Reissner's membrane** or *vestibular wall of the cochlear duct* (**B15**).

The scala vestibuli communicates with the perilymphatic space of the vestibule and turns into the scala tympani at the **helicotrema** (**AC16**). The scala tympani runs toward the cochlear window (p. 365, D)

which is closed by the *secondary tympanic membrane*. The connection between the two ducts is made possible by separation of the spiral lamina from the modiolus and formation of the hamulus. In this way, the helicotrema is created medially. Only the scala vestibuli and the cochlear duct ascend to the uppermost tip of the cochlea, the cupula (**A17**). In contrast to the rest of the cochlea, the cupula thus contains only two membranous spaces.

Frequency analysis in the cochlea. The oscillations of sound waves are transmitted to the perilymph through the vestibular window via eardrum and auditory ossicles. The resulting movements of the fluid ascend in the scala vestibuli and descend in the scala tympani to the cochlear window, where the waves of movement are absorbed (**C**). Movements of the fluid lead to oscillations of the basilar membrane (traveling waves). The site of maximal displacement of the basilar membrane (and hence stimulation of the receptor cells in the organ of Corti) depends on the frequency of the traveling wave or stimulating sound. High tonal frequencies cause maximal displacement of the basilar membrane in the basal convolutions (where the basilar membrane is narrow), middle frequencies in the middle of the cochlea, and low frequencies in the uppermost convolutions (where the basilar membrane is wide). Hence, different frequencies are registered in different parts of the cochlea, namely, frequencies of 20 000 Hz in the basal convolutions and frequencies of 20 Hz in the uppermost convolutions. This local arrangement provides the basis of the tonotopic organization of the acoustic system (p. 381, C).

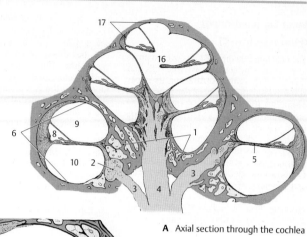

A Axial section through the cochlea

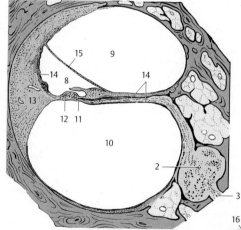

B Spiral canal of cochlea

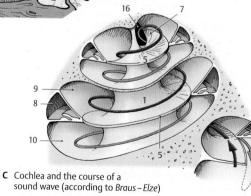

C Cochlea and the course of a
sound wave (according to *Braus–Elze*)

Inner Ear (continued)

Organ of Corti (A–C)

The **basilar membrane** (**AB1**) is covered on its lower surface by connective tissue cells that line the scala tympani (**AB2**); on its upper surface rests the *organ of Corti*, or **spiral organ** (**A3, B**). Lateral to the spiral organ, the epithelium continues as the *vascular stria* (**A4**), which contains numerous intra-epithelial capillaries. Medial to the spiral organ, at the margin of the bony spiral lamina, lies a thickened tissue layer derived from the inner periosteum, the *limbus of the spiral lamina* (**A5**); it is covered by epithelial cells. The limbus tapers off into two lips, the *tympanic lip of the limbus* (**A6**) and the *vestibular lip of the limbus* (**A7**), which enclose the *internal spiral sulcus* (**A8**).

The organ of Corti extends in a spiral from the basal convolution to the cupula of the cochlea. The figures (**A, B**) show cross sections through the organ; it consists of sensory cells and a variety of supporting cells. The center is occupied by the *inner tunnel* (**B9**), which contains a perilymph-like fluid, the *cortilymph*. Its medial wall is formed by the **inner pillar cells** (**B10**) and its lateral wall by the obliquely arranged **outer pillar cells** (**B11**). The pillar cells have a wide base (**B12**) containing the cell nucleus, a narrow middle part, and a head. They contain long bundles of supporting tonofilaments. The inner pillar cell forms the head plate (**B13**), while the outer pillar cell forms a round head part (**B14**), which approaches the head plate from below, as well as a flat process, the *phalangeal process* (**B15**). Laterally follows the group of **Deiters' supporting cells** (outer phalangeal cells) (**B16**); these carry the sensory cells (**B17, C**) on a protrusion of their lower part. Their tonofilament bundles branch below the sensory cells to form a *support basket* (**B18**). From each Deiters' cell, a narrow process ascends between the sensory cells and ends with a flat phalangeal process (**C19**). The phalangeal processes together form a perforated superficial membrane, the *reticular membrane*, and the apical ends of the sensory cells are firmly secured in the openings of the mem-

brane. Between the outer pillar cells and the Deiters' cells lies the *inner tunnel*, or *Corti's space* (**B20**), and laterally to the Deiters' cells lies the small *outer tunnel*, or *Nuel's space* (**B21**). This is followed by a group of simple, elongated supporting cells that merge into the epithelium of the vascular stria to form the medial wall of the *external spiral sulcus* (**A22**). The inner phalangeal cells border onto the inner pillar cells.

The sensory cells (**C**) include the **inner hair cells** (**C23**), which form only one row, and the **outer hair cells** (**C24**); the latter form three rows in the basal convolution of the cochlea, four rows in the middle convolution, and five rows in the upper convolution. All hair cells have below their apical surface a dense terminal web of microfilaments (**C25**) through which the sensory hairs (**C26**) are firmly secured. The sensory hairs are stiff, specialized microvilli (stereovilli) which are arranged in a semicircle, usually in three rows of different lengths. Nerve fibers with synapse–like contacts (**C27**) terminate at the base of the hair cells.

A nonoscillating gelatinous mass, the **tectorial membrane** (**AB28**), lies on top of the hair cells; it covers the spiral limbus and extends beyond its vestibular lip. It is not clear whether or not the stereovilli of the sensory cells are attached to the tectorial membrane and their tangential deflection during oscillation of the basilar membrane is brought about by a shift against the tectorial membrane. It is also possible that the stereovilli do not touch the tectorial membrane at all and are merely moved by the flow of endolymph.

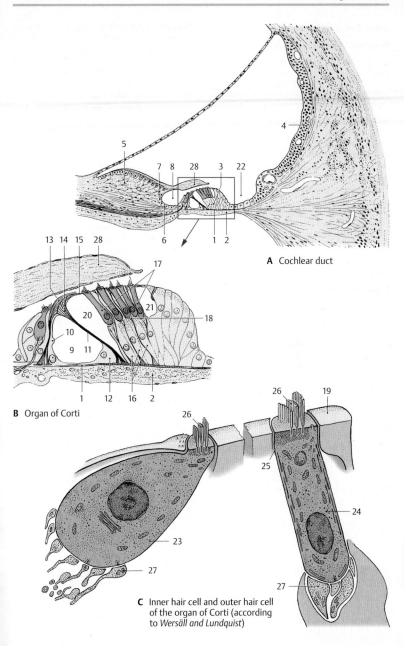

A Cochlear duct

B Organ of Corti

C Inner hair cell and outer hair cell
of the organ of Corti (according
to *Wersäll and Lundquist*)

Inner Ear (continued)

Vestibular Apperatus (A–D)

Saccule, utricle, and the three *semicircular ducts* emanating from the utricle form the organ of balance, the **vestibular apparatus**. It contains several sensory fields, namely, the two acoustic maculae, **macula of saccule** and **macula of utricle**, and the three **ampullary crests**. They all register acceleration and positional changes and, therefore, serve spatial orientation. The maculae react to linear acceleration in different directions, the crests react to rotational acceleration. The maculae occupy specific positions in space (p. 369, B8, B9); the macula of the utricle lies roughly horizontally on the floor of the utricle, and the macula of the saccule lies vertically on the anterior wall of the saccule. Thus, both are arranged at right angles to each other. (For positions of the semicircular ducts, see p. 369.)

Acoustic maculae (A). The epithelium lining the endolymphatic space increases in height in the oval areas of the maculae and differentiates into supporting cells and sensory cells. The *supporting cells* (**A1**) carry and surround the *sensory cells* (**A2**). Each sensory cell is shaped like a flask or ampoule and bears 70–80 stereovilli on its apical surface (**A3**). The sensory epithelium is surmounted by a gelatinous membrane, the *statoconic (otolithic) membrane* (**A4**), which carries crystalline particles of calcium carbonate, the ear crystals, or *statocones (otoliths)* (**A5**). The stereovilli of the sensory cells do not directly project into the statoconic membrane but are surrounded by a narrow space containing endolymph.

Function of the maculae. The proper stimulus for the stereovilli is a shearing force affecting the macula; with increasing *acceleration* there is a tangential shift between sensory epithelium and statolithic membrane. The resulting deflection of the stereovilli leads to stimulation of the sensory cell and to induction of a nerve impulse.

Ampullary crest (B, C). The crest (**BC6**) is formed by a protrusion in the ampulla and is oriented transversely to the course of the semicircular duct (**C**). Its surface is covered by *supporting cells* (**B7**) and *sensory cells* (**B8**). Each sensory cell bears approximately 50 stereovilli (**B9**) that are considerably longer than those of the macular cells. The ampullary crest occupies about one third of the height of the ampulla. It is surmounted by a gelatinous cap, the *ampullary cupula* (**B–D10**), which reaches to the roof of the ampulla. The cupula is traversed by long channels into which the hair bundles of the sensory cells protrude. The bases of the sensory cells are innervated by nerve endings (**A–C11**).

Function of the semicircular canals (D). The semicircular canals respond to *rotational acceleration* which sets the endolymph in motion. The resulting deflection of the ampullary cupula bends the stereovilli of the sensory cells and acts as the triggering stimulus. For example, if the head is turned to the right (red arrows), the endolymph of the lateral semicircular duct initially remains in place due to its inertia; this results in a relative movement in the opposite direction (hydrodynamic inertia, black arrows) so that both cupulae are deflected toward the left (**D12**). The endolymph then slowly follows the rotation of the head. However, once the rotation has stopped (broken, arrested arrows), it continues to flow for a certain distance in the same direction so that the cupulae are deflected to the right (**D13**). The function of the semicircular ducts serves primarily the reflex eye movements. Rapid eye movements caused by rotation of the head (*rotatory nystagmus*) depend on cupular deflection. The slow component of nystagmus always follows the direction of the cupular deflection.

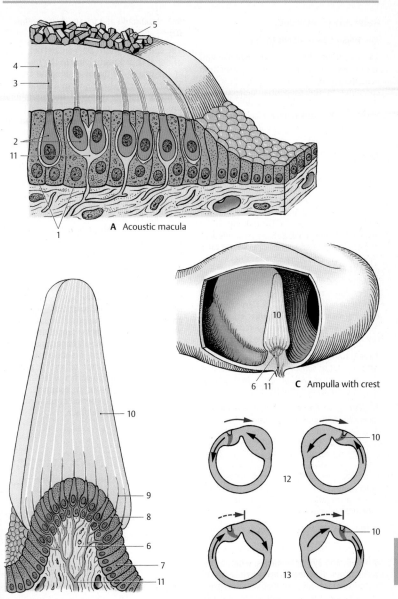

A Acoustic macula

B Ampullary crest

C Ampulla with crest

D Function of the semicircular canals (according to *Trincker*)

Inner Ear (continued)

Vestibular Sensory Cells (A–C)

The hair cells of the acoustic maculae and ampullary crests share the same structural principle. They are mechanoreceptors responding to the tangential deflection of their sensory hairs. There are two types of vestibular hair cells: *type I cells* have the shape of a flask, while *type II cells* have the shape of a cylinder.

Type I hair cells (**A1**) have a round cell body and a narrow neck; below their apical surface lies a dense terminal web (**A2**). The apical cell surface is differentiated into approximately 60 *stereovilli* (**A3**) of graduated lengths and a single, very long motile *kinocilium* (**A4**) with a basal body at its origin. Each type I hair cell is surrounded on its lateral and basal surfaces by a **nerve calyx** (**A5**), which is formed by a thick nerve fiber. The upper part of the calyx contains vesicles and closely adjoins the hair cell; this region is therefore regarded as the synaptic part of the hair cell. The nerve calyx, in turn, is contacted by heavily granulated nerve endings (**A6**), possibly representing the terminals of efferent nerve fibers.

Type II hair cells (**A7**) are equipped with an identical set of sensory hairs. At the base of the cell lie more (**A8**) or less heavily granulated terminals of nerve fibers.

All hair cells of a sensory field show a uniform orientation of their kinocilia (**B**). Electrophysiological studies have shown that bending of the stereovilli toward the kinocilium results in stimulation (green arrow), while bending in the opposite direction leads to inhibition (red arrow) (**C**). Movements in intermediate directions cause stimulation or inhibition below the threshold of sensation.

In this way, the vestibular apparatus is able to register each movement precisely. The *semicircular ducts* (**kinetic labyrinth**) control primarily the eye movements, whereas the *acoustic maculae* (**tonic labyrinth**) directly affect muscular tone, in particular, the tension of extensor muscles and neck muscles (p. 383 C).

Spiral Ganglion and Vestibular Ganglion (D)

The **spiral ganglion** (**D9**) consists of a chain of nerve cell clusters lying in the modiolus at the margin of the osseous spiral lamina. Together these clusters form a spiral. They contain bipolar neurons, the peripheral processes (dendrites) of which extend to the hair cells of the organ of Corti. Their central processes (axons) run as *foraminous spiral tract* (**D10**) to the axis of the modiolus, where they unite to form the *radix cochlearis* (root of the cochlear nerve) (**D11**).

The **vestibular ganglion** (**D12**) lies at the floor of the internal acoustic meatus. It consists of a superior part and an inferior part. The bipolar neurons of the *superior part* (**D13**) send their peripheral processes to the ampullary crests of the anterior semicircular duct (**D14**) (anterior ampullary nerve) and the lateral semicircular duct (**D15**) (lateral ampullary nerve), to the macula of the utricle (**D16**) (utricular nerve) and to part of the macula of the saccule (**D17**). The neurons of the *inferior part* (**D18**) supply the ampullary crest of the posterior semicircular duct (**D19**) (posterior ampullary nerve) and part of the macula of the saccule (saccular nerve). The central processes form the *radix vestibularis* (root of the vestibular nerve) (**D20**), which together with the radix cochlearis extends in a common nerve sheath through the *internal acoustic meatus* into the middle cranial fossa.

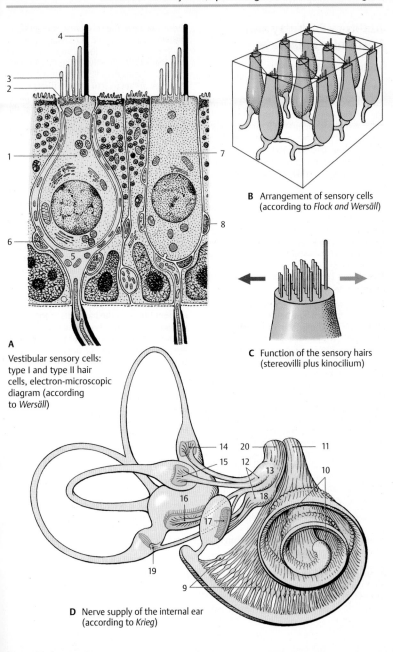

B Arrangement of sensory cells (according to *Flock and Wersäll*)

A
Vestibular sensory cells: type I and type II hair cells, electron-microscopic diagram (according to *Wersäll*)

C Function of the sensory hairs (stereovilli plus kinocilium)

D Nerve supply of the internal ear (according to *Krieg*)

The Ear

Auditory Pathway and Vestibular Pathways

Auditory Pathway

Cochlear Nuclei (A, B)

The fibers of the **radix cochlearis** (**A1**) enter the medulla oblongata at the level of the **anterior cochlear nucleus** (**AB2**) and bifurcate. The ascending branches extend to the **posterior cochlear nucleus** (**AB3**), and the descending branches to the anterior cochlear nucleus. The projection of the cochlea onto the nuclear complex is highly organized; fibers from the basal cochlear convolutions terminate in the dorsomedial parts of the nuclei, while fibers from the uppermost convolutions terminate in the ventrolateral parts. This regular distribution of the afferent fibers is the basis for the subdivision of the cochlear nuclei according to tonal frequencies (see p. 370, frequency analysis in the cochlea).

Such a *tonotopic organization* of the cochlear complex can be demonstrated by electrical recordings in an experimental animal (cat) (**B**). Recordings from individual neurons and simultaneous sonication with different sounds can detect the frequency at which an individual cell responds best. The frontal section through the oral region of the cochlear nuclei shows how the electrode, which scans from top to bottom, registers in the posterior nucleus (**B3**) the recorded points in a precise sequence from high to low frequencies. Here lie the neurons for specific tonal frequencies in a regular sequence. When the electrode enters the anterior nucleus (**B2**), this sequence suddenly stops and frequencies oscillate only in a specific range.

The secondary fibers of the auditory pathway originate from the neurons of the cochlear nuclei. Fiber bundles from the anterior cochlear nucleus cross to the opposite side as a broad fiber plate mixed with nerve cells, the **trapezoid body** (**A4**) (p. 111, AB15), and then ascend as the **lateral lemniscus** (**A5**) (p. 133, D22) to the inferior colliculus (**A6**). Fibers originating from the pos-

terior cochlear nucleus cross obliquely as *posterior acoustic striae* (**A7**). A large portion of the lemniscus fibers run from the cochlear nuclei directly to the inferior colliculus. A considerable number of fibers, however, are relayed to tertiary fibers in the intermediate nuclei of the auditory pathway, namely, in the **posterior nucleus of the trapezoid body (superior olive)** (**A8**), in the **anterior nucleus of the trapezoid body** (**A9**), and in the **nuclei of the lateral lemniscus**. A tonotopic organization has been demonstrated in the nucleus of the posterior trapezoid body. The accessory nucleus lying medially to it, the **medial nucleus of the superior olive** (**A10**), receives fibers from the cochlear nuclei of both sides and is interposed into a fiber system serving *directional hearing*. The fiber connections from the posterior nucleus of the trapezoid body to the abducens nucleus (**A11**) (*reflex eye movements resulting from the sensation of sound*) are still disputed. The fibers are thought to extend beyond the abducens nucleus to the contralateral cochlear nucleus and terminate as efferent fibers on the hair cells of the organ of Corti. They probably control the inflow of stimuli. The lateral nuclei of the lemniscus are scattered clusters of nerve cells interposed in the course of the lateral lemniscus. From the *posterior nucleus of the lateral lemniscus* (**A12**), fibers cross to the contralateral lemniscus (Probst 's commissure) (**A13**).

Inferior Colliculus (A)

The predominating portion of the lateral lemniscus terminates in the principal nucleus of the inferior colliculus in a topical pattern. Electrophysiological studies have demonstrated a tonotopic organization of this nucleus. It is a relay station for acoustic reflexes, from where acoustico-optic fibers extend to the superior colliculi and tectocerebellar fibers to the cerebellum. The inferior colliculi are interconnected by the *commissure of the inferior colliculi* (**A14**).

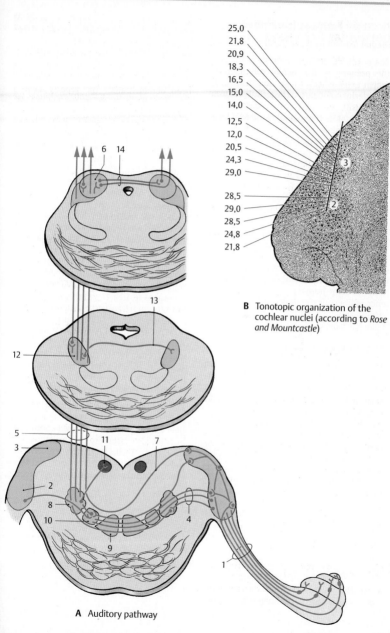

B Tonotopic organization of the cochlear nuclei (according to *Rose and Mountcastle*)

A Auditory pathway

Auditory Pathway (continued)

Medial Geniculate Body (A, B)

The next section of the auditory pathway is the **peduncle of the inferior colliculi** (**A1**), which extends as a strong fiber tract from the inferior colliculus to the **medial geniculate body** (**AB2**), from where the *acoustic radiation* originates. The medial geniculate body is thought also to contain somatosensory fibers from the spinal cord as well as cerebellar fibers. Obviously, it is not solely a relay station of the acoustic system but is also involved in other systems. Some fiber bundles of the peduncle of the inferior colliculi are derived from the trapezoid nuclei and reach the medial geniculate body without synapsing in the inferior colliculus. The two medial geniculate bodies are thought to be interconnected by crossing fibers running through the *inferior supraoptic commissure* (Gudden's commissure) (**A3**). It is uncertain whether such connections exist in humans. However, the presence of fibers descending from the auditory cortex and terminating in the geniculate body has been confirmed.

Acoustic Radiation (A, B)

The fibers of the **acoustic radiation** (**AB4**) run from the medial geniculate body transversely through the inferior posterior part of the internal capsule and ascend vertically in the temporal lobe to the auditory cortex. The fibers show a topical organization with individual sections of the medial geniculate body projecting to specific regions of the auditory cortex. Within the acoustic radiation, the fibers undergo a spiral rotation so that the rostral parts of the geniculate body project to the caudal cortical areas and the caudal parts of the geniculate body project to rostral cortical areas (**B**). This rotation has been demonstrated experimentally in monkeys and during maturation of the myelin sheath in humans.

Auditory Cortex (A–C)

Electrical recordings from the cortex of various experimental animals (cat, monkey) under simultaneous exposure to sounds of different frequencies have uncovered the *tonotopic organization* of the auditory cortex (**AB5**, **C**), where the uncoiled cochlea is represented from the basal convolution to the cupula. Three regions of hearing have been demonstrated: the **primary auditory area** (**AI**) (**C6**), the **secondary auditory area** (**AII**) (**C7**), and the area of the **posterior ectosylvian gyrus** (**Ep**) (**C8**). In the primary auditory area, the neurons responding best to high frequencies (large blue dots) lie rostrally; the neurons responding best to low frequencies (small blue dots) lie caudally. In the secondary auditory area, the frequency sensitivity is arranged in the reverse order. Auditory area AI is the *primary terminal of the acoustic radiation*, while auditory areas AII and Ep are regarded as *secondary auditory areas*. The relationships may be compared to those of the visual cortex, where area 17 is the terminal of the optic radiation while areas 18 and 19 are secondary integration areas. Auditory area AI corresponds in humans to area 41, which covers Heschl's convolutions (transverse temporal gyri) and is the terminal of the acoustic radiation (p. 252, C1). Areas 42 and 22 are secondary auditory areas and include Wernicke's speech center for the comprehension of languages (p. 262, A1). Hence, the auditory cortex must be regarded as a region much larger than Heschl's convolutions.

The auditory pathway has several commissural systems along its course that provide fiber exchanges at various levels. Some fiber bundles also ascend to the ipsilateral auditory cortex. The latter therefore receives impulses from both organs of Corti, and this is especially important for directional hearing.

The Ear

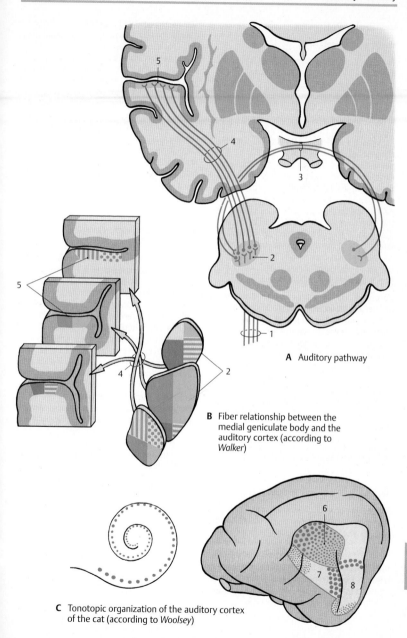

A Auditory pathway

B Fiber relationship between the medial geniculate body and the auditory cortex (according to *Walker*)

C Tonotopic organization of the auditory cortex of the cat (according to *Woolsey*)

Vestibular Pathways

Vestibular Nuclei (A, B)

The fibers of the **radix vestibularis** (**A1**) enter the medulla oblongata at the level of the **lateral vestibular nucleus** (Deiters' nucleus) (**AB2**) and bifurcate into ascending and descending branches that terminate in the **superior vestibular nucleus** (Bechterew's nucleus) (**AB3**), in the **medial vestibular nucleus** (Schwalbe's nucleus) (**AB4**), and in the **inferior vestibular nucleus** (**AB5**) (p. 120, B). The nerve fibers for different parts of the labyrinth extend to specific regions of the nuclear complex. Fiber bundles for the *macula of the saccule* (**B6**) terminate in the lateral part of the inferior nucleus, while fibers for the *macula of the utricle* (**B7**) end in the medial part of the inferior nucleus and in the lateral part of the medial nucleus. Fibers for the *ampullary crests* (**B8**) terminate primarily in the superior nucleus and in the upper part of the medial nucleus.

Certain groups of neurons respond to linear acceleration and others to rotational acceleration. Some neurons respond to ipsilateral rotation, others to contralateral rotation. The vestibular complexes of both sides are interconnected by commissural fibers, through which some nerve cell clusters are stimulated by the labyrinth of the contralateral side. In addition to labyrinthine fibers, cerebellar fibers from the vermis and the fastigial nuclei (p. 164, B) as well as spinal fibers transmitting impulses from the joint receptors terminate in the nuclear complex. Efferent fibers for central control run from the vestibular nuclei back to the sensory epithelia.

Secondary Vestibular Pathways (A, C)

These are connections to the spinal cord, to the reticular formation, to the cerebellum, and to the oculomotor nuclei. The **vestibulospinal tract** (**A9**) originates from the neurons of Deiters' nucleus (lateral vestibular nucleus) and reaches into the sacral spinal cord. Its fibers terminate at the spinal interneurons and activate the α- and γ-motoneurons of the extensor muscles.

The numerous fibers extending to the reticular formation stem from all vestibular nuclei. To the cerebellum run direct fibers from the vestibular ganglion as well as fiber bundles from the medial and inferior vestibular nuclei. They terminate in the nodulus and flocculus (**A10**) and in parts of the uvula (vestibulocerebellum, p. 152, A6; p. 164, B). The fibers ascending to the oculomotor nuclei (**AC11**) stem primarily from the medial and superior vestibular nuclei; they form part of the medial longitudinal fasciculus (**A12**). There is also a vestibulocortical connection through the thalamus (ventral intermediate nucleus, p. 184, B13). Electrophysiological studies resulted in a projection of vestibular impulses to a small region in the anterior postcentral area near the facial region of the sensory homunculus (p. 251, C).

Interaction of eye muscles, neck muscles, and organ of balance (C). The connection between vestibular complex and oculomotor nuclei brings circumscribed groups of neurons into contact with each other. Nerve cell clusters receiving impulses from a specific semicircular duct are probably connected with nerve cell clusters innervating a specific eye muscle. This would explain the exceptionally precise interaction of vestibular apparatus, eye muscles, and neck muscles, thus permitting fixation of an object even during movements of the head. We always perceive a stationary, vertical image of our surroundings despite our head movements. To guarantee such a constant visual impression, each head movement is compensated for by rotation of the eyeballs. The finely tuned interaction of neck muscles and ocular muscles is controlled by the vestibular apparatus via γ-motoneurons (**C13**).

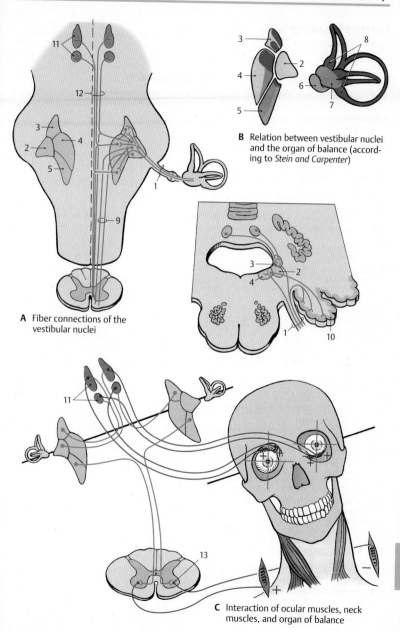

B Relation between vestibular nuclei and the organ of balance (according to *Stein and Carpenter*)

A Fiber connections of the vestibular nuclei

C Interaction of ocular muscles, neck muscles, and organ of balance

Further Reading

Textbooks, general

Ariëns Kappers, C. U., G. C. Huber, E. C. Crosby: The Comparative Anatomy of the Nervous System of Vertebrats, Including Man. Hafner, New York 1936, Reprint 1960

Brodal, A.: Neurological Anatomy. Oxford University Press, Oxford 1981

Carpenter, M. B.: Core Text of Neuroanatomy. Williams & Wilkins, Baltimore 1978

Clara, M.: Das Nervensystem des Menschen. Barth, Leipzig 1959

Creutzfeld, O. D.: Cortex Cerebri. Springer, Berlin, Heidelberg, New York, Tokyo 1983

Curtis, B. A., S. Jakobson, E. M. Marcus: An Introduction to the Neurosciences. Saunders, Philadelphia 1972

Dejerine, J.: Anatomie des centres nerveux. Rueff, Paris 1895–1901

Eccles, J. C.: Das Gehirn des Menschen. Piper, München 1973

Friede, R. L.: Topographic Brain Chemistry. Academic Press, New York 1966

Greger, R., U. Windhorst: Comprehensive Human Physiology. From Cellular Mechanisms to Integration. Vol. I und II. Springer, Berlin, Heidelberg, New York 1996

Kandel, E. R., J. H. Schwartz, T. M. Jessell: Principles of Neural Science. 3rd Edition. Appleton & Lange 1991

Ludwig, E., J. Klingler: Atlas cerebri humani. Karger, Basel 1956

Nieuwenhuys, R., J. Voogd, Chr. van Huizen: The Human Nervous System: A Synopsis and Atlas, 3rd ed. Springer, Berlin 1988

Retzius, G.: Das Menschenhirn. Norstedt, Stockholm 1896

Schaltenbrand, G., P. Bailey: Einführung in die stereotaktischen Operationen mit einem Atlas des menschlichen Gehirns. Thieme, Stuttgart 1959

Schmidt, R. F., G. Thews (Hrsg.): Physiologie des Menschen, 26. Aufl. Springer, Berlin 1995

Sidman, R. L., M. Sidmann: Neuroanatomie programmiert. Springer, Berlin 1971

Thompson, R. F.: Das Gehirn. Von der Nervenzelle zur Verhaltenssteuerung. 2. Auflage. Spektrum, Heidelberg, Berlin, Oxford 1994

Villiger, E., E. Ludwig: Gehirn und Rückenmark. Schwabe, Basel 1946

Zigmond, M. J., F. E. Bloom, S. C. Landis, J. L. Roberts, L. R. Squire: Fundamental Neuroscience. Academic Press, San Diego, London, Boston 1999

Introduction

Bullock, Th. H.: Introduction to Nervous System. San Francisco 1977

Bullock, Th. H., G. A. Horridge: Structure and Function in the Nervous System of Invertebrates. University Chicago Press, Chicago 1955

Drescher, U., A. Faissner, R. Klein, F. G. Rathjen, C. Stürmer (eds.): Molecular bases of axonal growth and pathfinding. Cell Tissue Res. Special Issue 290 (1997) 187–470

Eccles, J. C.: How the Self Controls the Brain. Springer, Berlin 1994

Edelman, G. M., W. E. Gall, W. M. Cowan (eds.): Molecular Bases of Neural Development. John Wiley & Sons, New York, Chichester, Brisbane 1985

Goodman, C. S., C. J. Shatz: Developmental mechanisms that generate precise patterns of neuronal connectivity. Cell 72/Neuron 10 (Suppl.) (1993) 77–98

Hamburger, V.: The Heritage of Experimental Embryology. Hans Spemann and the Organizer. Oxford University Press, New York 1988

Herrick, J. C.: Brains of Rats and Men. University Chicago Press, Chicago 1926

Herrick, J. C.: The Evolution of Human Nature. University Texas Press, Austin 1956

Kahle, W.: Die Entwicklung der menschlichen Großhirnhemisphäre. Springer, Berlin, Heidelberg 1969

Kölliker, A.: Entwicklungsgeschichte des Menschen und der höheren Tiere. 2. Ausgabe. W. Engelmann, Leipzig 1879

Le Gros Clark, W. E.: Fossil Evidence for Human Evolution. University Chicago Press, Chicago 1955

Le Gros Clark, W. E.: The Antecedents of Man. Edinburgh University Press. Edinburgh 1959

Sherrington, Sir Charles: Körper und Geist – Der Mensch über seine Natur. Schünemann, Bremen 1964

Sperry, R. W.: Chemoaffinity in the orderly growth of nerve fiber patterns and connections. Proc. Natl. Acad. Sci. USA 50 (1963) 703–710

Tessier-Lavigne, M., C. S. Goodman: The molecular biology of axon guidance. Science 274 (1996) 1123–1133

Tobias, P. V.: The Brain in Hominid Evolution. Columbia University Press, New York 1971

Basic Elements of the Nervous System

Akert, K., P. G. Waser: Mechanisms of Synaptic Transmission. Elsevier, Amsterdam 1969

Babel, J., A. Bischoff, H. Spoendlin: Ultrastructure of the Peripheral Nervous System and Sense Organs. Thieme, Stuttgart 1970

Barker, J. L.: The Role of Peptides in Neuronal Function. Dekker, Basel 1980

de Belleroche, J., G. J. Dockray: Cholecystokinin in the Nervous System. Verlag Chemie, Weinheim 1984

Björklund, A., T. Hökfeld: Classical Neurotransmitters in the CNS. Handbook of Chemical Neuroanatomy. Elsevier, Amsterdam 1984

Bloom, F. E.: The functional significance of neurotransmitter diversity. Am. J. Physiol. 246 (1984) C184–C194

Cajal, S. R.: Histologie du système nerveux de l'homme et des vertébrés. Maloine, Paris 1909–1911

Causey, G.: The Cell of Schwann. Livingstone, Edinburgh 1960

Cold Spring Harbour Symposia 40: The Synapse. Cold Spring Harbour Laboratory, New York 1976

Cottrell, G. A., P. N. R. Usherwood: Synapses. Blackie, Glasgow 1977

Cowan, W. M., M. Cuenod: The Use of Axonal Transport for Studies of Neuronal Connectivity. Elsevier, Amsterdam 1975

De Robertis, E. D. P., R. Carrea: Biology of Neuroglia. Elsevier, Amsterdam 1965

Eccles, J. C.: The physiology of synapses. Springer, Berlin, Göttingen, Heidelberg, New York 1964

Emson, P. C.: Chemical Neuroanatomy. Raven, New York 1984

Eränkö, O.: Histochemistry of Nervous Transmission. Elsevier, Amsterdam 1969

Fedoroff, S., A. Vernadakis: Astrocytes. Academic Press, London 1986

Frotscher, M., U. Misgeld (eds.): Central Cholinergic Synaptic Transmission. Birkhäuser, Basel, Boston, Berlin 1989

Fuxe, K., M. Goldstein, B. Hökfeld, T. Hökfeld: Central Adrenalin Neurons. Pergamon, Oxford 1980

Gray, E. G.: Axo-somatic and axo-dendritic synapses of the cerebral cortex: an electron microscope study. J. Anat. 93 (1959) 420–433

Heimer, L., L. Záborszky (eds.): Neuroanatomical Tract-Tracing Methods 2. Recent Progress. Plenum Press, New York, London 1989

Heuser, J. E., T. S. Reese : Structure of the synapse. In: Handbook of Physiology, Section 1: The Nervous System. Vol. I: Cellular Biology of Neurons, Part 1 (E. R. Kandel, Ed.), American Physiological Society, Bethesda, Md., 261–294, 1977

Jonas, P., H. Monyer (eds.): Ionotropic Glutamate Receptors in the CNS. Handbook of Experimental Pharmacology. Vol. 141. Springer, Berlin, Heidelberg, New York 1999

Jones, D. G.: Synapses and Synaptosomes. Chapmann & Hall, London 1975

Landon, D. N.: The Peripheral Nerve. Chapmann & Hall, London 1976

Loewi, O.: Über humorale Übertragbarkeit der Herznervenwirkung. Pflügers Arch. Gesamte Physiol. 189 (1921) 239–242

Nakai, J.: Morphology of Neuroglia. Igaku-Shoin, Osaka 1963

Neher, E.: Ion channels for communication between and within cells. Neuron 8 (1992) 606–612

Pappas, G. D., D. P. Purpura: Structure and Function of Synapses. Raven, New York 1972

Penfield, W.: Cytology and Cellular Pathology of the Nervous System. Hoeber, New York 1932

Peters, A., S. L. Palay, H. F. Webster: The Fine Structure of the Nervous System. Oxford University. Press, New York 1991

Pfenninger, K. H.: Synaptic morphology and cytochemistry. Prog. Histochem. Cytochem. 5 (1973) 1–86

Ransom, B., H. Kettenmann (eds.): Neuroglia. Oxford University Press, Oxford 1995

Rapoport, St. J.: Blood-Brain Barrier in Physiology and Medicine. Raven, New York 1976

Roberts, E., T. N. Chase, D. B. Tower: GABA in Nervous System Function. Raven, New York 1976

Sakmann, B.: Elementary steps in synaptic transmission revealed by currents through single ion channels. Neuron 8 (1992) 613–629

Sakmann, B., E. Neher: Single-channel recording. Plenum, New York 1983

Schoffeniels, E., G. Franck, L. Hertz: Dynamic Properties of Glia Cells. Pergamon, Oxford 1978

Stjaerne, L.: Chemical Neurotransmission. Academic Press, London 1981

Szentágothai, J.: Neuron Concept Today. Adakémiai Kiadó, Budapest 1977

Uchizono, K.: Excitation and inhibition. Elsevier, Amsterdam 1975

Unwin, N.: Neurotransmitter action: opening of ligand-gated ion channels. Cell 72/Neuron 10 (Suppl.) (1993) 31–41

Usdin, E., W. E. Burney, J. M. Davis: Neuroreceptors. Chichester, New York 1979

Watson, W. E.: Cell Biology of Brain. Chapman & Hall, London 1976

Windle, W. F.: Biology of Neuroglia. Thomas, Springfield 1958

Weiss, D. G.: Axoplasmic Transport. Springer, Berlin 1982

Yamamura, H. J., S. J. Enna, M. J. Kuhar: Neurotransmitter Receptor Binding. Raven, New York 1985

Zimmermann, H.: Synaptic transmission : Cellular and Molecular Basis. Thieme, Stuttgart 1993

Spinal Cord and Spinal Nerves

Brown, A. G.: Organization of the Spinal Cord. Springer, Berlin 1981

Dyck, P. J., P. K. Thomas, E. H. Lambert: Peripheral Neuropathy, Vol. I. Biology of the Peripheral System. Saunders, Philadelphia 1975

Foerster, O.: Spezielle Anatomie und Physiologie der peripheren Nerven. In: Handbuch der Neurologie, Bd. II/I, hrsg. von O. Bumke, O. Foerster. Springer, Berlin 1928

Foerster, O.: Symptomatologie der Erkrankungen des Rückenmarks und seiner Wurzeln. In: Handbuch der Neurologie, Bd. V, hrsg. von O. Bumke, O. Foerster. Springer, Berlin 1936

Hubbard, J. I.: The Peripheral Nervous System. Plenum, New York 1974

Kadyi, H.: Über die Blutgefäße des menschlichen Rückenmarkes. Gubrynowicz & Schmidt, Lemberg 1886

Keegan, J. J., F. D. Garrett: The segmental distribution of the cutaneous nerves in the limbs of man. Anat. Rec. 102 (1948) 409–437

v. Lanz, T., W. Wachsmuth: Praktische Anatomie, Bd. I/2.–4. Springer, Berlin 1955–1972

Murnenthaler, M., H. Schliack: Läsionen peripherer Nerven. 5. Aufl. Thieme, Stuttgart 1987

Noback, Ch. N., J. K. Harting: Spinal cord. In: Primatologia, Bd. II/1, hrsg. von H. Hofer, A. H. Schultz, D. Starck. Karger, Basel 1971

Villiger, E.: Die periphere Innervation. Schwabe, Basel 1964

Willis, W. D., R. E. Coggeshall: Sensory mechanisms of the spinal cord. 2nd ed. Plenum, New York 1991

Brain Stem and Cranial Nerves

Brodal, A.: The Cranial Nerves. Blackwell, Oxford 1954

Brodal, A.: The Reticular Formation of the Brain Stem. Oliver & Boyd, Edinburgh 1957

Clemente, C. D., H. W. Magoun: Der bulbäre Hirnstamm. In: Einführung in die stereotaktischen Operationen mit einem Atlas des menschlichen Gehirns, hrsg. von G. Schaltenbrand, P. Bailey. Thieme, Stuttgart 1959

Crosby, E. C., E. W. Lauer: Anatomie des Mittelhirns. In: Einführung in die stereotaktischen Operationen mit einem Atlas des menschlichen Gehirns, hrsg. von G. Schaltenbrand, P. Bailey. Thieme, Stuttgart 1959

Delafresnaye, J. F.: Brain-Mechanisms and Consciousness. Blackwell, Oxford 1954

Duvernoy, H. M.: Human Brain Stem Vessels. Springer, Berlin 1978

Jasper, H., L. D. Proctor, R. S. Knighton, W. C. Noshay, R. T. Costello: Reticular Formation of the Brain. Churchill, Oxford 1958

Mingazzini, G.: Medulla oblongata und Brücke. In: Handbuch der mikroskopischen Anatomie, Bd. IV. Springer, Berlin 1928

Olszewski, J., D. Baxter: Cytoarchitecture of the Human Brain Stem. Karger, Basel 1954

Pollak, E.: Anatomie des Rückenmarks, der Medulla oblongata und der Brücke. In: Handbuch der Neurologie, Bd. I, hrsg. von O. Bumke, O. Foerster. Springer, Berlin 1935

Riley, H. A.: An Atlas of the Basal Ganglia, Brain Stem and Spinal Cord. Williams & Wilkins, Baltimore 1943

Spatz, H.: Anatomie des Mittelhirns. In: Handbuch der Neurologie, Bd. I, hrsg. von O. Bumke, O. Foerster, Springer, Berlin 1935

Cerebellum

Angevine, jr. J. B., E. L. Mancall, P. I. Yakovlev: The Human Cerebellum. Little, Brown, Boston 1961

Chain-Palay, V.: Cerebellar Dentate Nucleus. Springer, Berlin 1977

Dichgans, J., J. Bloedel, W. Precht: Cerebellar Functions. Springer, Berlin 1984

Dow, R. S., G. Moruzzi: The Physiology and Pathology of the Cerebellum. University Minnesota Press, Minneapolis 1958

Eccles, J. C., M. Ito, J. Szentágothai: The Cerebellum as a Neuronal Machine. Springer, Berlin 1967

Fields, W. S., W. D. Willis: The Cerebellum in Health and Disease. Green, St. Louis 1970

Ito, M.: The Cerebellum and Neural Control. Raven Press, New York 1984

Jakob, A.: Das Kleinhirn. In: Handbuch der mikroskopischen Anatomie, Bd. IV, hrsg. von W. v. Moellendorff. Springer, Berlin 1928

Jansen, J., A. Brodal: Das Kleinhirn. In: Handbuch der mikroskopischen Anatomie, Bd. IV, hrsg. von W. Bargmann, Erg. zu Bd. IV/I. Springer, Berlin 1958

Larsell, O., J. Jansen: The Comparative Anatomy and Histology of the Cerebellum. University of Minnesota Press. Minneapolis 1972

Llinás, R.: Neurobiology of Cerebellar Evolution and Development. American Medical Association, Chicago 1969

Palay, S. L.: Cerebellar Cortex. Springer, Berlin 1974

Diencephalon

Akert, K.: Die Physiologie und Pathophysiologie des Hypothalamus. In: Einführung in die stereotaktischen Operationen mit einem Atlas des menschlichen Gehirns, hrsg. von G. Schaltenbrand, P. Bailey. Thieme, Stuttgart 1959

Ariëns Kappers, J., J. P. Schadé: Structure and Function of the Epiphysis Cerebri. Elsevier, Amsterdam 1965

Bargmann, W., J. P. Schadé: Lectures on the Diencephalon. Elsevier, Amsterdam 1964

De Wulf, A.: Anatomy of the Normal Human Thalamus. Elsevier, Amsterdam 1971

Diepen, R.: Der Hypothalamus. In: Handbuch der mikroskopischen Anatomie, Bd. IV/7, hrsg. von W. Bargmann. Springer, Berlin 1962

Emmers, R., R. R. Tasker: The Human Somesthetic Thalamus. Raven, New York 1975

Frigyesi, T. L., E. Rinvik, M. D. Yahr: Thalamus. Raven, New York 1972

Harris, G. W., B. T. Donovan: The Pituitary Gland, Bd. III. Pars Intermedia and Neurohypophysis. Butterworths, London 1966

Hassler, R.: Anatomie des Thalamus. In: Einführung in die stereotaktischen Operationen mit einem Atlas des menschlichen Gehirns, hrsg. von G. Schaltenbrand, P. Bailey. Thieme, Stuttgart 1959

Haymaker, W., E. Anderson, J. H. Nauta: Hypothalamus. Thomas, Springfield/Ill. 1969

Kuhlenbeck, H.: The human diencephalon. Confin, neurol. (Basel), Suppl. 14 (1954)

Macchi, G., A. Rustioni, R. Speafico: Somatosensory Integration in the Thalamus. Elsevier, Amsterdam 1983

Morgane, P. J.: Handbook of the Hypothalamus (3 Vol.) Dekker, Basel 1979–81

Nir, I., R. J. Reiter, R. J. Wurtman: The Pineal Gland. Springer, Wien 1977

Pallas, J. E.: La journée du thalamus, Marseille 1969

Purpura, D. P.: The Thalamus, Columbia University Press, New York 1966

Wahren, W.: Anatomie des Hypothalamus. In: Einführung in die stereotaktischen Operationen mit einem Atlas des menschlichen Gehirns, hrsg. von G. Schaltenbrand, P. Bailey. Thieme, Stuttgart 1959

Walker, A. E.: The Primate Thalamus. University of Chicago Press, Chicago 1938

Walker, A. E.: Normale und pathologische Physiologie des Thalamus. In: Einführung in die stereotaktischen Operationen mit einem Atlas des menschlichen Gehirns, hrsg. von G. Schaltenbrand, P. Bailey. Thieme, Stuttgart 1959

Wolstenholme, G. E. W., J. Knight: The Pineal Gland. Ciba Foundation Symposium Churchill-Livingstone, London 1971

Wurtman, R. J., J. A. Axelrod, D. E. Kelly: The Pineal. Academic Press, New York 1968

Telencephalon

Alajouanine, P. Th.: Les grandes activitées du lobe temporal. Masson, Paris 1955

v. Bonin, G.: Die Basalganglien. In: Einführung in die stereotaktischen Operationen mit einem Atlas des menschlichen Gehirns, hrsg. von G. Schaltenbrand, P. Bailey. Thieme, Stuttgart 1959

Braak, H.: Architectonics of the Human Telencephalic Cortex. Springer, Berlin 1980

Brazier, M. A. B., H. Petsche: Architectonics of the Cerebral Cortex. Raven Press, New York 1978

Brodmann, K.: Vergleichende Lokalisationslehre der Großhirnrinde. Barth, Leipzig 1925

Bucy, P. C.: The Precentral Motor Cortex. University of Illinois Press, Urbana 1949

Chan-Palay, V., C. Köhler (eds.): The Hippocampus - New Vistas. Neurology and Neurobiology. Vol. 52. Alan R. Liss, Inc., New York 1989

Ciba Foundation Symposium 58: Functions of the Septo-Hippocampal System. Elsevier, Amsterdam 1977

Creuzfeldt, O. D.: Cortex Cerebri. Springer, Berlin 1983

Critchley, M.: The Parietal Lobes. Arnold, London 1953

Denny-Brown, D.: The Basal Ganglia. Oxford University Press, London 1962

Descarries, L., T. R. Reader, H. H. Jasper: Monoamin Innervation of the Cerebral Cortex. Alan R. Riss, New York 1984

Dimond, St.: The Double Brain. Churchill-Livingstone, Edinburgh 1972

Divac, I. R., G. E. Öberg: The Neostriatum. Pergamon, Oxford 1979

Eccles, J. C.: Brain and Conscious Experience. Springer, Berlin 1966

v. Economo, C., G. N. Koskinas: Die Cytoarchitektonik der Hirnrinde des erwachsenen Menschen. Springer, Berlin 1925

Eleftheriou, B. E.: The Neurobiology of the Amygdala. Plenum, New York 1972

Feremutsch, K.: Basalganglien. In: Primatologia, Bd. II/2, hrsg. von H. Hofer, A. H. Schultz, D. Starck. Karger, Basel 1961

Freund, T. F., G. Buzsáki (eds.): Interneurons of the Hippocampus. Hippocampus 6 (1996) 347–473

Frotscher, M., P. Kugler, U. Misgeld, K. Zilles: Neurotransmission in the Hippocampus. Advances in Anatomy, Embryology and Cell Biology, Vol. 111, Springer, Berlin, Heidelberg 1988

Fuster, J. M.: The Prefrontal Cortex. Raven, New York 1980

Gainotti, G., C. Caltagirone: Emotions and the Dual Brain. Springer, Berlin 1989

Gastaud, H., H. J. Lammers: Anatomie du rhinencéphale. Masson, Paris 1961

Goldman, P. S., W. J. Nauta: Columnar distribution of cortico-cortical fibers in the frontal association, limbic and motor cortex of the developing rhesus monkey. Brain Res. 122, 393–413 (1977)

Goodwin, A. W., J. Darian-Smith: Handfunction and the Neocortex. Springer, Berlin 1985

Hubel, D. H., T. N. Wiesel: Anatomical demonstration of columns in the monkey striate cortex. Nature (Lond.) 221 (1969) 747–750

Isaacson, R. L., K. H. Pribram: The Hippocampus. Plenum, New York 1975

Jones, E. G., A. Peters: Cerebral Cortex. Vol. 1–6. Plenum, New York 1984–1987

Kahle, W.: Die Entwicklung der menschlichen Großhirnhemisphäre. Springer, Berlin 1969

Kennedy, C., M. H. Des Rosiers, O. Sakurada, M. Shinohara, M. Reivich, J. W. Jehle, L. Sokoloff: Metabolic mapping of the primary visual system of the monkey by means of the autoradiographic C¹⁴deoxyglucose technique. Proc. nat. Acad. Sci. (Wash.) 73 (1976), 4230–4234

Kinsbourne, M., W. L. Smith: Hemispheric Disconnection and Cerebral Function. Thomas, Springfield 1974

Passouant, P.: Physiologie de le hippocampe. Edition du centre national de la recherche scientifique. Paris 1962

Penfield, W., H. Jasper: Epilepsy and the Functional Anatomy of the Human Brain. Little, Brown, Boston 1954

Penfield, W., T. Rasmussen: The Cerebral Cortex of Man. Macmillan, New York 1950

Penfield, W., L. Roberts: Speech and Brain Mechanisms. Princeton University Press. Princeton 1959

Ploog, D.: Die Sprache der Affen. In: Neue Anthropologie, hrsg. von H. G. Gadamer, P. Vogler. Thieme, Stuttgart 1972

Rose, M.: Cytoarchitektonik und Myeloarchitektonik der Großhirnrinde. In: Handbuch der Neurologie, Bd. I, hrsg. von O. Bumke, O. Foerster. Springer, Berlin 1935

Sanides, F.: Die Architektonik des menschlichen Stirnhirns. Springer, Berlin 1962

Schneider, J. S., T. J. Lidsky: Basal Ganglia and Behavior. Huber, Bern 1987

Schwerdtfeger, W. K.: Structure and Fiber Connections of the Hippocampus. Springer, Berlin 1984

Seifert, W.: Neurobiology of the Hippocampus. Academic Press, London 1983

Squire, L. R.: Memory and Brain. Oxford University Press, New York 1987

Stephan, H.: Allocortex. In: Handbuch der mikroskopischen Anatomie, Bd. IV/9, hrsg. von W. Bargmann. Springer, Berlin 1975

Valverde, F.: Studies on the Piriform Lobe. Harvard University Press. Cambridge/Mass. 1965

Weinstein, E. A., R. P. Friedland: Hemiinattention and Hemisphere Spezialization. Raven, New York 1977

Cerebrospinal Fluid System

Hofer, H.: Circumventrikuläre Organe des Zwischenhirns. In: Primatologia, Bd. II/2, hrsg. von H. Hofer, A. H. Schultz, D. Starck. Karger, Basel 1965

Lajtha, A., D. H. Ford: Brain Barrier System. Elsevier, Amsterdam 1968

Millen, J. W. M., D. H. M. Woollam: The Anatomy of the Cerebrospinal Fluid. Oxford University Press, London 1962

Schaltenbrand, G.: Plexus und Meningen. In: Handbuch der mikroskopischen Anatomie, Bd IV/2, hrsg. von W. Bargmann. Springer, Berlin 1955

Sterba, G.: Zirkumventrikuläre Organe und Liquor. VEB Fischer, Jena 1969

Cerebrovascular System

Dommisee, G. F.: The Arteries and Veins of the Human Spinal Cord from Birth. Churchill-Livingstone, Edinburgh 1975

Hiller, F.: Die Zirkulationsstörungen des Rückenmarks und Gehirns. In: Handbuch der Neurologie, Bd. III/11, hrsg. von O. Bumke, H. Foerster. Springer, Berlin 1936

Kaplan, H. A., D. H. Ford: The Brain Vascular System. Elsevier, Amsterdam 1966

Krayenbühl, H., M. G. Yasargil: Die zerebrale Angiographie, 2. Aufl. Thieme, Stuttgart 1965; 3. Aufl. 1979

Luyendijk; W.: Cerebral Circulation. Elsevier, Amsterdam 1968

Szilka, G., G. Bouvier, T. Hovi, V. Petrov: Angiography of the Human Brain Cortex. Springer, Berlin 1977

Autonomic Nervous System

Burnstock, G., M. Costa: Adrenergic Neurons, Chapman & Hall, London 1975

Csillik, B., S. Ariens Kappers: Neurovegetative Transmission Mechanisms. Springer, Berlin 1974

Furness, J. B., M. Costa: The enteric nervous system. Churchill Livingstone, Edinburgh 1987

Gabella, G.: Structure of the Autonomous Nervous System. Chapman & Hall, London 1976

Kuntz, A.: The Autonomic Nervous System. Lea & Febiger, Philadelphia 1947

Langley, J. N.: The autonomic nervous system, part 1. Heffer, Cambridge 1921

Mitchell, G. A. G.: Anatomy of the Autonomic Nervous System. Livingstone, Edinburgh 1953

Newman, P. P.: Visceral Afferent Functions of the Nervous System. Arnold, London 1974

Pick, J.: The Autonomic Nervous System. Lippincott, Philadelphia 1970

Thews, G., G. Vaupel.: Vegetative Physiologie. Springer, Berlin 1990

White, J. C., R. H. Smithwick: The Autonomic Nervous System. Macmillan, New York 1948

Functional Systems

Adey, W. R., T. Tokizane: Structure and Function of the Limbic System. Elsevier, Amsterdam 1967

Andres, K. H., M. v. Dühring: Morphology of cutaneous receptors. In: Handbook of Sensory Physiology, Bd. II, hrsg. von H. Autrum, R. Jung, W. R. Loewenstein, D. M. MacKay, H. L. Teuber. Springer, Berlin 1973

Barker, D.: The morphology of muscle receptors. In: Handbook of Sensory Physiology. Bd. III/2, hrsg. von H. Autrum, R. Jung, W. R. Loewenstein, D. M. MacKay, H. L. Teuber. Springer, Berlin 1974

Campbell, H. J.: The Pleasure Areas. Eyre Methuen, London 1973

Couteaux, R.: Motor endplate structure. In: The Structure and Function of Muscle, hrsg. von G. H. Bourne. Academic Press, New York 1973

Douek, E.: The Sense of Smell and Its Abnormalities. Churchill-Livingstone, London 1974

Gardner, H.: Dem Denken auf der Spur, der Weg der Kognitionswissenschaft. Klett-Cotta, Stuttgart 1989

Halata, Z.: The Mechanoreceptors of the Mammalian Skin, Advances in Anatomy, Embryology and Cell Biology, Bd. 50/5. Springer, Berlin 1975

Isaacson, R. L.: The Limbic System. Plenum Press, New York 1974

Janzen, R., W. D. Keidel, G. Herz, C. Steichele: Schmerz. Thieme, Stuttgart 1972

Jung, R., R. Hassler: The extrapyramidal motor system. In: Handbook of Physiology, Section 1, Bd. 2, hrsg. von J. Field, H. W. Magoun, V. E. Hall. American Physiological Society Washington 1960

Knight, J.: Mechanisms of Taste and Smell in Vertebrates. Ciba Foundation Symposium. Churchill, London 1970

Lassek, A. M.: The Pyramidal Tract. Thomas, Springfield/Ill. 1954

Monnier, M.: Functions of the Nervous System, Bd. 3: Sensory Functions and Perception. Elsevier, Amsterdam 1975

Munger, B. L.: Patterns of organization of peripheral sensory receptors. In: Handbook of Sensory Physiology, Bd. I/1, hrsg. von H. Autrum, R. Jung, W. R. Loewenstein, D. M. MacKay, H. L. Teuber. Springer, Berlin 1971

de Reuk, A. V., S. J. Knight: Touch, Heat and Pain. Ciba Foundation Symposium. Churchill, London 1966

Sezntágothai, J., J. Hámori, M. Palkovits: Regulatory Functions of the CNS: Motion and Organization Principles. Pergamon, Oxford 1981

Wiesendanger, M.: The pyramidal tract. In: Ergebnisse der Physiologie, Bd. 61. Springer, Berlin 1969

Zacks, S. J.: The Motor Endplate. Saunders, Philadelphia 1964

Zotterman, Y.: Olfaction and Taste. Pergamon, Oxford 1963

Zotterman, Y.: Sensory Mechanisms. Elsevier, Amsterdam 1966

Zotterman, Y.: Sensory Functions of the Skin. Pergamon, Oxford 1977

The Eye

Carpenter, R. H. S.: Movements of the Eyes. Pion, London 1977

Fine, B. S., M. Yanoff: Ocular histology. Harper & Row, New York 1972

Hollyfield, J. G.: The Structure of the Eye. Elsevier, Amsterdam 1982

Hubel, D. H., T. N. Wiesel: Die Verarbeitung visueller Informationen. Spektrum der Wissenschaft, November 1979

Livingston, M., D. Hubel: Segregation of form, color, movement, and depth: Anatomy, physiology and perception. Science 240 (1988) 740–749

Masland, R. H.: Die funktionelle Architektur der Netzhaut. Spektrum der Wissenschaft, 66–75, Februar 1989

Marr, D.: Vision. Freeman, San Francisco 1982

Polyak, St.: The Vertebrate Visual System. University Chicago Press, Chicago 1957

Rodieck, R. W.: Vertebrate Retina. Freeman, San Francisco 1973

Straatsma, B. R., M. O. Hall, R. A. Allen, F. Crescitelli: The Retina Morphology, Function and Clinical Characteristics. University California Press, Berkeley 1969

Wässle, H., B. B. Boycott: Functional architecture of the mammalian retina. Physiol. Rev. 71 (1991) 447–480

Walsh, F. W., W. F. Hoyt: Clinical Neuroophthalmology. Williams & Wilkins, Baltimore 1969

Warwick, R.: Wolff's Anatomy of the Eye and Orbit. Lewis, London 1976

Zeki, S.: A Vision of the Brain. Blackwell, London 1993

Organ of Hearing and Balance

Ades, H. W., H. Engström: Anatomy of the inner ear. In: Handbook of Sensory Physiology, Bd. V/1, hrsg. von H. Autrum, R. Jung, W. R. Loewenstein, D. M. MacKay, H. L. Teuber. Springer, Berlin 1974

Brodal, A., O. Pompeiano, F. Walberg: The Vestibular Nuclei and Their Connexions. Oliver & Boyd, Edinburgh 1962

Gualtierotti, T.: The Vestibular System. Springer, Berlin 1981

Kolmer, W.: Gehörorgan. In: Handbuch der mikroskopischen Anatomie, Bd. III/1, hrsg. von W. v. Moellendorff. Springer, Berlin 1927

Precht, W.: Neuronal Operations in the Vestibular Systems. Springer, Berlin 1978

Rasmussen, G. L., W. F. Windle: Neural Mechanisms of the Auditory and Vestibular Systems. Thomas, Springfield/Ill. 1960

deReuck, A. V. S., J. Knight: Hearing Mechanisms in Vertebrates. Ciba Foundation Symposium, Churchill, London 1968

Whitfield, I. C.: The Auditory Pathway. Arnold, London 1960

Index

Page numbers in **bold** indicate extensive coverage of the subject